SCHOPENHAUER'S BROKEN WORLD-VIEW

Science and Philosophy

VOLUME 10

This series has been established as a forum for contemporary analysis of philosophical problems which arise in connection with the construction of theories in the physical and the biological sciences. Contributions will not place particular emphasis on any one school of philosophical thought. However, they will reflect the belief that the philosophy of science must be firmly rooted in an examination of actual scientific practice. Thus, the volumes in this series will include or depend significantly upon an analysis of the history of science, recent or past. The Editors welcome contributions from scientists as well as from philosophers and historians of science.

The titles published in this series are listed at the end of this volume.

SCHOPENHAUER'S BROKEN WORLD-VIEW

Colours and Ethics between Kant and Goethe

by

PAUL F.H. LAUXTERMANN

University of Twente, Enschede, The Netherlands

KLUWER ACADEMIC PUBLISHERS

DORDRECHT / BOSTON / LONDON

A C.I.P. Ctalogue record for this book is available from the Library of Congress.

ISBN 0-7923-6585-2

Published by Kluwer Academic Publishers,
P.O. Box 17, 3300 AA Dordrecht, The Netherlands.

Sold and distributed in North, Central and South America
by Kluwer Academic Publishers,
101 Philip Drive, Norwell, MA 02061, U.S.A.

In all other countries, sold and distributed
by Kluwer Academic Publishers,
P.O. Box 322, 3300 AH Dordrecht, The Netherlands.

Printed on acid-free paper

Printed in the Netherlands.

TABLE OF CONTENTS

Preface

Why care about Schopenhauer, and (if one does) what is there to say about him that has not been said already?

The principal thesis of the present book is that a major, unresolved question about Schopenhauer's work still remains. To anticipate the issue towards which a fairly lengthy argument is to carry us, we shall seek to show that its numerous contradictions compel us to ask in the end: 'What, after all, is here meant to be the *real* 'Thing-in-itself': 'Will', or Nirvana?' Or, to use somewhat less technical language, 'Is, in Schopenhauer's true view, that which transcends all empirical phenomena to be found in what he (rightly or wrongly) calls 'Will' or rather in a human Intellect capable in the end of extinguishing that Will?'

To be sure: the sheer fact that Schopenhauer's work is marked by many a contradiction has often enough been pointed out by his numerous, less and more recent, on the whole very able commentators. 'His commentators', *not* himself; on the contrary, he always vehemently denied ever having committed any. But precisely this circumstance presents us with something of a riddle. For Schopenhauer was an extraordinarily acute thinker, who knew how to express himself in literary prose of high standing and a clarity rarely surpassed, and whose utterances about his Idealist contemporaries in German philosophy display a temperamental impatience with muddle-headedness of any kind. Whence, then, those evident contradictions in his own work — contradictions that may be summed up in one fundamental antinomy (which we have dubbed 'can the world be in my head, yet my head be in the world?') and one fundamental paradox (similarly dubbed by us 'can God commit suicide?')?

The aim of the present book, in building upon the accomplishments of previous commentators, is to shed some new light on this problem. Its principal thesis has already been suggested by the superficially somewhat naive-looking question raised above. That thesis implies that there are in fact *two* ultimate principles in Schopenhauer's philosophy and that his, at first sight, beautifully rounded-off system reveals itself at second sight as the secret arena of two conflicting world-views.

Whence the conflict? Here our first auxiliary thesis is that an inquiry into the origins of the discord brings to light Schopenhauer's grappling in his own unique way with a problem that is still very much with us, to wit: how to acquire a solid grasp of the world at large (the eternal longing of all metaphysicians of all times) in a manner compatible with all that modern science has validly to say about the constitution of our world? This has become such a problem because, ever since the 'mechanization of the world-picture' in the 17th century, modern science may lay claim to valid knowledge of nature without, however, being able to provide a satisfactory world-view all by itself.

Our second (closely related) auxiliary thesis concerns the intellectual guides under whose joint patronage Schopenhauer's pertinent effort took place. We shall try to show how his critical allegiance to *Kant* (who drew dualist conclusions in the domains of epistemology and ethics from the birth of modern science) as well as to *Goethe* (who sought to point out basic shortcomings of modern science from a monist point of view) carried him towards a philosophical vision of the world as enticing as it is problematic.

Our programme, then, is to follow Schopenhauer along his chosen road — to find out how he interpreted the legacies of Kant and Goethe (sometimes, but always revealingly so, misinterpreting them); what in these legacies he adopted and what he rejected (often portions of one in view of the other); and what doctrines of his own he went on to construct out of them. We shall follow three lines in particular:

a. his transformation (in view of what he held to be Kant's epistemology) of Goethe's doctrine on the nature of colours and how his pertinent effort compares with a somewhat similar exercise undertaken by his perennial foe, Hegel;
b. his transformation, in the spirit of his own 'Will'-metaphysics, of Kant's ethical doctrine;
c. his construction — undertaken in parallel with an improved (or so at least he believed) Kantian epistemology — of an essentially though not explicitly Goethean epistemology, meant to serve as a kind of propaedeutic to the apotheosis of his entire philosophical enterprise: the gospel of the final redemption from the bondage of 'Will'.

We carry out this programme of ours by blending, in a perhaps unusual sort of way, the history of philosophical currents with history of science, history of ideas generally, and (in order to elucidate relevant portions of Schopenhauer's biography and intellectual and social environment) German history as well.

A more detailed exposition of what all this is going to entail follows presently in an introductory chapter. For now, we wish to express our gratitude to the present and former members of the Department of History at the University of Twente: Petra Bruulsema for ever-ready secretarial help, and Christiaan Boudri, Floris Cohen, Fokko Jan Dijksterhuis, Marius Engelbrecht, the late Casper Hakfoort, Nancy Nersessian, Hans Sparnaay, Tomas Vanheste and Péter Várdy for their vigilant assessment of ideas in the course of our regular ongoing exchanges (Floris Cohen also revised with care the entire typescript, ensured its camera-ready preparation, and came up with the title). Further, Joachim Aul is to be mentioned, with gratitude, for his workmanlike criticisms based on a shared predilection for Schopenhauer, and Helen Ure for carrying out the 'light editing' recommended by a generous referee. Finally, on a more informal level, I gladly avail myself of the opportunity to thank, above all, Brigitte Mazohl for many stimulating conversations in the course of long years of unabated friendship; among other names that might be mentioned I choose those of Ralph Melville and Claus Scharf (of the Institut für europäische Geschichte in Mainz) for their lively interest in the author's scholarly development.

Quotations are given in English in the main text; unless mentioned otherwise, the translations are all ours. If the text quoted is a translation of an original text in German or French, the latter is given in a footnote. The footnotes also offer substantial excursions beyond the course of our argument, whereas references to pertinent sources and to the literature are provided in the endnotes.

INTRODUCTION

Bertrand Russell, in the Schopenhauer chapter of his well-known *History of Western Philosophy*, charges the protagonist of the present book — apart from a list of other un-British vices like habitually dining well — with having been "extremely quarrelsome".[1] And so much is certain that Arthur Schopenhauer was a philosopher with a remarkable propensity for being provoked to anger. For example, when a certain Adolph Cornill published a book on his philosophy which seemed to have been written with the specific aim of showing up an impressive number of contradictions inherent in it, Schopenhauer's reaction was as one might expect. In a letter to his faithful 'evangelist' Frauenstädt of 11 July 1856 he castigated such an attempt as "a stupid, bad trick, such as is often used by knaves", and which was especially ludicrous "with me, the most consistent and monolithic of all philosophers".[*2] And in a letter of 31 March 1854 to the 'apostle' Johann August Becker we read (this time about worthy old Frauenstädt himself!): "To seek for contradictions in my work is wholly in vain; it is all of a piece."[†3] Cornill's book has fallen into, probably well-deserved, oblivion; but, in spite of the master's warning, commentators have continued (unless they belong to some Schopenhauer-orthodoxy) to be puzzled, rightly or wrongly, by the same phenomenon. This is true not only of such rather unsympathetic commentators as Bertrand Russell, who naturally did not fail in his turn to point out inconsistencies, and who went on to sum up his judgment in the verdict that "Schopenhauer's gospel of resignation is not very consistent nor very sincere".[4] It is also true of those who were impressed by the fact that under Schopenhauer's hands, out of such heterogeneous and sometimes (at least apparently) contradictory elements such a philosophically as well as artistically beautiful edifice could have arisen. *This* kind of commentary tended to show admiration for the wide range of philosophical topics and currents that came together in Schopenhauer's mind rather than to scorn its alleged shallowness (which then should have expressed itself in the 'contradictions' which, *mirabile dictu*, Schopenhauer had not even noticed himself); and they sought to determine the nature and origins of the varied contributions to the great symphony with which Schopenhauer's system strikingly has been compared.[5] Such a search brought his first biographer, Wilhelm Gwinner, to the conclusion that "it was Schopenhauer's vocation to complement not only the physical realism of the sensualists but also the aesthetic realism of Goethe with the

---

[*] "... ein dummer, schlechter Kunstgriff, dessen sich die Lumpe oft bedienen"; "... bei mir, dem konsequentesten und einheitlichsten aller Philosophen".

[†] "Bei mir Widersprüche zu suchen ist ganz eitel: Alles ist aus einem Guß."

ethical realism of Kant."*[6] Rather a mouthful; and one might imagine that the master, had he lived to see his admirer's book, would have agreed as to the crucial importance of these three currents but would have objected to the term 'ethical *realism*' as far as Kant was concerned. ''Realism', he might have said, 'is precisely what *distinguishes* my own moral philosophy favourably from Kant's, which — for all its great merits — remained far too abstract to be convincing. Immortal, indeed, is Kant's great doctrine of the difference between the empirical and the intelligible I, but this is precisely the consequence of Kant's equally great and immortal doctrine of the ideal nature of space and time. So you might rather say that it was my vocation to complement this Kantian *idealism* — which, however epochal, was also one-sided — with, indeed, the *realism* of both Goethe and those French sensualists whose outstanding representative was Cabanis.'

It is indeed true that one encounters this — at first sight — quite remarkable hodgepodge of influences in several places in Schopenhauer's work, as we shall have ample opportunity to show in the course of the present book. His epistemology, for example, centres on the problem of how to reconcile Kant's allegedly 'subjective' approach with the 'objective' one of Cabanis; paradoxically, however, this very effort led, under Schopenhauer's hands, (as we shall see in our fourth chapter), to a new edition of the *really* 'subjective' idealism of Berkeley! That he had never been, nor had even believed himself to be, a fully orthodox Kant adept appears already from a revealing remark in his notebooks of the year 1810, when he was still a student in Göttingen. He had enrolled there as a *studiosus medicinae* taking classes in the sciences of nature, especially physiology — 'the physical realism of the sensualists'! Yet in his heart he was already devoted to philosophy, into which discipline he was introduced by his teacher Schulze-Aenesidemus, who recommended to him readings of Kant, in particular. In that notebook of 1810, then, young Schopenhauer pointed to the need to complement Kant's results with Goethe's. We first read there: "Kant's shortcomings are perhaps best expressed by saying that he has not known contemplation",† which expression is to be understood in the given context as that serene, intuitively synthetic vision which so strikingly contrasted with Kant's painstakingly analytical frame of mind — the very vision Gwinner was later to characterize as 'aesthetic realism'. This is followed by the remark that Germany is lucky in simultaneously having a Goethe in its ranks, since otherwise the development of the German spirit might become too one-sided. But the two giants together may perhaps elevate the German spirit to a height that will even surpass Antiquity.[7] Conversely, in a comment which he gave much later on the episode of his collaboration with Goethe in the domain of the theory of colours, he came to the conclusion that the amazing objectivity of Goethe's mind (the hallmark of artistic genius!)

---

* "Die Ergänzung des ethischen Realismus Kants zu dem physischen der Sensualisten nicht nur, sondern auch zu dem ästhetischen Goethes ... Dies war seine Aufgabe."

† "Es ist vielleicht der beste Ausdruck für Kants Mängel, wenn man sagt: er hat die Kontemplation nicht gekannt."

unfortunately prevented the great man from appreciating the subjective approach of his younger collaborator who had also gone through the school of Kant!

However, with these three names — crucially important though they are — the survey of influences that have contributed to Schopenhauer's ultimate philosophical world-view is certainly not exhausted. In a conversation with a French visitor he explained that there had been no more than three philosophers of the truly first rank: the Buddha, Plato, and Kant.[8] Of these, the first-mentioned was important for Schopenhauer as a confirmation rather than as a source of inspiration for his own philosophy; but the *Upanishads* (in the tradition of which the Buddha, in however heretical a way, stands) had been such an inspiration, as Schopenhauer never tired of assuring. Equally, he never ceased to point out that Schulze had directed his attention not only to the "amazing" Kant, but also to the "divine" Plato. However, this time we are inclined to agree with Russell when he writes: "I do not think that he [Schopenhauer] owes as much to Plato as he thinks he does."[9] Indeed, we believe that Plato served him as an antidote to Kant in more or less the same way as Goethe did; and this in spite of Plato's indubitable 'idealism' which, however, was an *objective* idealism, the Platonic Ideas being interpreted moreover by Schopenhauer in a way that makes them look rather like Goethe's *Urphänomene* ('primordial phenomena'), as we shall seek to make plausible in our concluding chapter. Considering further that 'physiological realism' served him, as it did the Buddha,[*10] to some extent as an appropriate training in fathoming the misery of the human condition and the inexorable *bellum omnium contra omnes* which is the core of nature, one might put forward the thesis that the originality as well as the inevitable problems of Schopenhauer's philosophy have their ultimate roots in the *rendez-vous* of the twin heritages of Kant and Goethe in the mind of a German philosopher whose pessimistic temperament made him especially prone to the exotic consolations of Indian wisdom (which at that time gradually became somewhat better known to the West) — as well as to a grim biologically tinged realism which was soon to acquire systematic status in Darwinism (although the latter was criticized as "shallow" by him from his metaphysical backgrounds).

This 'temperament' is indeed crucial, for it was hardly unique in itself for a young philosopher at that time to undergo the influence of both Kant and Goethe. It is not by chance, after all, that the entire period from the later Enlightenment up to and including Romanticism is wont to be called the 'Goethe era' (*Goethezeit*) which, as a matter of fact, coincides roughly with Goethe's life-span covering the second half of the 18th and the first third of the 19th century: No one could escape from the giant shadow cast by Goethe, not even those who, like the Romantic poets, opposed his authority. Nor, for that matter, could anyone, as far as

---

[*] To the training of a Buddhist monk belongs contemplation of his body in which are to be found 32 more or less impure parts and nine openings from which continually stream dirty and disgusting substances; in the words of a 20th-century commentator "the confrontation of the image of a tempting woman with this vision is surely to have a devastating influence on any erotic passion."

philosophy was concerned, escape from Kant's shadow, not even those who, like the German Idealists, believed to have overcome him. It has therefore been said with justice that "precisely the conjunction of Goethe's impulses where one's vision of life is concerned with Kant's critical philosophy urged exceedingly gifted youngsters towards a productivity as richly variegated as it was feverish."[*11] Indeed, the mere fact that the philosophy of Hegel, whom Schopenhauer so heartily despised, is quite as inconceivable as Schopenhauer's own without the preceding 'Copernican Revolution' brought about by Kant, and the further fact that Hegel, again like Schopenhauer (though in a very different way, as we shall see), publicly championed Goethe's much-contested theory of colours, seems to suggest that the characteristic given in the above of Schopenhauer's philosophical system as a *rendez-vous* of the twin legacies of Kant and Goethe might be applied to Hegel as well. But of course Schopenhauer saw in Hegel not a legitimate heir of Kantian philosophy, but rather its corrupter; and, whatever one may think of Schopenhauer's claim to be the only true and legitimate heir of Kantian philosophy, this image of himself is nevertheless crucial for Schopenhauer's position in the German intellectual climate of the time.

Schopenhauer's route to Kant has been characterized by Safranski in his brilliant biography as a *Krebsgang* (i.e., as directed backwards) in that, belonging by the date of his birth to the rearguard of the *Goethezeit*, he came to Kant via the detour of Romanticism, and, "having arrived there, he will proceed to a revision of the trial that the successors had started against Kant."[†12] Indeed, Schopenhauer, notwithstanding his explicit critique of Kant's philosophy (in its most systematic guise to be found in the 'Critique of Kantian Philosophy' ('Kritik der Kantischen Philosophie') he wrote as a supplement to the first volume of his own main work '*The World as Will and Representation*' (*Die Welt als Wille und Vorstellung*)), always acknowledged his profound debt to Kant; and it is precisely the proclaimed need of a return to Kant which distinguishes his position from that of German Idealism (at least as far as Hegel is concerned). His attitude, not only towards Kant but, at bottom, towards Goethe as well, may therefore best be characterized as one of *ite ad fontes*. And it is precisely this desire for a return to the sources which, combined with the unmistakable 'materialist' streak in his work (which was itself the fruit of his life-long interest in natural science, especially physiology) and with its no less unmistakable 'pessimistic' flavour (which came quite naturally to him and had therefore not to be 'learned' from the Indians who only confirmed it and provided it with a theoretical fundament) made appear him 'modern' after 1850 when the time of German Idealism had passed. Comparable to Stendhal in this respect, who was likewise a stranger in his time in that he was in many ways a man of the 18th and at the same time of the 20th century who happened to live in the

---

[*] In fact Schiller, who had deeply undergone the influence of Kant before he became Goethe's companion, was the first in whom this conjunction took place.

[†] "Im Krebsgang, über die Romantik, wird er zu Kant kommen, und dort angelangt, wird er eine Revision des Prozesses vornehmen, den die Nachfolger gegen ihn angestrengt hatten."

first half of the 19th century, (a man, moreover, whose predilection for the music of Rossini, though not that of mathematics, was shared by our philosopher), Schopenhauer was convinced that he would be understood later on in the century; and, moreover, he had the good fortune to live to see the first dawn of his fame.

To be sure, this fame was bound to see its twilight, too. In the 20th century, Schopenhauer — although he never fell into complete oblivion — moved to the background of the philosophical stage; and, in spite of a certain revival during the last decades (for which an increased number of, partly excellent, publications on him bear witness) he seems unlikely to regain the place he occupied in the former *fin de siècle*. The ideals of the Enlightenment especially (to which Schopenhauer firmly stood) are not fashionable nowadays; there is rather, perhaps, an inclination to mysticism which is equally characteristic of him and which made Thomas Mann define "the mixture of Voltaire and Jakob Böhme in him" as the most fascinating aspect of his philosophy.[*13] In response to this postmodern trend one might very well remind its spokesmen of the fact that, like it or not, we are all heirs to the Enlightenment: not only the agnosticists among us but also (apart from a few fundamentalists) contemporary Christian believers as far as their interpretation of traditional dogma and morals and their acceptance of the separation of Church and State are concerned (just as we are all heirs to its inimical brother Romanticism, the influence of which we discern not only in the contemporary flirtation with all kinds of esotericism we have already hinted at, but above all in modern man's sentiment-dominated conception of marriage and of personal life in general). But there is another, more important consideration. Philosophers and poets really do not corrupt the youth; even those people who see in the spirit of the Enlightenment the root of all the evils of the 20th century (we shall return to this in our final chapter) ought to be aware that such chapters from the history of ideas like Enlightenment and Romanticism, important as they are, were, in the last resort, Enlightened reflections upon and conclusions from, or Romantic reactions against, a much more fundamental development which had started much earlier: the Scientific Revolution — a revolution of which it has been remarked that it abandoned the ideal of contemplation,[14] the very ideal, that is, of which we saw Schopenhauer deplore the lack in Kant and praise the presence in Goethe. Indeed, if Schopenhauer's philosophy may to a large degree be considered as an effort to unite the joint legacies of Kant and Goethe, the work of these two protagonists can in its turn be seen as two fundamentally different modes of coming to terms with that decisive event in the history of the West, which set it apart for good from humanity's (up to then) customary course — the rise of modern science. With this in mind we come to the following outline of our book:

In Chapter One we analyse the 'mechanization of the world-picture' from a philosophical, a cultural–psychological, and a historical point of view. This is followed by an interpretation of Kant's and Goethe's respective responses to it, so as to lead us into a broad sketch of the intellectual climate of the *Goethezeit*. We

* "... die Mischung aus Voltaire und Jakob Böhme in ihm ..."

then characterize Schopenhauer as a latecomer to this epoch — a position which (in tandem with his personal background, to be sure) predestined him for the highly non-conformist role he was to play upon the philosophical stage.

In Chapter Two his first appearance there will be characterized in terms of an analysis of his dissertation '*On the Fourfold Root of the Principle of Sufficient Reason*' (*Über die vierfache Wurzel des Satzes vom zureichenden Grunde*) and a comparison of the original with the later version of that book. We then point to the acquaintance he made with the *Upanishads* and to his short-lived collaboration with Goethe on the theory of colours as the events responsible for the crucial development in his thought revealed by that comparison.

In Chapter Three we outline Goethe's controversy with Newton as laid down in the former's '*Theory of Colours*' (*Zur Farbenlehre)*. Our subsequent analysis of Schopenhauer's own treatise '*On Vision and Colours*' (*Vom Sehn und den Farben*) brings to light a fundamental difference between his frame of mind and Goethe's, in spite of their shared rejection of Newton's *Opticks*. Since the main service his theory of colours was to render Schopenhauer resided in the more general theory of perception it gave rise to in his mind, this will lead us

In Chapter Four to a closer investigation into his physiology-based epistemology and to a critical comparison with the position taken on that score by Kant (to whose theory of knowledge Schopenhauer's own epistemology was meant to give its much-needed, final perfection in the same way as his own theory of colours to Goethe's). All this culminates in our effort to lay bare the roots of the first great 'contradiction', to wit, the Schopenhauerian antinomy (an antinomy as such recognized by him, but indeed as an *antinomy*, i.e., as an *inevitable* contradiction) 'can the world be in my head, yet my head be in the world?'. Here we conclude that Schopenhauer's heretical allegiance to Goethe's theory of colours and his equally heretical allegiance to Kant's critical philosophy have their common root in what we (adopting a term used by Schopenhauer himself in his critique of Kantian philosophy) choose to call his πρωτον ψευδος (*prôton pseudos*), or primal fallacy. This virtually made him lapse from Kantian 'transcendental' (or 'critical') idealism into a neo-Berkeleyan psychological (or subjective) idealism, thus creating at the same time an unbridgeable gap between Goethe's general outlook and his own. The latter circumstance makes it illuminating to compare

In Chapter Five the mostly different way in which (of all conceivable couples) both Hegel and Schopenhauer defended Goethe's theory of colours. Our comparison will yield a characterization of both men as 'positivist metaphysicians', which resemblance, however, cannot obscure their fundamental difference in the ultimate questions of philosophy. This brings us

In Chapter Six to Schopenhauer's ethics, the outline of which is to be found already in the fourth book of his main work '*The World as Will and Representation*', but which is the special topic of the two essays he brought together in '*The two basic problems of ethics*' (*Die beiden Grundprobleme der Ethik*). A faithful exposition of its chief content reveals Schopenhauer once more as a heretical disciple of Kant, whose own moral philosophy, as laid down in the '*Groundwork to a Metaphysics of Morals*' (*Grundlegung zur Metaphysik der Sitten*) as well as in the second great '*Critique*' and in the '*Metaphysics of Morals*' (*Metaphysik der Sitten*), will equally be summarized with care before (following upon an excursion into more recent currents of ethical thought) we proceed

In Chapter Seven to a critical comparison between Kant's and Schopenhauer's moral philosophies, followed by an attempt to deduce the differences between them from their respective metaphysical backgrounds. Thus we arrive at the intrinsic problems of Schopenhauer's 'Will'-metaphysics in connection with the problem of moral freedom; and this in its turn will provide a good occasion to tackle

In Chapter Eight the second great 'contradiction', to wit, the paradox 'can God commit suicide?'. This eventually confronts us with the big question: 'What, then, is the real 'Thing-in-itself': 'Will' or 'Nirvana'?'; a question which in its turn stimulates us to undertake an effort to unveil at long last the hidden core of Schopenhauer's philosophy. This finally brings us

In Chapter Nine to the influence that Schopenhauer's philosophy has exercised on later generations. Here our attention is directed mostly at the third book of '*The World as Will and Representation*'. There we find Schopenhauer's aesthetic doctrine in the first place. Yet we shall mainly seek to demonstrate the *ethical* side of the alternative epistemology he developed there to shore up his aesthetic doctrine. This more hidden epistemology of his, alternative to the modified 'Kantian' theory of knowledge set forth in the first book like previously in his dissertation, will appear to be really at least as 'Goethean' as it is allegedly 'Platonic'. We end the book by considering what Schopenhauer still has to say that is of value for our own time.

And now we finally commence!

I

Kant, Goethe, and the Mechanization of the World-Picture

a. The mechanization of the world-picture

"Modern man lacks a unified conception of the world. He lives in a dual world: in his environment, which is naturally given to him, and, at the same time, in the world which since the beginning of the modern era has been created for him by sciences founded upon the principle that the laws of nature are, in essence, mathematical. The non-unity which has thus come to penetrate our entire life is the true source of the spiritual crisis we are going through today," says the Czech philosopher Jan Patočka.*[15] Patočka refers here to that loss of a pristine unity in man's experience of the world which has been lamented so often. He appears to link it closely to the 17th-century Scientific Revolution — an event that has also been described as the "mechanization of the world-picture",[16] and interpreted as the point of no return in the process of European civilization's "deviation from the general human pattern".[17] What was this Scientific Revolution all about?†

Patočka himself, in the passage quoted, points to the principle that "the laws of nature are, in essence, mathematical". This is not, to be sure, a novel idea; Patočka finds himself in agreement here with such pioneers in shaping the concept of the 'Scientific Revolution' and in seeking to define its essence as E.A. Burtt and Alexandre Koyré. The former spoke of the "metaphysical foundations of modern physical science" which, in his view, amount to the following. Reality is not

---

* "L'homme moderne n'a pas du monde une conception unifiée. Il vit dans un monde double: dans son environnement, qui lui est naturellement donné et, en même temps, dans le monde que depuis l'ère moderne créent pour lui des sciences fondées sur ce principe que les lois de la nature sont d'essence mathématique. Le non-unité qui a pénétré ainsi toute notre vie est la source propre de la crise spirituelle par laquelle nous passons actuellement."

† We are aware that the concept of the 'Scientific Revolution' has in our time become a disputed one. This is so, in part, no doubt, because of the general postmodernist distrust of 'grand narratives', but also because the word 'revolution' has really become subject to a deplorable process of word-inflation over time. Nevertheless, in view of the indisputable fact that with the emergence of modern science something arrived upon the European stage which had never been there before (nor upon any non-European stage), and in view of the tremendous consequences of this arrival, the admittedly somewhat worn-out term 'revolution' seems in this case to us still useful and even indispensable.

understood as the world of qualitative things that intuition presents to us, but as mathematical entities and relations; finalist motives are excluded from operative action so that causality becomes purely efficient; qualities, except insofar as these are reducible to geometrization, are banished to subjectivity. Mathematical idealizations are the only 'true' world, the only 'objective' world.[18] So far Burtt in 1924, and it seems indeed evident that modern science [*19] is not simply a further development of the naive and 'natural' world-picture suggested by the *sensus communis*, but rather its radical overthrowing. This goes for heliocentrism as well as for the new conception of motion, as it is much more natural to assume that bodies keep moving because a force presses upon them or draws them, than to assume, in accordance with the principle of inertia, that bodies keep moving until something intervenes to stop them or deflect their course. Viewed in this perspective, we have here just another manifestation of the eternal human desire to *know*, i.e., of that distrust of simple sensual evidence which is the prerogative of an animal species characterized by, among other things, the ability to put itself at a certain distance from the world that surrounds it. It is not therefore so strange as it might seem that Patočka seeks to link modern science to Eleatism, [†20] or that Burtt and Koyré are eager to stress Platonic influences in the creators of the modern scientific world-view — not only in Copernicus and Kepler (which is not so surprising), but even in Galileo.

Of course they are aware that we are dealing here with a 'Platonism of sorts', so to speak, precisely because "geometry becomes real", i.e., because modern science is characterized by the fact that "celestial precision has descended on earth" ("la précision du ciel est descendu sur la Terre"), in Koyré's beautiful phrase.[‡21] But is not the implication that Platonic influences are being somewhat overemphasized here? After all, in another passage we find Koyré summing up the "destruction of the cosmos" (i.e., of the idea of a finite and hierarchically ordered world) and the "geometrization of space" (i.e., the substitution of the homogeneous and abstract space of Euclidean geometry for the qualitatively differentiated and concrete cosmic space of pre-Galilean physics) as the two principal characteristics of the

[*] The term 'modern science' is used here as an overall label for the whole development which starts with the Scientific Revolution. For our purpose this seems justified. Otherwise it is indeed desirable to distinguish between the science of Kepler, Galileo, and Newton, and science since Einstein and Bohr, which may be called with Dijksterhuis 'classical' (to be distinguished from the 'ancient' of the Greeks) and 'modern', respectively; or, with Cohen, 'early modern' and 'modern', respectively.

[†] So did Dijksterhuis, in spite of his own acknowledgement that it is almost unfeasible for a philosophy which so radically denies the reality of the empirical world ever to provide any inspiration for a science concerned precisely with this empirical world. Yet in the conservation laws and in the idea of causality, which Dijksterhuis (just like Meyerson) took to be, in the last resort, reduction to identity, he recognized the same desire to trace the unchangeable behind all change.

[‡] To the Greeks, only in the celestial sphere was absolutely and perfectly regular motion possible, so that only mathematical *astronomy* was possible, not mathematical *physics*.

intellectual revolution of the 17th century. This, in its turn, means that, henceforth, all things in the universe are on the same ontological level; there is no longer an opposition between Heaven and Earth, and astronomy and physics are united.[22] Hence, as Koyré puts it, "all considerations of value, perfect harmony, meaning and design are banished from science", which is the same conclusion we have seen Burtt arrive at in somewhat other words. Now this is precisely the fundamental break in man's experience of the world of which Patočka spoke; and it looks rather un-Platonic.

Kepler's indubitable, and Galileo's possible, conceptions of themselves as spiritual heirs of Plato may well therefore be less relevant than they seem. Man is a conservative being — even the perpetrators of a genuine revolution tend to feel assured about the legitimacy of their enterprise only if they can present it as a *return* to a position still older than that of their immediate predecessors and which, unfortunately, has become submerged over time. One decisive difference between the Platonist position and Galileo's enterprise is, after all, that the latter is about a new conception of *motion*, which means that in Platonic terms we are concerned here with an analysis of Becoming rather than with a contemplation of Being (just as, in mathematics, new forms like analytical geometry and the infinitesimal calculus were soon to be developed, by Descartes, Newton, and Leibniz). For another, no less decisive difference, the claim that the ideal world postulated has some bearing upon our everyday world demands *experiment*, in the sense of creating an artificial situation in which abstract, mathematically formulated law becomes empirically true (or nearly so). Koyré surely recognized the role of experiment as an essential component of modern science (i.e., of experiment as opposed to the spontaneous experience of common sense characteristic of Aristotelianism). However, he was always anxious to stress its secondary nature, in that experiment presupposes a new *language* in which to phrase questions and a new *dictionary* enabling us to read and interpret the answers. This language is the one of mathematics or, more precisely, of geometry. For, although Pythagoras already had declared number to be the essence of things, and although Scripture says that God founded the world upon number, weight, and measure, this (so Koyré insists) had never been taken seriously before Galileo.[*23] Prior to Galileo, then, humanity lived in the "world of the more or less", entering from then the "universe of precision" first revealed to us by Galileo's mathematical physics. All this, in his view, is why Greek technics and even medieval technics, in spite of the admirable achievements of the latter especially, never transcended the boundary of

---

[*] It is perhaps interesting to note that Dijksterhuis, in an earlier work of his, used almost the same wording in order to point out that the *Archimedean* influence did not fully reveal itself until the advent of Galileo. Indeed, the Alexandrian school of Greek science (among which Archimedes must be reckoned in the intellectual, though not of course in the strictly geographical sense) may be said to have used mathematics very well as (in Koyré's words) a "method for the study of sublunar phenomena" and to provide therefore at least one counter-example to Koyré's thesis that it was only in modern times that geometry became *real*. Yet it may be said that only in modern science has this trend become *dominant*.

τεχνη (*technè.*). Adopting here Mumford's vocabulary,[24] he says that it is only in the wake of a turning of ´επιστημη (*epistèmè)* towards τεχνη that the "eotechnical" machine has been transformed into the modern, "paleotechnical" machine.* For example, the Middle Ages and the Renaissance era, and even Antiquity, possessed optical instruments, yet the microscope made its appearance only at the beginning of the 17th century, even though (at least according to Koyré) it had been technically possible, other than with the telescope, to construct it already in the 13th century. But, so Koyré continues, the *idea* behind it was lacking. For example, spectacles, invented in the 13th century, were only a "tool" (*outil)*, not an "optical instrument" (*instrument optique*).[25] Now all this may be perfectly true, yet the question remains: *why* was this idea lacking and *why* was the appropriate language not formulated? Not because people were too stupid, but evidently because they were content with "tools" and with "roughly" (*à-peu-près*), whereas Galileo's generation no longer was.

Why was that generation no longer content with it? A simple question at first sight, but in fact an intricate one, if only because it must be dealt with on three distinct levels. On the first level we must address a preliminary question — provided that it does indeed make sense to join Burtt in speaking of the "metaphysical foundations of modern physical science": what exactly do we mean in this context by 'metaphysics'? After all, in view of the indisputable fact that science 'works' and that, as a consequence, our mastery over nature has increased in a spectacular way — in view of the fact, in short, that we simply *know* more about nature than before — positivism may to a certain extent perhaps be forgiven its failure to recognize any metaphysical element in modern science at all. Why (a positivist might ask) suggest that, in spite of everything, science is under the spell of new metaphysical presuppositions, whereas in fact we are dealing here with nothing but a new *method* which, given its overwhelming success, has proven to be the right one?

That question may be put less crudely. Take the way in which the French historian of Western philosophy, Revel, has subjected Cartesian metaphysics (indeed: Cartesian *metaphysics*, for here, and with pre-Kantian aprioristic systems in general, no one will deny that we are dealing with metaphysics, even if, at least in intention, science-based metaphysics) to a rather devastating critique.[26] Revel conceives of Descartes' entire physics as an extension of his theology (a 'theology of sorts', no doubt, but still ...) in that it is on the basis of the assumption of divine immutability that Descartes proclaims the principle of the conservation of motion, and on the basis of the assumption that God has endowed matter with motion and rest and conserves as much of both in it as He has put into it at creation that he decrees everything in the world to happen "by figure and by motion". Moreover, this geometrism may renounce experiment, for, if matter is nothing but 'extension',

* "Eotechnical" refers to pre-industrial technics; "paleotechnical" to what is usually called the 'First Industrial Revolution'. The technology of the 'Second Industrial Revolution', which was the first to be truly science-based, is called by Mumford "neotechnical".

one can turn physics into a deductive *a priori* science like geometry. And since all reality that is not purely spiritual can be reduced to geometry, life is materialized (or rather, geometrized and mechanized) by Descartes — yet not out of 'materialism', but out of the old prejudice of the immense chasm between man and animal. In short: "Descartes, useless and uncertain" ("Descartes, inutile et incertain"), as Pascal had already noticed!*[27] Now it might be argued against Revel that Descartes (together and in rivalry with Beeckman and Gassendi, to be sure) was among the first to provide an elaborate mechanicist picture of the world, based on corpuscularism, which at least makes him important as a historical episode (apart, of course, from his very great merits as a mathematician). But here precisely is the crux. For, in spite of the circumstance that Descartes was a much more accomplished mathematician than Galileo, there was in fact little that was particularly mathematical in the way he presented his mechanicist universe beyond the generally geometrical nature of these 'primary properties' of figure, size, and motion. Descartes treated those corpuscles in motion which are the building-blocks of his universe in a wholly qualitative way; a circumstance which has recently provoked the nasty remark that "the greatest effective merit of his corpuscular view of the world would be to acquaint the world at large with the new science in the guise of a poor man's universe of precision".[28]

The moral of the story seems to be that one should distinguish between, on the one hand, what might be called 'surface metaphysics', i.e., philosophies of nature which try to restore in an up-to-date, would-be scientific fashion an all-embracing world-view and which are therefore almost by definition ephemeral;† and, on the other hand, the underlying and abiding 'depth metaphysics', which is, indeed, Koyré's 'universe of precision'. After all, science may no longer be 'mechanicist' in our day, but it is more mathematical than ever. But, so one might object, can this 'depth metaphysics' still be called 'metaphysics' at all? It must certainly be admitted that, at the very least, we are concerned here with a 'metaphysics of sorts' ... in view of the circumstance *that for the first time in the history of metaphysics the postulated 'true' world is no longer an object of awe and reverence*! Yet, as already hinted at in the above, we believe that it does make sense to speak of 'metaphysics' here. For, although the most striking aspect of modern science resides in its emancipation from philosophy, i.e., in its being no longer embedded in an all-encompassing world-view, it is nonetheless true that the 'material' world, the 'world of fact', to which it confines its explorations, has preciously little to do

* In his stressing the role of experience and experiment Pascal has more in common with the British empiricist school (which did not become fashionable in France until the 18th century under the belated influence of Newtonian science) than with continental rationalism; however, his dominant motive (namely, his religious preoccupations) of course separates him greatly from the former.

† Cartesianism, to confine ourselves for the moment to this example, was already made definitely obsolete by Newtonian physics with its reintroduction of the concept of force. But, as we shall see in Chapter VIII, p. 215, the same is true of later non-mechanicist philosophies of nature insofar as these pretend to be based upon science.

with the world of everyday experience.* Or, to use a distinction made long ago by the French philosopher and (not by chance) engineer, Georges Sorel: science does not present us with "natural nature"; it creates an "artificial nature" instead. This is a world, not of theoretical abstractions alone (even though these provide its basis) but also of scientific instruments, which ensure that, e.g., a feather falls down to earth in the same time — or almost — as a stone, thanks to the created vacuum which does not exist in "natural nature". And since this "artificial nature" pretends to reveal to us 'true reality' and thus shares with metaphysics (in the customary sense of the term) the depreciation of everyday sense-perception we addressed at the beginning of this chapter, one may rightly join Sorel in speaking of the "metaphysical preoccupations of modern physicists" (even though we are concerned here with metaphysical preoccupations that most often remain unconscious, as positivism involuntarily demonstrates).†[29]

'Tell me what your metaphysical preoccupations are ...' Indeed, we have by now arrived at our second question (with which we leave the strictly philosophical level and enter what might be called the 'cultural–psychological' level): 'On the soil of what kind of mentality could such metaphysical preoccupations arise and, eventually, become predominant?' Perhaps a clue is indeed to be found in the preoccupations of a 19th-century philosopher who, with his concept of "will-to-power", has been said by a 20th-century philosopher to have brought to completion (and therefore terminated) the history of Western metaphysics.‡[30] In

---

* Or, to adopt the expression Patočka derived from his phenomenologist intellectual environment, with the "life-world" (*Lebenswelt*).

† According to this view scientific laws are accounts of the actions of man-made tools into which a part of nature has been incorporated; and since there is no reason to suppose that such tools are models of nature's hidden mechanisms, Sorel arrives at the conclusion (surprising for naive positivists, who indeed were his *bêtes noires*) that there is no determinism (i.e. *order*) whatsoever in nature left to herself ("natural nature"). This seemed to be confirmed by the second law of thermodynamics with its announcement of a tendency towards increasing entropy. Sorel's entire philosophy of culture sees culture as the ceaseless human endeavour to create islands of order (doomed to perish in the long run) in the arbitrariness of nature. A thoroughgoing study of this interesting figure is to be found in work of the Dutch essayist J. de Kadt, to which I am much indebted.

‡ This is the conclusion drawn by Heidegger in his two-volume study of Nietzsche. Elsewhere, too, Heidegger points to the central position 'Will' occupies in Western philosophy, e.g., in his essay 'Wozu Dichter?' (an allusion to Hölderlin's elegy 'Brot und Wein', where the poet asks: "Wozu Dichter in dürftiger Zeit?"). Here it is stated that modern man may rightly be called "willing man". One might surely object (we shall return to it in our final chapter) that the whole philosophical development from Descartes to Nietzsche is all too easily brought together by Heidegger under the one denominator of a voluntarist subjectivism, in which no difference is made between 'thinking' and 'willing' subject, and 'will' is moreover identified with 'will-to-power' (which, according to Heidegger, is in fact "will-to-will", since power is the essence of will). What made Heidegger do so is evidently his general opposition to modern civilization. The dubious consequences of this generalizing attitude become clear when we remember that in the

this context we want to draw attention to the same 20th-century philosopher's famous lecture on "the query after technics" (an exposition that is, in many ways, fascinating until the moment — at around page 23 — when the 'profundity' that 'has never elucidated anything'* and the corresponding linguistic fornication definitely gain the upper hand). Here we meet again the likes of Sorel's 'instrumentalist' view of science as well as Koyré's conception of the Industrial Revolution as a turning of ʹεπιστημη (*epistèmè*) towards τεχνη (*technè*). Only, in the latter case Heidegger's conception is almost diametrically opposite.†[31] To Heidegger's mind the big question to be asked is: What is it in modern technics that has enabled it to make use of exact science? And the answer, according to Heidegger, must be that modern technics, just as science, is a form of "unhiding"; only, the unhiding that prevails in modern technics is not a "bringing forth" in the sense of ποιησις (*poièsis),* but a "claiming" (*Herausfordern*) compelling nature to supply energy which can be transformed, stored, and distributed as such.‡ Such specific human conduct towards nature is displayed for the first time in the rise of modern science, its mode of representing being directed towards nature as a calculable complex of forces. In other words, modern physics is experimental, not because it uses machines for the interrogation of nature, but rather the other way round: experimentation comes in because physics as pure theory compels nature to present itself as a complex of forces calculable in advance, and it does so in order to provide information on whether and how nature reports itself.

Now one might raise two objections here. First, although the 'universe of precision' certainly is about an 'artificial nature' (which indeed suggests that the 'depth metaphysics' we believe to underlie the Scientific Revolution has on the cultural–psychological level something to do with 'will-to-power'), this does not in the least exclude the fact that a disinterested search for truth was the foremost guiding motive of the pioneers of the Scientific Revolution; for, against a

---

political sphere it induced him to lump together in the same sweeping way Western liberalism and Russian communism as having the same metaphysical background (to say nothing of his, at least temporarily, hailing of Nazism as a sound alternative; cp. Victor Farias' well-known study).

* This is an allusion to Arthur Schnitzler's well-known distich: "Tiefsinn hat nie ein Ding erhellt / Klarsinn schaut tiefer in die Welt".

† Heidegger insists that the Greek word τεχνη extended beyond the strictly artisanal sphere, not only because it included equally the artistic field (and was therefore part of ποιησις) but also because already since Plato it had been connected with ʹεπιστημη. This could be so, because the essence of technics does not reside in making and handling, but in the "unhiding" of that which does not generate itself ("unhiding" — "Entbergen" — being Heidegger's favourite (and in fairness, I must add, quite beautiful, for once) etymological translation of Greek ʹα-ληθεια (*a-lètheia*)).

‡ In this connection Heidegger compares an old wooden bridge built into the river, with a modern power station. Here it is rather the river which is built into the power-station, for it owes its present nature as a river (namely, its being a supplier of hydraulic pressure) only to the nature of the power-station.

fashionable, contemporary relativism (of which Feyerabend may be regarded as the pioneer, and which assumes caricature-like guises in some forms of sociology of knowledge wherein science is refused any claim to objective truth) it must be repeated that, indeed, we *know* more of nature than before the rise of modern science!*[32] Second, there is a little chronological problem in Heidegger's account: for, after all, history shows that, once modern science was born, almost two hundred years still had to pass until modern, science-based technology, which transforms the big world of everyday life, came into being. To be sure, Heidegger anticipated this objection by insisting that that which is "*historisch*" later was nonetheless "*geschichtlich*" earlier, since the essence of modern technics is already implicit in modern physics. Now this subtle difference is hardly translatable into English. A possibly suitable alternative might be to fall back upon the medical metaphor of an 'incubation period' (which, as a metaphor, should certainly not be stretched too far, since — unlike Heidegger probably — we wish by no means to suggest that modern science is a contagious illness). In this view, the Industrial Revolution † appears as the eventual breakthrough of a 'will-to-power' attitude towards nature, already prefigured by the mathematical reductionism of the Scientific Revolution.

We may now pursue the metaphor in the other direction as well, by asking: 'When was it that the European organism became susceptible to this infection?' This, then, is our third question, and with it we have reached the historical level. The question, if taken in the strict sense, cannot be answered, as it is of course impossible (or arbitrary at the very least) to fix any specific date here; yet there may be said to exist something close to a consensus among historians that Europe's response to the challenge presented by its profound economic as well as spiritual crisis in the 14th century is to be regarded as a decisive turning-point.‡[33]

---

* It has been rightly observed that the quantitative method of modern science meant a decisive step on the road to ever greater objectivation, since in this new symbolism we are no longer concerned with *words* (with their always ambiguous meanings), but with numbers whose meaning is defined by the position they occupy in the full numerical system. *For scientific purposes* it is therefore superior to the symbolism of speech; in Ernst Cassirer's words: "... for what we find are no longer detached words but terms that proceed according to one and the same fundamental plan and that, therefore, show us a clear and definite structural law." Perhaps it is the most striking point of the adventure of modern science that, in it, two basic human tendencies, the will to power and the will to knowledge (indeed a 'will', too!), for the first time in history came together.

† To be more precise: the Second Industrial Revolution (Mumford's "neotechnical age"), since, in spite of the steam-engine and, with it, the beginning of the penetration of technics by science, the technics of the First Industrial Revolution (the "paleotechnical age") cannot yet be characterized as 'technology', i.e. as science-based technics.

‡ "Turning point"; for an inherent dynamism in European society and a resulting tendency towards expansion can certainly be traced back to the period *before* the 14th-century crisis, as has been demonstrated by (among others) Lynn White. He stresses the importance of the agricultural revolution of the early Middle Ages since, from the 10th century on, it permitted rapid urbanization and with it the advent of a new way of life which, in its turn,

This response took shape as a 'flight forward' towards an unprecedented expansion beyond its original boundaries, resulting in the rise of what Immanuel Wallerstein has called "the modern world-system".[34] It is true that, just as with science and technology, there is a time-lag here, too, even if it is in the opposite direction, in that the aspect of this process which speaks most to the imagination (namely, the great adventure of the Voyages of Discovery) certainly took place long before the Scientific Revolution.[*35] Yet these Voyages of Discovery were the first realization of the *plus ultra*, which was the motto of Francis Bacon's *Novum Organon*, on the frontispiece of which, not by chance, the pillars of Hercules are depicted.[†] In other words, this was the first great breakthrough of a certain direction of mind, or mentality, or spiritual climate, or whatever one wishes to call it, out of which the Scientific Revolution in its turn could arise and, above all, could be hailed for the promise it seemed to hold for a new world (even if it was to take some time before that promise could be realized). This is, of course, not to deny that this revolution was itself the *truly* decisive event in the history of the West (as we have, indeed, ourselves asserted at the beginning of this chapter). For, however spectacular the European penetration of the world was in terms of space covered, with the sole exception of the Americas that penetration really remained a rather marginal affair (in Asia and even more so in Africa) until the 19th century, which was the end of the 'incubation period'. But for the rise of modern science and, in its long wake, modern technology, the 'modern world-system' might well therefore have gone the way of all flesh.

One other qualification is needed when we speak of *European* expansion. 'Which Europe?' it may be asked. For Wallerstein's "long 16th century", ranging

---

germinated important medieval progress made in the mechanical exploitation of natural forces for human purposes. Likewise, the great pioneer in the study of Chinese science, Joseph Needham pointed to the inherent instability of Europe as compared to China's "spontaneous homeostasis". He ascribed this in the first place to Europe's geographical predicament, which tended from the beginning to turn it into "a culture of rovers". Fernand Braudel rather follows Marx in underscoring the 13th-century origins of capitalism in the Italian city states, which, according to him, already by then had given rise to a kind of European-centred world economy. At all events, however much Braudel sought (against Wallerstein) to play down the point, not even he could deny that 'capitalism' acquired a new dimension with the unprecedented European expansion which began in the last decade of the 15th century.

[*] The Dutch historian of science, R. Hooykaas, although of course recognizing that the Scientific Revolution took place much later than the Voyages of Discovery, was nevertheless eager to point to the importance of these sailors being confronted with a host of new phenomena considered impossible up until then, because this fostered a respect for empirical facts and a critical stance towards authorities which was to be essential for the later development.

[†] In this sense, Francis Bacon, who contributed nothing to science in the strict sense, and who quite misunderstood its mathematical nature, may nonetheless be regarded as its prophet. We come back to the 'Baconian' tradition in science in Chapter V, p. 122.

from about 1450 to 1650, equally saw the disappearance of the religious unity of Europe, as well as the rise of the modern state-system, i.e., of a plurality of modern 'sovereign' states which no longer, even in theory, recognized any superior or common tie, and were engaged in much mutual rivalry. A rivalry, to be more precise, between the states of Atlantic Europe, since the Mediterranean was no longer the 'world sea', and Germany (the remainder of the Holy Roman Empire, which had once been the embodiment of the idea of a universal monarchy) sank away in the devastations of the Thirty Years' War. This war, itself a result of the religious split which challenged the universal claims of Roman Catholicism, was the wicked fairy that sent Germany into its princess' sleep. For how long was that sleep to endure, and what prince was to kiss the German lands awake?

Late in the 18th century, Germany became an example of the Gospel's saying that "the last will be the first". With the rise of the Enlightenment, which was itself inspired by the 17th-century Scientific Revolution, a new universal idea arrived upon the European stage. With England wrapped up in the practical and utilitarian adventure of the Industrial Revolution, and with France similarly engaged in its radical attempt to create a new society inspired by the American Declaration of Independence, which it hailed as the first triumph of Enlightenment philosophy in practice, it remained for Germany, where the Enlightenment ideas could find no other 'practical' embodiment than 'enlightened despotism', to turn towards a deeper 'theoretical' reflection on the spirit and essence of the Enlightenment. This movement of reflection might well have led to a complete rupture with the Enlightenment rather than to its critical deepening, as indeed it was to do to a large extent with Romanticism and with some of its precursors like Herder. However, this is certainly *not* what happened with the two greatest representatives of the impressive intellectual upsurge in 18th-century Germany: Kant and Goethe. It is towards these twin stars that we shall now direct our attention for a while

b. Kant and Goethe

Kant's philosophy has often been described as the synthesis of continental rationalism and British empiricism; and there is nothing to be said against this for as long as one keeps in mind that the way in which this synthesis was attained by Kant was not some kind of gentleman's agreement, but a veritable revolution leading philosophy towards wholly new paths. What was the essence of this 'synthesis'?

This has often been said to reside in a turn towards the subject, but that is not very satisfactory. Descartes' *cogito*, after all, is nothing else. A better entry may be found by recalling that both rationalism and empiricism (each in its own way, to be sure) still clung to the traditional notion of truth as *adaequatio rei et intellectus* (the adequate correspondence between things and the intellect). The difference is of course that rationalism held the thinking subject capable of producing such harmony from within, by means of purely logical, deductive operations, whereas according to empiricism it enters the mind from the outside. The latter view may lead to a materialist 'copying'-theory (to be found in its most unsubtle form in Lenin's *Materialism and Empiriocriticism*), which, it is true, can be traced back to Locke. Yet, on the other hand, the Locke–Berkeley–Hume lineage shows that the empiricist version of *adaequatio* bears in itself "the seeds of its own decay". For empiricism, in taking seriously Descartes' prescription that one should proceed from the simplest given elements, and in identifying these with sense-data, was bound (at least in consistent minds) sooner or later to arrive at the painful question: 'What is it in sense-experience that justifies our belief in such notions as 'substance' — to which Berkeley still clung as far as 'spiritual' substances were concerned — or 'causality'?'

As is well known, it was Hume's scepticism over the latter issue which awakened Kant from his "dogmatic slumbers". Bertrand Russell once gave this Kantian statement an ironic, though wholly unjustified, twist in commenting that "the awakening was only temporary, for he soon invented a soporific which enabled him to sleep again".[36] Kant really meant those "dogmatic slumbers" of his to refer to his having taken for granted, so far, the continental–rationalist approach (not in its Cartesian, but rather in its Leibnizian–Wolffian guise, to be sure). Yet already in his 'pre-critical' period (i.e., in his studies written before c. 1770) there are signs of an increasing distrust of rationalist pretensions. If certainty could be attained that way, how then could it be explained that everybody had his own system? Mathematics and physical science (in which he was well-versed; one need only think of the Kant–Laplace theory on the coming into being of the physical universe) evidently had been more successful in becoming fully fledged sciences than metaphysics had; yet this very success seemed to argue against Hume's scepticism as well. Hence Kant's 'Copernican Revolution', the essence of which may be said to consist of its *constructivist* notion of truth. This means that truth is a property not of reality, but of our *judgements* about it, which are true if confirmed by *experiment*. To illustrate this, we quote at some length a deservedly famous passage in the preface to the second edition of the *Critique of Pure Reason*, which may at the same time be regarded as one of the classic attempts to define the essence of what is now (as it was not in Kant's time) called the Scientific Revolution:

"When *Galileo* made balls run down an oblique plane with self-chosen gravity, or when *Torricelli* made the air bear a weight preconceived by him as equal to that of a known column of water, or when, later still, *Stahl* changed metals into lime and these back again into metal by withdrawing something from them and then returning it: then a light flashed upon all students of nature. They understood that Reason conceives only of what it produces itself in accordance with its own project; that, using principles of its own judgement, it must proceed in accordance with constant laws, and that it must compel Nature to answer its questions instead of allowing itself to be kept by Nature on the leash. For if not, observations made accidentally and without a premeditated plan fail completely to be connected in a necessary law, such as Reason after all seeks for and is in need of. Reason must approach Nature having in one hand its own principles, which alone justify our turning corresponding phenomena into laws, and carrying in the other hand experiment, conceived after the latter. As a result, Reason certainly learns from Nature, but not as a schoolboy to whom the teacher dictates whatever he likes, but as an ordained judge, who compels the witnesses to answer the questions he puts before them ... This is how physics has for the first time been enabled to reach the highway of science, after so many centuries of having produced nothing but a mere groping in the dark".[*37]

On the one hand this implies an unconditional acceptance of the scientific world-picture such as had by then culminated in the Newtonian synthesis, but on the other hand it implies the recognition that this is, indeed, a *picture* — not in the sense of a 'copy', to be sure, but rather in the sense of a 'model'.[†] And this means that the 'true world' of mechanical–mathematical science is indeed a true world insofar as (unlike the world of immediate sense-experience) it is an *objective* world, which is to say that its propositions are valid universally, not just for me as an individual. Yet, paradoxically, it is precisely *because of this objectivity* (since there is no object without a subject, even if this 'subject' is a 'subject of sorts', as we shall see) that it is no longer the 'real' true world, so to speak. *This* world, the world of the "Things-in-themselves" (*Dinge an sich*) constitutes what Kant calls the "noumenal" realm; and this realm is not an object of theoretical knowledge, since the categories of Understanding (which constitute the *a priori* elements of science) find their legitimate application only in what Kant calls the "phenomenal" realm.

---

[*] "Als *Galilei* seine Kugeln die schiefe Fläche mit einer von ihm selbst gewählten Schwere herabrollen, oder *Torricelli* die Luft ein Gewicht, was er sich zum voraus dem einer ihm bekannten Wassersäule gleich gedacht hatte, tragen ließ, oder in noch späterer Zeit *Stahl* Metalle in Kalk und diese wiederum in Metall verwandelte, indem er ihnen etwas entzog und wiedergab; da ging allen Naturforschern ein Licht auf. Sie begriffen, daß die Vernunft nur das einsieht, was sie selbst nach ihrem Entwurfe hervorbringt, daß sie mit Prinzipien ihrer Urteile nach beständigen Gesetzen vorangehen und die Natur nötigen müsse auf ihre Fragen zu antworten, nicht aber sich von ihr allein gleichsam am Leitband gängeln lassen müsse; denn sonst hängen zufällige, nach keinem vorher entworfenen Plane gemachte Beobachtungen gar nicht in einem notwendigen Gesetze zusammen, welches doch die Vernunft sucht und bedarf. Die Vernunft muß mit ihren Prinzipien, nach denen allein übereinkommende Erscheinungen für Gesetze gelten können, in einer Hand, und mit dem Experiment, das sie nach jenen ausdachte, in der anderen, an die Natur gehen, zwar um von ihr belehrt zu werden, aber nicht in der Qualität eines Schülers, der sich alles vorsagen läßt, was der Lehrer will, sondern eines bestallten Richters, der die Zeugen nötigt, auf die Fragen zu antworten, die er ihnen vorlegt ... Hierdurch ist die Naturwissenschaft allererst in den sichern Gang einer Wissenschaft gebracht worden, da sie so viele Jahrhunderte durch nichts weiter als ein bloßes Herumtappen gewesen war."

[†] One might object that the primacy of a constructivist notion of truth belongs, not to Kant, but rather to Giambattista Vico with his *verum factum*. But for Vico this was precisely a reason to grant true scientific status, not to natural science, but to History (the latter being indeed a creation of man, 'nature', however, being a creation of God). Cf. on this subject also our Chapter V, p. 130.

Now *this* realm in its turn is marked by a dual aspect, which stands expressed in Kant's repeated assertion that our knowledge possesses two stems, namely, *receptivity* of impressions and *spontaneity* of concepts. For essential to Kant's constructivism is his recognition of "the given" as something in its own right (unlike traditional rationalism, which treats sense-knowledge as confused rational knowledge). At the same time, unlike in traditional empiricism, he distinguishes between "sensation" (*Empfindung*), "intuition" (*Anschauung*), and "experience" (*Erfahrung*). For the first is only the material aspect of the second, the formal aspect of which are the *a priori* forms of intuition (space and time); and between the second and the third the apparatus of the *a priori* categories of understanding is to be found. These categories presuppose indeed the synthesizing activity of the subject (no object without a subject); only, this subject is no longer Descartes' *metaphysical* "thinking substance", nor Hume's *psychological* "bundle of perceptions". It is a "*transcendental* I" or a "*transcendental* subject", i.e., it is nothing but the necessary correlate of the diverse "forms of judgement" from which, in their turn, those "categories" can be deduced ("judging" being the essence of "thinking").

So *Kant does not in fact deduce the possibility of science from the certain existence of a thinking subject; he rather deduces the existence of the latter from the existence of the former*; this being the true meaning of his 'Copernican Revolution'! The actual implication is a *cleavage* of the subject: a "transcendental I" perceives an "empirical I", the latter being what we usually call an individual.* But the former, which is the very condition of all objectivation, cannot itself be objectivated; hence Kant's critique of "rational psychology", and, at the same time, of two other pseudo-sciences: "rational cosmology" and "rational theology". For, as has been suggested already in the preceding remarks on the phenomenal and the noumenal realms, the transcendental presuppositions of science point equally towards its own *limits* and to the *boundaries* of "pure" reason in general.† Yet there is a fundamental tendency in human reason to trespass those boundaries; hence the "Ideas of Reason", which, however, are solely legitimate either in a *regulative* sense in the domain of Theoretical Reason, or as "postulates of Practical Reason". Thus "Reason", too, is split up, to wit, into "theoretical" and "practical"

---

* Kant may really be said to distinguish between *three* "I"s (the "intelligible I" being the third one). For, although the distinction — made in the *Critique of Practical Reason* — between "empirical character" and "intelligible character" runs closely parallel with that between the "empirical" and the "transcendental" subject of the *Critique of Pure Reason*, the "intelligible character" as the bearer of man's deepest nature as a *moral being* (and as such belonging to the "noumenal" realm), cannot be identified with the purely formal "transcendental unity of apperception", which, although (being the condition of any phenomenon) it cannot be a "phenomenon" itself, is not a "noumenon" either; in fact it 'is' nothing. We shall discuss the "intelligible I" at some length in chapters VII and VIII; for now it suffices to confine ourselves to the two mentioned first.

† More on the distinction made by Kant between 'limits' (*Schranken*) and 'boundaries' (*Grenzen*) in our Chapter VI, p. 136.

Reason, the first being the domain of *necessity* and the second of *freedom*. Primacy belongs to the second, for it should never be forgotten that, for Kant, the problem of freedom was the central problem of philosophy — *his fundamental preoccupation was an ethical one*!

Still, in both cases Reason appears as *lawgiver*. This goes without saying for 'practical' reason, which is the domain of freedom, precisely because Reason is lawgiver to the *inner* senses: to passions, desires, inclinations, to which Man is not therefore, like the animals, helplessly exposed. But it is also true of 'theoretical' reason, which is the domain of necessity, although in this case indirectly in that theoretical reason is split up once more into "Understanding" (*Verstand*) and "Reason" proper (*Vernunft*). It is the former which (though under the ultimate inspiration of the Principle of Reason) is the direct lawgiver to the *outer* senses, i.e., to the "turmoil of the senses" ("das Gewühl der Sinne") which is processed by the concepts of Understanding into true, i.e., scientific, cognition.* For, although it may be said (as we did in the above) that Kant vindicated sensual knowledge, and that his 'receptivity' meant a good deal of respect for 'the given', it is nonetheless clear that for his constructivist mode of thought the relationship between rational concepts and sensual intuitions was, in the last resort, one of subordination, with Understanding playing the role of the "ordained judge"! This in spite of his assurance that neither of the two is to be preferred to the other, as intuitions without concepts are just as "blind" as concepts without intuitions are "empty".[38] Further, although it may likewise be said that Kant's 'Transcendental Aesthetic', with its doctrine of 'pure intuition', saved the unity of the senses from the atomism of British empiricism, it did so by linking it to the transcendental conditions of pure mathematics which, by themselves, serve only as an introduction to the transcendental investigation of the conditions of natural science carried out in the 'Transcendental Analytic'. So one might well ask whether the senses have no other job than to provide raw materials, fit only to be ground in the mill of Newtonian science?

This, then, was the fundamental problem for Goethe who, about the same time that Kant entered the stage as the discoverer of 'artificial nature', came forward as the champion of 'natural nature' and of what Bergson was later to call "expérience intégral" (i.e., a kind of empiricism which does not — like traditional empiricism — proceed by 'association' but rather by 'dissociation' since it sees the world as it presents itself to our *Anschauung* not as a jig-saw-puzzle composed of atomic ultimate elements — 'sense data' — but rather as the unfurling into colourful

---

* The 'categories' of Understanding are thus *constitutive* for science (in that they may indeed be said to prescribe laws to nature), whereas the 'ideas' of Reason have (in the domain of 'theoretical' reason!) only the *regulative* function of inspiring the search for utmost unity and totality in the knowledge thus obtained. In our Chapter IV we shall see that one of the essential differences between Kant and Schopenhauer consists of the latter's fundamentally different usage of the terms 'Reason' and 'Understanding'.

diversity of an original unity). His position, as distinct from Kant's, implies a rejection of the 'mechanization of the world-picture' and a call for an alternative science. This becomes particularly clear in his theory of colours, with its outspoken stance against Newton, as we shall see in some detail in Chapter III. Here, already, we point to his illuminating complaint that Newton, in his *experimentum crucis*, truly crucified nature. He had put Nature on the rack to compel her to confess to what the investigator had already decided upon beforehand. "Only, Nature is a steadfast and noble person who, under whatever pains, stands by Truth. If it is stated otherwise in the protocol, then the inquisitor has misheard or the clerk has miswritten".[*39]

This reminds us of Kant's "ordained judge", even though his role is here being judged somewhat differently ... It was not indeed by chance that young Schopenhauer, as we quoted him in the Introduction, praised Goethe for the "contemplation" he found wanting in Kant. For, according to Goethe, it is not by violating Nature through experiment, or (which amounts to the same thing) through the strait-jacket of mathematical constructions that we are able to approach Nature's inner core. Rather, we do so by intuitively grasping the 'primordial phenomena' (*Urphänomene*) which provide for unity amidst all variability, and for eternal recurrence amidst the myriads of changing circumstances. In the Introduction to his *Theory of Colours*, Goethe "believe[s] to have deserved the gratitude of the philosophers for striving to pursue the phenomena up to their ultimate sources, up to the point where they simply appear and are there — where nothing in them is susceptible to further explanation".[†40]

In the organic sphere Goethe prefers to speak of "prototypes" (*Urtype*) — the ideal types to which the anatomical particularities of each species are to be compared. Originally he believed that the primordial plant (the *Urpflanze*), of which all existing plants are "metamorphoses", was still to be found in nature; he even searched for it in Sicily. In his famous conversation with Schiller on the occasion of a meeting of the Society for Natural Studies at Jena, Goethe gave an exposition of his theory of metamorphosis and, "with some characteristic strokes of the pen, conjured up a symbolic plant before his eyes".[‡] But Schiller, who was an infinitely lesser poet but had a better philosophical training than the older literary *confrater* he envied (and wooed) so much, shook his head and said: "This is not experience, this is an Idea";[§] upon which Goethe retorted that he was very

[*] "Allein die Natur gleicht einer standhaften und edelmüthigen Person, welche selbst unter allen Qualen bei der Wahrheit verharrt. Steht es anders im Protocoll, so hat der Inquisitor falsch gehört, der Schreiber falsch niedergeschrieben."

[†] "Vom Philosophen glauben wir Dank zu verdienen, daß wir gesucht die Phänomene bis ihren Urquellen zu verfolgen, bis dorthin, wo sie bloß erscheinen und sind, und wo sich nichts weiter an ihnen erklären läßt."

[‡] "... und ließ, mit manchen charakteristischen Federstrichen, eine symbolische Pflanze vor seinen Augen entstehen."

[§] "Das ist keine Erfahrung, das ist eine Idee."

pleased to have ideas without knowing it, and even to be able to see them with his own eyes.[41] In the following period of intensive cooperation with Schiller, Goethe learned indeed to correct his original, naive realism and to understand that real 'experience' (i.e., experience with a claim to objective validity) cannot do without 'ideas'; so the *Urpflanze* (to stick to the example) must be seen as a 'model' not as something to be found in nature. This seems to be rather like Kant's position; yet Goethe remained convinced that his *Urphänomene* are at the same time ideal *and* real, intelligible *and* sensible. He trusted his eyes, because he was convinced that every sense-organ is analogous to the phenomenon it perceives. Light and the eye, for example, are complementary entities. As he put it (once again, in the Introduction to his *Theory of Colours*): "The eye owes its existence to Light. Out of a neutral animal auxiliary organ Light calls into being an organ that is to become its equal counterpart; thus the eye forms itself *by* Light *for* Light, in order that the inner light may meet the external light,"*[42] a thesis which he illustrated by means of a German adaptation of Plotinus' well-known words:

Wär' nicht das Auge sonnenhaft,	"If the eye were not sun-like,
Wie könnten wir das Licht erblicken?	how could we ever behold Light?
Lebt' nicht in uns des Gottes eigne Kraft,	And if God's own power did not live in us,
Wie könnt' uns Göttliches entzücken?	how could the Divine ever delight us?"

All this implied a much more radical vindication of sense intuition (*Anschauung*) than Kant had offered. It was nourished by Goethe's pantheist belief in Nature as a living whole, and we may well ask what stood at the origin of this conception of his.

The name Plotinus, just cited, may well offer a clue. It points, of course, towards the Platonic legacy, i.e., to the very body of ideas that Koyré (to mention him once again) persuasively argued has been made use of in two quite distinct guises: the "productive" Platonism of the Scientific Revolution, and its "unproductive" counterpart which led only to speculation.[43] Now Goethe as a youth had already been influenced by Neo-Platonic speculation, or, to be more precise, by what has been called its "magical" variety. This, as distinct from the "mystical" variety assimilated in the end by Christian orthodoxy, "resulted in the heterodoxy of alchemists, hermeticists, and kabbalists."[44] Goethe, in his autobiography *'Fiction and Truth'* (*Dichtung und Wahrheit*), describes the spiritual crisis he underwent after his return from Leipzig in 1768, and his contact with the Pietist circle around Suzanna von Klettenberg, who, incidentally, was to serve as the model of the "beautiful soul" in *Wilhelm Meister*. In this circle alchemy was highly appreciated (which may seem somewhat suspicious from a Christian point of view, but we should not forget that we are concerned here with a mystical, strongly subjectivist, and therefore rather heterodox form of Christianity). Here Goethe indulged for a time in alchemical experiments.[45] Yet

* "Das Auge hat sein Dasein dem Licht zu danken. Aus gleichgültigen thierischen Hülfsorganen ruft sich das Licht ein Organ hervor, das seines Gleichen werde; so bildet sich das Auge am Lichte fürs Licht, damit das innere Licht dem äußeren entgegentrete."

later on, once he had learnt self-restraint, at first from his immersion in practical activities in Weimar, and then from his discovery of the classical ideal of clarity and objectivity in Italy, he became well aware of the danger of the magical "pact with the Devil", which he was to depict so impressively in his *Faust*. Indeed, that is how he moved "from alchemy to science", as Ronald Gray remarks in his instructive study; yet his longing for an *unbroken* world-view ("That I may detect the inmost force / Which binds the world, and guides its course / Its germs, productive powers explore / And rummage in empty words no more") always remained alive in him.[*46] To quote Gray: "Here the distinction between Goethe's science and both alchemy and modern science becomes clear. Modern science has taken over from alchemy only its concern with practical experiment, and has rejected the mystical philosophy on which it was based. Goethe accepted the philosophy and began to develop the respect for empiricism, which, though theoretically essential to alchemy, had been entirely neglected."[47]

"The philosophy", then, was, first and foremost, the Neo-Platonic, Hermeticism-imbued legacy which had been so conspicuous in Renaissance thought but had subsequently been pushed to the background owing to the 'mechanization of the world-picture' (even though it was always to remain alive as an undercurrent).

Indeed, Goethe had already tried to read Plotinus' *Enneads* at the age of fifteen, without, of course, understanding much of it; and, as we have just seen, this theoretical approach had soon given way to the more practical one of alchemical experimentation. But in 1805 he read the *Enneads* once again, with more profit; and he was always to remain in sympathy with it, as he was with the writings of Giordano Bruno. By then, however, the Neo-Platonic influence had for a long time been enriched with the influence exerted by Spinoza, who no less than Bruno believed in one divine substance, and who likewise identified it with Nature ... but who had done so in a much more abstract, 'mechanicist' way. Spinoza was, in fact, a typical representative of 17th century Continental rationalism; small wonder, then, that Goethe's reception of Spinoza was in fact a highly conditional one: "It must not be thought that I endorsed his writings and adhered to their letter," he warns us in his autobiography; for it was of course impossible for the author of *Werther* and *Faust* fully to understand a man "who, as a disciple of Descartes, had reached the summit of thought by means of mathematical and rabbinical erudition".[†48] He nonetheless praised Spinoza in that this philosopher, from his firm conviction that Nature everywhere acts in accordance with the same eternal, necessary, and divine laws, teaches us neither to be bewildered when we observe something reason-like in animals or even in plants (who are not therefore to be

---

[*] "Daß ich erkenne was die Welt /Im Innersten zusammenhält / Schau' alle Wirkenskraft und Samen / Und tu' nicht mehr in Worten kramen."

[†] "Denke man aber nicht, daß ich seine Schriften hätte unterschreiben und mich dazu buchstäblich bekennen mögen ..."; "...der als Schüler von Descartes durch mathematische und rabbinische Cultur sich zu dem Gipfel des Denkens hervorgehoben".

considered as machines!), nor when, by contrast, we see human beings act unreasonably against generally acknowledged moral laws or unwisely against their own interests. Spinoza's work thus exudes an atmosphere of serenity and quiet comprehension of the world; this can help us accept something with which our physical as well as our social life unceasingly confronts us, namely that in the end we must *resign.* True, most people are capable of *partial* resignation; but they resign only in order at once to replace one passion with another, thus in the end arriving at the false, nay blasphemous, maxim that 'all is in vain'. Only those few who are fully alive to the eternal, necessary, and law-determined (*gesetzmäßige*) dimension of life and of the world are capable of gaining insights that are not annulled, but rather confirmed, by the contemplation of transitoriness, and thus of arriving at true resignation.[49] And elsewhere, following upon his assertion that what had attracted him more than anything else towards Spinoza was the boundless unselfishness which radiates from every sentence of his and which made him forge his often misunderstood saying: "Who loves God need not desire that God loves him in turn", Goethe adds: "Here, too, it should not be forgotten that the closest connections result only from contrasts. Spinoza's all-reconciling quietude contrasted with my all-exciting strivings; and his mathematical method was the opposite of my poetic way of reflecting and representing. It was precisely this orderly treatment — which many people did not think appropriate to moral subjects — which turned me into his most passionate disciple and his most determined admirer".[*50]

So much for Goethe's "philosophy". As to his "empiricism", this was not indeed, as we have seen, the variety of empiricism which, by means of artificial experiment, seeks to corroborate some preconceived theory. For this is coercive empiricism in the manner of Newton and Kant, of which Goethe so strongly disapproved. Rather, it is what he called "tender empiricism", which, on the contrary, rests upon a patient and unprejudiced *empathy* with the phenomena Nature presents to us,[†51] in order to arrive in the end at the *Urphänomen*, at which point the human intellect cannot but resign. In a variation upon Gray one might say that, whereas modern science has accepted the 'Faustian' element of the hermeticist tradition but not its organic world-view, with Goethe it was exactly the other way round.

---

[*] "Übrigens möge auch hier nicht verkannt werden, daß eigentlich die innigsten Verbindungen nur aus dem Entgegengesetzten folgen. Die alles ausgleichende Ruhe Spinozas kontrastierte mit meinem alles aufregenden Streben, seine mathematische Methode war das Widerspiel meiner poetischen Sinnes- und Darstellungsweise, und eben jene geregelte Behandlungsart, die man sittlichen Gegenständen nicht angemessen finden wollte, machte mich zu seinem leidenschaftlichsten Schüler, zu seinem entschiedensten Verehrer."

[†] "Es gibt eine zarte Empirie, welche sich mit dem Gegenstand innigst identisch macht und dadurch zur eigentlichen Theorie wird." ("There is a tender empiricism which identifies itself, as intensely as can be, with the object, and which turns thereby into theory proper").

c. The legacy of Kant and Goethe, and how Schopenhauer received it

The preceding account may perhaps have given rise to the expectation that there existed a sharp antagonism between the two "giant brothers" *(Riesenbrüder)* of the mind, as Schopenhauer used to call them. But that is not so. Kant — one generation older than Goethe, but always an interested observer of the intellectual scene of his time — had accepted in principle the invitation Schiller extended to him in 1794 to contribute to his periodical *Die Horen*. Probably it was nothing but a certain mental fatigue which prevented the, by then, 70-year-old philosopher from keeping his promise, unlike Goethe, who indeed became an active contributor. That is how, much later, Goethe had to admit, to his faithful Eckermann, that Kant never took any notice of him. The same conversation gave rise to Goethe calling Kant the most eminent German philosopher, for "this man has penetrated into German culture in the most profound manner, and he has made an impact upon one even if one has never read him."[*52] Yes, Goethe went even further in suggesting that he himself had always been a Kantian without being aware of it. He refers in this context to his essay on 'experiment as a mediator between subject and object' ('Der Versuch als Vermittler von Subjekt und Objekt'), but this is not too convincing for anyone acquainted with that particular piece of writing.[†53] Reading it rather confirms the truth of a confession made by Goethe elsewhere that "he lacked the organ for philosophy proper".[54]

Nevertheless, in the very same essay that opens with this confession, Goethe gives an interesting account of his road to Kant. Kant's *Critique of Pure Reason*, we read there, for a long time after it came out in 1781 still fell completely beyond his own sphere of interest. Conversations gave him the impression that the book was just the latest instalment in the hoary dispute over the respective contributions made by ourselves and by the external world to our mental existence (two aspects he himself had never felt a need to separate from one other). Thus he was inclined to agree with those who, in accordance with Kant, claimed that, although all the knowledge we have starts with experience, it does not therefore arise wholly from experience. He equally liked the synthetic judgements *a priori*, since, throughout his own life as a poet as well as an observer of nature, he had proceeded now

---

* "*Kant* (sagte er) ist der vorzüglichste, ohne allen Zweifel. Er ist auch derjenige, dessen Lehre sich fortwirkend erwiesen hat, und die in unsere deutsche Cultur am tieffsten eingedrungen ist. Er hat auch auf Sie gewirkt, ohne daß Sie ihn gelesen haben."

† Man is customarily inclined, so Goethe writes here, to look at the objects around him from a subjective point of view ('do I like them or not?'; 'do they profit or harm me?'). The man of science, instead, must look at them in a detached and objective way; he should therefore be cautious not to draw conclusions too quickly from one single experiment (nor, for that matter, from two or more mutually isolated experiments). More especially, he should not use them rashly as arguments for some favourite hypothesis or theory (guess at whom Goethe is aiming here ...?). The reader is likely to nod his approval, yet not to be unduly impressed with these rather trivial observations, nor to discern any particularly 'Kantian' flavour about them.

synthetically, now analytically. Although it might therefore be said that the entrance-gate to Kantian theory strongly attracted him, he nevertheless could not bring himself to venture into the labyrinth itself. But later he made the acquaintance of the *Critique of Judgement*, and "to this I owe a most joyful period of my life. Here I saw my most disparate occupations brought together, with products of art and of nature being treated the same way and aesthetic and teleological judgement being made mutually illuminating."[*55] Still he did not manage to establish a meaningful contact with the Kantians, who used to smile when he said something he believed to be in accordance with Kant's mode of thought (not surprisingly, we would add ...). All this changed, however, once his relationship with Schiller had become more intense. There certainly remained a difference in the points of departure between the two ("he preached the gospel of Freedom; I did not want to see the rights of Nature curtailed"),[†56] but these appeared now to supplement rather than to exclude one another. This, as Goethe ends his account, is how he gradually became accustomed to that Kantian language that had hitherto been wholly alien to him.

Yet how selective and, in truth, high-handed his relationship with Kantian philosophy remained, becomes clear when, in still another essay, we find Goethe delivering himself of the following opinion. Kant had strictly confined thinking to discursive judgement, out of a justified rejection of the human inclination unhindered by much experience to apply all kinds of unreflected intellectual whims to objects. How, then, is it possible for Kant to allow us some freedom in this respect? Here Goethe pointed to the well-known assertion in the *Critique of Judgement* that we can imagine an Understanding which, since it is not discursive like our own, but intuitive, goes from the intuition of a Whole to the particular, or from the whole to the parts. It is true, Goethe half-naively, half-ironically concedes, that Kant seems to reserve this *intellectus archetypus* (as against our own, discursive *intellectus ectypus*) for a divine Understanding. Yet, Goethe goes on, if we are able in the moral sphere to elevate ourselves — through our belief in God, Virtue, and Immortality — on to a higher realm and thus to approach the First Being (as Kant taught), then (attention now!) this might equally be the case in the intellectual sphere insofar as, by intuiting an ever-creative Nature, we make ourselves worthy of spiritual participation in her productions![57] Thus Goethe, by means of this rather daring interpretation, felt himself legitimized by "the old man from Königsberg" to continue his own studies in his own way.

All this is to say that Goethe fully recognized Kant's greatness, and took from him what he could assimilate, but in his heart always remained aware that he was

---

[*] "Nun aber kam die Kritik der Urteilskraft mir zuhanden, und dieser bin ich eine höchst frohe Lebensepoche schuldig. Hier sah ich meine disparatesten Beschäftigungen nebeneinander gestellt, Kunst- und Naturerzeugnisse eins behandelt wie das andere; ästhetische und teleologische Urteilskraft erleuchteten sich wechselsweise."

[†] "Er predigte das Evangelium der Freiheit, ich wollte die Rechte der Natur nicht verkürzt wissen." Cf. our remarks in Chapter IX, p. 250.

himself of a different mind-set. With Schiller the case was both similar and somewhat different. While never showing great interest in Kant's epistemology, he had occupied himself intensively with his ethics and aesthetics. This study had the effect of making him happily turn away from his own original *Sturm und Drang* period, which Goethe had already abandoned much earlier; that is why it took rather a long time until he and Schiller found each other. Well, Goethe must have thought, let those who have not been in a position to attain the classical ideal of self-restrained, harmonious humanity by getting acquainted under Italy's sky with the beauties of the Mediterranean landscape and of Graeco–Roman art (which Goethe, of course, still interpreted in the naively devoted way of Winckelmann) attain it in their own way! In short, Goethe made room for a certain Schillerian–Kantian influence upon his artistic and intellectual development, while always remaining true to his own nature; and it is very significant that at the same time the influence of *Herder* declined.

This particular influence had been very strong for a while; one need only recall Goethe's stay in Strasbourg in 1770–71, where, under Herder's guidance, he learned to admire Gothic architecture (considered wrongly to be particularly German) and allowed his own poetry to be inspired by the native originality (*Urwüchsigkeit*, to use the plastic and almost untranslatable German expression) of the *Volkslied* (folk song). Their friendship persisted, until it was broken off because of a nasty remark by Herder. As is well known, Herder, when Goethe published his drama *Die natürliche Tochter* ('The natural daughter'), allowed himself to joke: "I prefer your natural son" ("Dein natürlicher Sohn ist mir lieber"), subtly alluding to Goethe's son August, begotten with Christiane Vulpius to whom Goethe was not yet officially married at that time. *Geheimrat* von Goethe was not amused. Still, personal irritation can hardly have been decisive.

The point is rather this. In Herder, we find the classic example of a mind in which the possibility to which we pointed in the previous section of a fundamental rupture between the German spirit and the spirit of the Enlightenment (the latter taken here for the spirit of the West, and of France in particular) was indeed in danger of being realized. This can already be seen in Herder's *Auch eine Philosophie der Geschichte* ('*Another Philosophy of History*'), written in 1774 as a reply to Voltaire's *Essai sur les moeurs*. Now Voltaire had likewise been made a target in *Lessing*'s critique of the servile imitation of classicist French drama by authors like Gottsched, to whom Lessing recommended Shakespeare as a better source of inspiration for the German theatre; yet Lessing is rightly considered one of the great representatives of the German Enlightenment. But what in Lessing had simply been healthy, non-conformist opposition to a worn-out literary canon acquires, with Herder, a different slant. In his conception of history, 18th-century individualism is given a characteristic bent: *Nations* are seen as contrasting cultural individualities in which every now and then mankind expresses one defining aspect of itself, whereas individual man tends to be reduced to nothing but the partaker of a fixed cultural identity. True, all cultures are basically endowed with equal rights: German culture is not proclaimed by Herder to be intrinsically superior to others, and it is only fair to add that in his later *Ideen zur Philosophie der Geschichte der Menschheit* ('*Ideas for the Philosophy of the History of Mankind*'), written between 1784 and 1791, Herder sought to reconcile Enlightenment universalism with his discovery of the mysterious and unique national soul. He did so by proclaiming his notion of "humanity" (*Humanität*) to be the ultimate goal of history, to which each nation in its heyday makes its own contribution. Yet it remains the case that human history is conceived by him as a part of natural history (conceived in its turn pantheistically as a revelation of the Deity) — that, therefore, like nature, it obeys laws, and that different cultures are likewise to be considered as organic entities. It was this divinization and, at the same time, naturalization of human history that lay at the root of Kant's rather unfavourable review of the first and the second volume of the *Ideen* in 1784 and 1785, respectively (more on this in our chapter V, p. 131). The review led to a lasting estrangement between Kant and his former pupil, who until then had always spoken with great warmth of his former teacher. In 1799 Herder, by way of a belated reply to Kant's review, wrote his *Metakritik*, in which, singling out the *Critique of Pure Reason*, he subjected Kant's philosophy to a severe critique.

We have already noted that Herder's belief in the uniqueness of cultures in the sense of collective personalities was not necessarily bound to lead to national hatred, as indeed in himself it did not; the Slav peoples, in particular, enjoyed his lively and sympathetic interest. Yet the period of the Napoleonic wars that followed was all too liable to give a more aggressive bent to the hoary German complex of inferiority towards France — a complex due to those many decades of national sleep we have talked about. It did so in a dual sense. The French revolutionary nation-state called up a sense of resistance in that it was felt to be an alien oppressor; at the same time, its demonstrated capacity for power-deployment was envied.*[58] Glorification of the collective identity easily came to serve as compensation for the military defeat of Prussia at Jena in 1806 and for the humiliating peace concluded in its aftermath at Tilsit, which marked the nadir of

* This national will-to-power indeed added a decisively new element to the German concept of 'nation', which is not yet to be found in Herder, but was to bear fruit in the further course of German history. I cannot therefore quite agree with Finkielkraut, when, in distinguishing between two types of racism (namely, one that proclaims one scale of values for all peoples and thus establishes a hierarchy between them in accordance with the extent to which they conform to it; and another one that is based on each people's unrepeatable singularity, thus denying any unity of mankind), he arrives at the conclusion that the former ends in colonialism, whereas the latter (which he rightly traces back to Herder) reaches its apogee in Hitlerism. It is true that Herder fits fairly well Finkielkraut's description of his second category, but only because he was still content with cultivating his own national uniqueness without wishing to subjugate other nations to an inherently superior *Deutschtum*. It was quite otherwise with later, aggressive and 'racist' nationalism culminating in Hitler, the origins of which can be traced back to the anti-Napoleonic wars of liberation. *This* particular racism may rather be said to be a very explosive *synthesis* of both types!

Germany's subjugation. Fichte's '*Speeches to the German Nation*' (*Reden an die deutsche Nation*) of 1807 are symptomatic in this respect. A hierarchy between peoples is proclaimed here: each period of history has a special chosen people who act as the representative of humanity for that period. Now is the German nation's turn; its claim is legitimized by its being the most creative and "original" nation of Europe — "original" in that, unlike the other great European nations, it has preserved its original language (*Ursprache*).

There followed the Wars of Liberation, from which Goethe, to the annoyance of many a German patriot, kept himself aloof. This was due in part, no doubt, to his admiration for Napoleon (the ribbon of whose *Légion d'honneur* he ostentatiously continued to wear), but in no smaller measure to his 18th-century cosmopolitanism. Here is what he thought of national hatred: "How could I, to whom culture and barbarism are the only things that matter, have hated a nation which belongs to the most cultivated on earth, and to which I myself owe such a large part of my own culture? More generally speaking, there is something peculiar about national hatred. You will always find it at its strongest and most virulent at the lowest cultural level. Yet there is a level where it disappears completely and where, in a sense, one stands *above* the nations, such that one feels the weal or woe of one's neighbour as if it had happened to one's own nation. This cultural level was appropriate to my nature, and I had confirmed it to myself before I had reached my sixtieth year."[*59] This is from a conversation with Eckermann. As to that "national literature" which he, too, had glorified in his youth under Herder's influence, in another conversation he pointed to a surprising resemblance between a Chinese novel he had recently read and his own *Hermann und Dorothea.* Eckermann suggested that, for its very Chineseness, the novel was likely to have been rather alien to him. But such appeared not to be the case, for interesting reasons. True, Goethe had an eye, and much regard, for its "specific" features, especially the ongoing presence of natural scenery (gold fish one almost hears splashing in the pond, or birds on their tree twigs singing on and on) amidst which its human protagonists act; also the description of the interior of the houses, which he had found "just as pretty and graceful as Chinese paintings",[†] and the classic moral legends interwoven with the story proper by way of illustrative proverbs. Yet what is really important, he insists, is not this exotic aspect of the book (precisely because such things are always historically determined), but the fact that its characters think, act and feel like us, and that, consequently, one feels oneself

[*] "Wie hätte ich, dem nur Cultur und Barbarei Dinge von Bedeutung sind, eine Nation hassen können, die zu den cultiviertesten der Erde gehört, und der ich selbst einen so großen Teil meiner eignen Cultur verdanke? Überhaupt ist es mit dem Nationalhaß ein eigenes Ding. Auf der untersten Stufe der Cultur werden Sie ihn immer am stärksten und heftigsten finden. Es gibt aber eine Stufe wo er ganz verschwindet und wo man gewissermaßen *über* den Nationen steht und ein Glück oder eine Wehe seines Nachbars empfindet als wäre es der eigenen begegnet. Diese Culturstufe war meiner Natur gemäß und ich hatte mich darin befestigt, ehe ich mein sechzigstes Jahr erreicht hatte."

[†] "... und das Innere der Häuser so nett und zierlich wie ihre Bilder."

very soon like one of them, except that everything happens with them in a more clear, clean and moral way.[60] All his scepticism towards the "gospel of freedom", and all his insistence on those "rights of nature" so dear to his heart notwithstanding, Goethe in his mature age appears here as a soulmate of Kant rather than as the spiritual kinsman of Herder he had once been in his youth.

Speaking about freedom, meanwhile, it is significant that Fichte, before those *Speeches* of his which radicalize *Herder*, had sought to radicalize *Kant* in his '*Attempt at a Critique of All Revelation*' (*Versuch einer Kritik aller Offenbarung*). This was the book with which, in 1792, he entered the philosophical stage, doing so anonymously so that the work was taken by the public to be by Kant himself until the latter published a denial in which he disclosed the true author's name. Here, as slightly later in his main work '*Foundation of the Whole Theory of Science*' (*Grundlage der gesamten Wissenschaftslehre*), he made a heroic attempt to overcome Kant's distinction between theoretical and practical reason. The 'Thing-in-itself' is dropped, and not only the 'form' but also the 'stuff' of the empirical world is said to have been created by the 'I', which first 'posits' itself and then goes on to posit the 'non-I' as its necessary field of action — not only of scientific activity, but primarily activity in the moral sense. This, to be sure, is a universal, supra-individual "I" — which also, however, in a more popular variety could be taken for a no longer universal, yet still supra-individual *Volksgeist*, or national spirit. Had not Fichte himself taken this popular course in his *Speeches*? All this might well serve as some kind of legitimation for the Romantic sense of life — the sense of life of a literary and artistic movement Goethe strongly disapproved of,[*61] not only because of its inclination towards Teutomania, nor solely for its medieval Catholicizing tendencies, but above all in view of its unmitigated subjectivism. Now a Byronian rebellion against the conventions of society, joined to a boundless drive for self-expression and self-realization, may very well go together with the desire for absorption into some Greater Whole (whether in Nature or in the *Volksgemeinschaft* (national community)), since in

[*] This disapproval found its expression in the well-known verdict: "Das Klassische ist das Gesunde, das Romantische das Kranke". At times, however, he allowed for some differentiation in his pertinent utterances. For example, in another conversation he credited Romanticism with having achieved, not only greater freedom in matters of form, but also greater diversity in content such that no subject matter whatsoever in the wide world or in the multifariousness of life is excluded any longer from poetry, going on to express his confidence that the excesses will over time disappear of themselves. Moreover, in 1814–15, through the brothers Boisseré in Heidelberg, he had made the acquaintance of early Rhenish and Flemish painting (notably that done by the Van Eycks) and had renewed his interest in medieval architecture (in particular, the cathedral of Cologne, the completion of which he advocated). Just as with the acquaintance he made with Oriental poetry (which served as a source of inspiration for his own *West-östlicher Divan* of 1819), this led to a certain softening of his previously strict classicism, which had made him regard Greek art as the only true art. None of this is to deny that the Greeks never ceased to serve as the ultimate model for him.

such a *unio mystica* the outer world ceases to be 'outer', i.e., a permanent and painful reminder of the subject's finitude! In all this Goethe must have recognized his own fascination, during his *Sturm und Drang* period (which as a movement was in fact a prelude to Romanticism), for such noble *Kraftmenschen* ('strongmen') as he had portrayed in *Götz von Berlichingen.* And in Fichte's doctrine of the world-positing 'I' he may have recognized the same titanic tendency, parading now in the stately mantle of Philosophy.

In this, to be sure, he would have been wrong insofar as Fichte's 'I' does not cease to be *Reason* in action, i.e. certainly 'Will', but then regarded as the will to realize the *universal Good of moral law*, as it was in Kant.* To that extent Fichte must still be regarded as a representative of the Enlightenment, not of Romanticism. For there is of course a fundamental difference between Enlightenment subjectivism and Romantic subjectivism in that for Romanticism the artistic genius (its hero, much more than the great scientific investigator) is a genius precisely because of his particular openness to the sources of inspiration that are provided by the forces of the Unconscious. Romanticism is therefore rightly considered as a reaction against the Enlightenment in that it rejected the latter's abstract, rootless conception of humanity; on the other hand, in spite of the above-mentioned Catholizing tendencies of many Romantics, it did not in fact abandon the Enlightenment's emancipation of Western man from the old hierarchical order. For, although it stresses man's need for rootedness, this is sought in the 'deeper' rather than in the 'higher' sphere; thus, when, as Heine mocked, "young people queued up as it were before the Roman Catholic Church",[†62] this had to do less with strictly religious motives (a pantheist worshipping of Nature is in fact more congenial to the Romantic temperament) than with aesthetic motives and with fascination for those Middle Ages that were seen as a period of German national greatness.[‡] Conversely, even the sober, not to say arid and pedantic and, thus, far from 'romantic' Kant reserved in his third great *Critique* a special place for artistic genius. In short, the dividing lines cannot always be drawn sharply, which becomes still more clear when we consider the

---

* Kant may indeed be regarded, with Heidegger, as already a philosopher of 'Will', but we are concerned here with an ethical will (indeed, a 'will', too, just as 'will to knowledge'; cf. footnote on p. 15)! It is otherwise with Schopenhauer, as we shall see in Chapter VII; on the other hand Schopenhauer's "Will" is in its turn different from Nietzsche's, as we shall see in Chapter VIII.

† "Wenn man nun sah, wie diese jungen Leute vor der römisch-katholischen Kirche gleichsam Queue machten ..."

‡ It should certainly be kept in mind that Romanticism is a general European phenomenon; already the name Byron which was mentioned by us in passing reminds us that there was also an English Romantic movement, just as there was a French one. But in those countries Romanticism is one period among others in the history of their national culture, whereas in the latecomer Germany it has decisively contributed to shaping the profile of its national culture in the first place; and only here did it become not only a literary but also a philosophical movement, especially with Schelling's *Naturphilosophie*.

two other protagonists of German Idealism besides Fichte, who with his *Theory of Science* had set the tone for the entire movement — namely, Schelling first and Hegel later, and their relationship with Goethe. In a sense they may both be said to have striven, though in different ways, to restore the balance Fichte had disturbed, in focusing their attention back upon the objective world (on 'Nature' and on 'History', respectively). But they did so by making reality as a whole subject-like; for we are concerned here with objectivations of Spirit, i.e., *of the hypostasized transcendental subject*, which, therefore, remains the ultimate reality for them, too. This new kind of metaphysics would certainly not have been feasible without Kant's 'Copernican Revolution'; but one may well ask what, in these heretical disciples of Kant, remains of the *critical* core of Kantianism with its sober and severe charting of the possibilities as well as the inexorable boundaries of human reason?[*] As compared to the *hybris* of German Idealism, Goethe, with his maxim: "The finest fortune of thinking man is to have investigated what can be investigated and quietly to revere what cannot",[†63] or with his "Doest thou want to proceed into the Infinite / then just move into the Finite in all directions",[‡64] paradoxically seems more of a true Kantian! It is no less true for all this that Schelling's romantic *Naturphilosophie* in particular (which by itself was only one, albeit important, stage in his Protaean philosophical development) was in line with Goethe's rejection of the mechanization of the world-picture. Goethe recognized the kinship, although his realistic mind remained critical of the metaphysical constructions which already characterized Schelling's *naturphilosophische* period (the later, Christianizing Schelling was wholly rejected by him).[§65] Further, the even more abstract and speculative Hegel, whose divinization of History cannot at all have been to Goethe's taste, in his philosophy of nature joined in with Goethe's rejection of Newton, and even defended the former's theory of colours, as we shall see in Chapter V!

Indeed, it is not by chance that the entire period from the later Enlightenment up to and including Romanticism is wont to be called the *Goethezeit*, as we remarked in our Introduction (p. 5). For all its acknowledged variety, the era as a

---

[*] As is well-known, Hegel had no theory of knowledge. More than that, he denied the very possibility of such a thing, using the argument that one can only learn to swim by going into the water. Indeed it is a topic among Hegelians that Kant's attempt at a critique of our faculty of knowledge is to be compared with the famous attempt of Baron Münchhausen to draw himself out of the swamp by his own wig. In our opinion this would count as an argument only if Kant had proceeded by way of introspection, rather than by way of a comparison between the publicly available results of positive science on the one hand, and those of metaphysics on the other.

[†] "Das schönste Glück des denkenden Menschen ist, das Erforschliche erforscht zu haben und das Unerforschliche ruhig zu verehren."

[‡] "Willst du ins Unendliche schreiten / Geh nur im Endlichen nach allen Seiten."

[§] One wonders what Goethe would have said about the esoteric speculations of Rudolf Steiner, the founder of Anthroposophy (a movement which — quite wrongly in our opinion — claims Goethe as its patron saint), had he lived to know them ...

whole may usefully be thought of as being held together by the common denominator of a final, large-scale effort to restore that unified world-view of which, at the beginning of the present chapter, we saw Jan Patočka (but he has really not been the only one) signalize the loss. This goes almost without saying for Goethe, whose life-long aspiration had been to achieve 'harmony', not only in his personal life, between the demands of individualist genius and those of society, but also in the domain of science, between the demands of analytical, abstract reason and those of immediate sensual intuition. The former aspiration accounts for what made him differ from Romanticism; the latter is responsible for what linked him to it and also to at least the Schellingian brand of German Idealism (which — more than the Fichtean and the Hegelian brand — may rightly be called the spirit of Romanticism brought into a philosophical system). Still, although it must be repeated that the dividing-lines cannot always be drawn sharply, we may not forget that there never ceased to be a fundamental difference between Goethe's position and that of German Idealism even in the guise Schelling had given to it in his *Naturphilosophie*. For, unlike German Idealism, Goethe accepted in principle Kant's resignation over 'ultimate questions';[*66] therefore, he also accepted in his own way Kant's confining genuine knowledge to the domain of 'experience' ... that is to say, of experience ordered by the intellect!

True, there is a difference here, too, which resides in the *nature* of this intellectual ordering. In our imagination we may watch Goethe seeking to impress upon Kant his own awareness that there are more things in the realm of experience than are dreamt of in Kant's philosophy! Indeed, Kant virtually identified this realm with Newtonian science, which implies that its theoretical component consists of causal chains expressed in the language of algebraic equations. But this implies in its turn that the individual phenomena which are connected into a general theory are in fact not those of sense perception but data yielded by experiment (Sorel's "artificial nature"!). In other words, Goethe knew as well as Kant did that human understanding cannot content itself with recording individual phenomena, nor with just categorizing them (as Linnaeus had done for botanical phenomena), but that it should search for the ways in which they are interconnected, for the conditions under which they change their appearances, and for

[*] In a conversation dated 1 September 1829, Eckermann tells Goethe about a visitor who had taken lectures with Hegel, and who reported that once more a proof of God's existence had been served up. Goethe thought this outdated. Philosophers do not in any way increase our knowledge of such problems as immortality, or the nature of the divine, or the nature of our soul and its connection with the body; rather, Goethe argued, it was precisely the great merit of Kant to have drawn the boundaries of possible human knowledge and to have left insoluble problems aside. He goes on to mock the Germans who torment themselves with attempts at the solution of metaphysical problems, while meantime the English with their practical wisdom laugh at them and conquer the world. (Goethe did not fail to add that the English, in so doing, knew how to serve up 'cant' (an expression not literally used by him), for example, in their apparently idealist advocacy of the abolition of the slave trade, which served their own interests).

the laws in accordance with which the one follows from the other. To Goethe, however, the single elements of the final synthesis remain those which are given in immediate empirical intuition, whereas in mathematical physics (Kant's model) these must first be transformed into quantitative values.[67] Goethe's urge to stick to a *qualitative* nature is without doubt connected closely to his immanent conception of the Divine:[68]

Was wär' ein Gott,	"What would be a God
der nur von außen stieße, /	who pushes only from without,
Im Kreis das All	and makes the universe turn
am Finger laufen ließe! /	around his finger!
Ihm ziemt's die Welt	It is proper to him
im Innern zu bewegen, /	to move the world from within,
Natur in Sich,	to keep nature into himself
Sich in Natur zu hegen, /	and himself into nature,
So daß was in Ihm lebt	so that all that lives within him,
und webt und ist, /	and moves and is
Nie Seine Kraft,	never lacks his force
nie Seinen Geist vermißt. /	or his spirit."

Kant's transcendent God, who is not to be sought in the natural, but rather in the ethical sphere, is of course much more compatible with a quantitatively reduced nature than Goethe's immanent one. But it is one thing to save, *against* Kant and mathematical–mechanical science, the full qualitative richness of the phenomenal world while at the same time, *just like* Kant, confining one's investigations to this phenomenal world (which is what Goethe did); it is quite something else to absorb the phenomenal world into the noumenal world by means of an all-encompassing metaphysical system (which is what German Idealism did). To use a metaphor, Goethe took the phenomenal world (the world of integral, not of quantitatively reduced, experience) as his starting-point, recognizing it as God's garment, while abstaining from further speculation about this God, whereas German Idealism took God as its point of departure, going on to give a detailed account of His adventures at the successive stages of His self-realization in Nature and/or in History. Alas, after Kant's destruction of traditional metaphysics and its attendant 'natural knowledge of God', this "God of the philosophers" (the *Dieu des philosophes* scorned by Pascal) could only be a hypostasis of Kant's 'transcendental I'. Hence, in the final analysis we have here, indeed, the apotheosis of the human subject, which, as cynical people have been eager to add, was, in its actual historical guise in Restoration Germany, so pitifully lacking in power ....

Hegel died in the autumn of 1831, Goethe in the spring of 1832. These dates are almost symbolic, as they followed closely upon the July revolution in 1830 Paris which, although it did not spread to Germany, in representatives of a younger generation aroused a sense of an approaching spring-time of liberty. Whereas Fichte, as a professor at the university of Jena, and once again as the Rector of the university of Berlin, had still got into conflict with the establishment, German

Idealism had subsequently, with Schelling and above all with Hegel, made its peace with society as Goethe had; both therefore seemed outdated now. The most eloquent spokesman of what proudly called itself "young Germany" (*Das Junge Deutschland*), Heinrich Heine, moved to Paris, commenting from there upon the German state of affairs with his sharp wit. His acquaintance with the ideas of early socialism (notably, those of Saint-Simon) made him, in his "winter tale" *Deutschland. Ein Wintermärchen*, replace "the old song of resignation" ("das alte Entsagungslied") with the "new and better song" ("ein neues Lied, ein besseres Lied") of establishing the Kingdom of Heaven here on Earth already while leaving Heaven "to the angels and the sparrows".[*69] Against the Christian churches, the joys of the senses are praised by him, and claimed for all men. As to the theoretical cobwebs of speculative philosophy, these are mocked somewhat further on in the same poem:[70]

Franzosen und Russen gehört das Land,	"To French and Russians is the land;
Das Meer gehört den Briten.	the sea is to the British;
Wir aber besitzen im Luftreich des Traums	but we are the undisputed rulers
Die Herrschaft, unbestritten.	of the airy realm of the Dream."

Yet there should be no misunderstanding: Heine always remained under the spell of Hegel. So did Ludwig Feuerbach, for all his extolling of the concrete 'I' of flesh and blood with its sensual intuition against the Idealists' apotheosis of the thinking 'I'. And so did young Karl Marx, for all his claims to surpass both Feuerbach — who allegedly had not arrived at conceiving sensual intuition as a "practical, social activity" and therefore still belonged to those philosophers who had only "interpreted the world in various ways instead of changing it" — and Hegel, whom he had allegedly "put onto his feet". For what united these and other 'Hegelians of the left' was a belief that behind the master's assertion that "all that is real is also rational" lay hidden the secret message that "all that is rational is also real" (i.e., the rational should, and also can, be made real).

Indeed, how 'idealist' and 'romantic' a period the *Vormärz* still really was! This label refers to the period between 1830 and March 1848, which is when the revolutionary wave that in February crested once again in Paris reached Central Europe. All these revolutions misfired, and yet 1848 marks the beginning of a new era — that of the final breakthrough not of socialism (as the *Communist Manifesto* had proclaimed) but on the contrary of industrial capitalism. The intellectual climate changed, too. Around the middle of the 19th century the impressive speculative castles built in the air by German Idealism finally collapsed under the impact of experimental science, which meanwhile, in England and France, had pursued its seemingly inexorable course. Philosophically the hour had struck for materialism and positivism, and this happened in Germany, too, where many people now began to blame Romantic *Naturphilosophie* for Germany's scientific

---

* "Wir wollen hier auf Erden schon / Das Himmelreich errichten"; "Den Himmel überlassen wir / Den Engeln und den Spatzen".

backwardness as compared to Western Europe. Justus Liebig, the famous chemist, left no doubt about his sentiments on this score. Having undergone, in his youth, the influence of Romantic *Naturphilosophie*, he later felt bound to pronounce the harsh judgment that this had been "a truly miserable period for chemistry in Germany"; and, in a personal review, to arrive at the bitter conclusion that this period ("so rich in words and ideas and so poor in real knowledge and profound study") had cost him two precious years of his life. "I cannot depict my fright and dismay when I was roused from this intoxication into consciousness", he tartly remarked.[71] But there were also those who, while inclined to assent to such a judgement in general, could not, for all that, find satisfaction in the shallowness of a 'philosophy' that reduced everything to *Kraft und Stoff* ("force and matter") — to quote the title of Ludwig Büchner's sensational book — and that culminated in Feuerbach's profound observation that man "is what he eats" ("Der Mensch ist was er ißt"). For people of such a bent the time had become ripe to discover a previously neglected philosophical outsider.

This outsider had been far too deeply rooted in the *Goethezeit* ever to accept the now fashionable positivism and materialism. On the other hand, born in 1788, he belonged to the rearguard of the *Goethezeit*, and this had made it easier for him right from the start to keep his critical distance from both Romanticism and German Idealism, the joint influence of which he had nonetheless undergone. Thus, the apotheosis of art (of music, especially) celebrated by Romantic poets like Wackenroder, Tieck, and Novalis was to leave its traces in the third book of his own *magnum opus*. Also, as a student at the university of Berlin he had not only heard Fichte — who had failed to impress him — but had also read Schelling, whose book '*On the Essence of Human Freedom*' (*Über das Wesen der menschlichen Freiheit*) is likely to have influenced him more than he was prepared to admit.

Still, much in this man's Hanseatic–patrician background and corresponding education had worked to set him somewhat apart from his German contemporaries. When his native Dantzig was occupied by Prussia in 1793 as a result of the second Polish partition, his father, out of attachment to the traditions of a free city-state, had moved to Hamburg, which was by then the most important maritime–commercial centre of Germany. His father, who had intended him originally to become a similarly large-scale merchant as he himself was, induced him 'to make the acquaintance of the world', so already as a child he had spent two years at the home of a business-friend of his father's in the French port Le Hâvre, and as an adolescent he joined his parents on an extended trip to the Netherlands, England, France, and Switzerland. All this allowed him to adopt a more gentlemanly and urbane attitude towards life and the world than used to be the case with those dressing-gown-clad German philosophers who so rarely left their studies. (It also allowed him, as the same cynical people cited before would promptly have added, to live "for, not from, philosophy" — as he prided himself — owing to a modest amount of capital he had inherited and had known how to increase over time through his own financial skill ...).

Finally, this man's somewhat peculiar psychological make-up (reinforced by some early, traumatic experiences, such as, notably, the probable suicide of his father) made its own contribution towards predisposing him to the role of an outsider.[72]

The reader is no doubt aware by now of whom we are talking. It was to become Arthur Schopenhauer's destiny to adopt, by way of a return to the sources, the dual legacy of Kant and Goethe, thus bypassing its perversion (as he saw it) in German Idealism, and to turn that legacy into another metaphysical system to be sure, yet one claimed by him with some justice to differ fundamentally from those of German Idealism in that, paradoxically, it was rooted in *experience*. Schopenhauer accepted the Kantian dualism of 'appearance' (*Erscheinung*) and 'Thing-in-itself', and was therefore prepared to leave the former (that is, the domain of *outer* experience) to materialist and positivist science. The right to provide an interpretation of life from the data of *inner* experience, however, Schopenhauer reserved for metaphysics, to which, unlike Kant, he attributed the capacity to pronounce upon the 'Thing-in-itself', though certainly not by way of a causal explanation of the empirical world (this had to remain the task of science). The revolutionary implication was that the world was no longer held to be grounded in the 'higher' regions of a hypostatized 'thinking I', but rather in the dark depths of the *irrational*, a glimpse of which is allowed us by means of our 'willing I'. This 'willing I', then, is nothing but the inner side of such a vulgar thing — as compared to the lofty heights of German Idealism —as our own *body*. This physiological approach made him define Kant's categorical apparatus (reduced by him to the category of causality alone) as a mere function of the brain (*Gehirnfunktion*). Herein resides Schopenhauer's link with the new materialism, just as in his (quite un-Kantian as well as un-Fichtean!) interpretation of "Will" as a dark unconscious force, he thereby showed himself as an heir of Romanticism (while, however, at the same time showing himself an heir of the Enlightenment by insisting that man by his intellect can detach himself from the bondage of "Will").

But we anticipate; let us return to Schopenhauer's early development. Upon his father's death young Arthur gained permission from his mother to drop the commercial career he had promised his father to follow in exchange for being allowed to partake in his parents' long voyage, and turn to the academy. For four semesters, from 1809 to 1811, he studied in Göttingen. Here he made the acquaintance of Kantian philosophy, thanks to his teacher in philosophy, Gottlob Ernst Schulze, who, under the name 'Aenesidemus', had published a critical commentary of Kant's main theses (especially the *a priori* nature of the law of causality). But, having originally matriculated as a *studiosus medicinae*, he also acquired a solid scientific knowledge, especially in the field of physiology, through the lectures of Johann Friedrich Blumenbach — a stimulus later to be deepened by his own reading of the great French physiologist Cabanis. His stay in Göttingen was followed by the period in Berlin (1811–1813) we have mentioned already. This came to an end owing to the political situation, i.e., the Wars of Liberation, towards which he adopted a similarly aloof attitude as was taken by the venerated

Goethe (also, for that matter, by the despised Hegel!). He fled from Berlin to Weimar, where his mother had moved after her husband's death, and which town was also, of course, the residence of Goethe. After a short stay there he moved to nearby Rudolstadt, where, in the quiet seclusion of a village guesthouse, he wrote in the summer of 1813 his doctor's dissertation — an epistemological investigation which revealed him to be a faithful, yet independent, disciple of the other venerated master, Kant. This booklet, which marks his entrance upon the philosophical stage, bore the scholarly title '*On the Fourfold Root of the Principle of Sufficient Reason*' (*Über die vierfache Wurzel des Satzes vom zureichenden Grunde*). An analysis of this juvenile work, which we shall show in one sense to present the clue to Schopenhauer's entire philosophy, will form the subject of our next chapter.

II

Five Decisive Years

Schopenhauer's dissertation is customarily read solely in the second, partly rewritten and enlarged edition of 1847; the original version of 1813 may with justice be called a forgotten work. This is much to be regretted. True, it is written in a much less brilliant, much more pedantic style than Schopenhauer's later work — clearly it still has an eye on the academy, as Safranski has rightly observed.[73] But the point is that this juvenile work throws a very interesting light upon Schopenhauer's intellectual development, such that a thorough acquaintance with it may well be called indispensable for a proper understanding of its author's place in the history of philosophy.

The most remarkable aspect of this epistemological study is that it *is* an epistemological study and that (in the original version, to be sure) it does not pretend to be anything more — all this in an era when Germany was sprouting all-embracing metaphysical systems galore! Nevertheless, the attentive reader, if aware of the author's later development, will already discern some faint pointers in the direction Schopenhauer's thought was about to take.

For example, in the penultimate chapter, in a passage meant to remind the reader of the conscious absence from this work of all ethical and aesthetic contemplation, Schopenhauer adds: "But perhaps this might some day become for me the subject of a larger piece of writing, the contents of which would compare to those of the present one, as being awake to dreaming."[*74] Moreover, the same chapter makes it clear that, in this as yet unborn *magnum opus*, Reason (*Vernunft*) is not to play the central part assigned to it by German Idealism, for here we read: "Further I have ... presented and defined *Reason* in a way unlikely to satisfy those who consider Reason the best thing in man in that, fitted out with the predicate of *practical* Reason, it is to them what turns one into a virtuous man and a saint; still less [will it satisfy] those who call Reason an "absolute faculty of knowledge" and bestow upon it an apotheosis which does not fit in at all with the way I have presented it."[†75]

---

[*] "Vielleicht aber könnte mir jenes ein Mal Gegenstand einer größern Schrift werden, deren Inhalt zu dem der gegenwärtigen sich verhalten würde wie Wachen zum Traum."

[†] "Ferner habe ich ... die Vernunft auf eine Weise dargestellt und bezeichnet, welche denen wohl nicht genügen kann, die die *Vernunft* für das Beste im Menschen halten, indem solche mit dem Prädikat der *praktischen* dasjenige sey, was den Tugendhaften, den Heiligen mache; noch weniger denen, die die Vernunft ein "absolutes Erkenntnisvermögen" nennen,

The second part of this sentence clearly aims at German Idealism, whereas its first part is evidently directed at Kant (mentioned by name in the same context a few sentences further down). This may at first sight seem rather astonishing in view of the great respect the author professes to feel for Kant, who is hailed as the "amazing" Kant in the opening sentence of the book (in one breath, it is true, with the "divine" Plato).[76] Yet it is in fact characteristic of the independent attitude the young philosopher took even towards his most revered masters — Goethe should have picked up the warning when he read the dissertation with so much apparent pleasure! True, it is the Kant of 'practical reason' who is concerned here, but even the Kant of 'pure reason' is not spared some gentle criticism, even though a mostly implicit one. His complicated mechanism of the twelve categories, while not yet rejected at this stage in so many words (as Schopenhauer was to do for the first time in his 'Critique of Kantian Philosophy'), is in practice reduced to the sole category of causality which, as the title of the book already suggests, he takes to be one of the four basic guises in which the principle of sufficient reason reveals itself.

This principle, then, for the wording of which Schopenhauer chooses Christian Wolff's formula: *nihil est sine ratione cur potius sit quam non sit* (i.e., 'nothing lacks a reason for why it is rather than is not') is called by him "the basis of all science". This is so because the principle allows us everywhere to ask 'Why?' — the question which is the mother of all sciences since we are concerned here with a system of concepts, that is to say, with a whole of *related* cognitions, not just with a simple aggregate of these.[77] Yet it should be kept in mind that this principle is nothing but the common expression of what in fact are *several* insights given *a priori*; and it is at this point that Schopenhauer's investigation sets in.

Leibniz — the first, according to Wolff, formally to draft the idea as a general principle — had distinguished between two different applications of the principle of sufficient reason. It applies to *judgements* (each of which needs a *reason* in order to be valid); it also applies to *changes in the outer world* (each of which needs a *cause* in order to be able to occur). Schopenhauer adds two more. He calls Leibniz' two applications the "principle of sufficient reason of cognition" (*principium rationis sufficientis cognoscendi*) and the "principle of sufficient reason of becoming" *(principium rationis sufficientis fiendi*), respectively, and goes on to supplement these with the "principle of sufficient reason of being" (*principium rationis sufficientis essendi*) and the "principle of sufficient reason of acting" (*principium rationis sufficientis agendi*). The 'being' principle applies to the domain of pure geometry and arithmetic, whereas the 'acting' principle asks for the *motives* why all that is done by human beings is done in the way it is done. The feature shared by these four applications is that we are concerned in every case with a *subject–object polarity*. This can be so because, in a general sense, our consciousness covers these two halves and contains nothing outside it. What, by

---

und ihr eine Apotheose widerfahren lassen, der meine Darstellung derselben keineswegs entspricht."

contrast, distinguishes one application from another are the different classes of objects that present themselves to our consciousness, i.e., the four different classes of the objects for the subject. These are:

a. The representations (*Vorstellungen)* that constitute the whole of our experience of the outer world, the data of which are provided by our sensuality (*Sinnlichkeit)*, and the connections between them by our understanding (*Verstand*). [See footnote * for a discussion of various renditions in English of the German expression *Vorstellungen*.][78]
b. The representations of these representations (*Vorstellungen von Vorstellungen*). These are concepts (*Begriffe*), which our reason (*Vernunft*) connects so as to form judgements (*Urteile)*. The faculty of reason is exclusively human, whereas we share sensuality as well as understanding with the animals.
c. The formal portion of our representations. These are the intuitions (*Anschauungen*), given *a priori*, of the forms of outer and inner sense, i.e., of space and time, respectively. [Footnote † is about 'intuition' as a translation of *Anschauung*.][79]

* It is always somewhat difficult to give a wholly adequate English equivalent for *Vorstellung*. Early translations have 'Idea'. Given the way in which the term appears in the great British empiricist philosophers, it indeed corresponds fairly well to Schopenhauer's *Vorstellung*. Only, it suffers from the disadvantage of using the same word for *Vorstellung* and for (Platonic) *Idee*, which also plays an important role in Schopenhauer's philosophy. Recent translations generally use 'Representation'. This can hardly be considered an improvement. Whereas 'Idea' may not be quite adequate, 'Representation' is still less so. Stephen Toulmin has already observed that 'representation', in view of its 'public' character, corresponds to German *Darstellung* rather than to *Vorstellung*; Christopher Janaway, for his part, objects to the realist connotation of 'representation' (since it suggests a representation *of* something else, like a portrait) and thinks that 'presentation' would be a more adequate translation. Personally, we would prefer 'Idea' after all, since its one drawback might easily be met by using the word 'Form' (or even better — in order to avoid confusion with the *immanent* Aristotelian 'forms' that are inseparable from their 'matter' — 'Transcendent' or 'Ideal' Form) for the Platonic 'Idea', which is rather common English usage in any case. Still, in order to avoid confusion I have preferred (like Janaway, for that matter) to conform to the contemporary habit and thus, even if reluctantly, to use 'Representation' for *Vorstellung*.

† Again, it is somewhat difficult to give a wholly adequate English equivalent here. 'Intuition' is generally used; we would perhaps prefer 'perception', if not Kant (more so than Schopenhauer) had distinguished between *Anschauung* and *Wahrnehmung*, the latter being the *awareness* of an empirical intuition. It seems better, therefore, to reserve 'perception' for those cases when in German *Wahrnehmung* is used, and to use 'intuition' for *Anschauung*. However, it should be kept in mind that this 'intuition' has nothing to do with some mystical illumination (a connotation unfortunately carried by the term). *This* kind of intuition is rather what Kant called "intellectual intuition" (*intuitus originarius*), and this, according to him, could only be attributed to God as the "Primeval Being" (*Urwesen*); in human knowledge, on the contrary, we are dealing solely with "empirical intuition" (*intuitus derivativus*). By contrast, German Idealism (Schelling especially) was

d. The immediate object of our inner sense. This is the *willing* subject which, as such, is an *object* to the *knowing* subject.

To follow, in detail, Schopenhauer's analysis of these four classes would lead us much too far astray. Suffice it here to point to those three aspects of his argument that are essential for our investigation.

There is, *first*, the general framework of the *Fourfold Root*, with its insistence on the *correlativity* of subject and object. We find this expressed in its most outspoken form in the crucially important Section 42, which bears the title 'Subject of knowledge, and object' ('Subjekt des Erkennens und Objekt'; Section 41 in the second version). Here Schopenhauer asserts that the subject can only be known insofar as it is a *willing* subject. This is so because the *knowing* subject, which, as the necessary correlate of all representations, forms the condition of the latter, can never become representation or object itself. To the possible objection, 'If the subject is not known, how, then, is it possible for its different faculties (sensuality, understanding, reason) to be known to us?', Schopenhauer responds that these are known to us only by way of the different classes of the representations for which they stand. "Just as, along with the subject, the object is immediately postulated as well (since the very word is meaningless without the latter) and also in the same way the subject along with the object, such that 'being a subject' means nothing else than 'having an object' and 'being an object' nothing else than 'being known by a subject' — in just the same way, along with an object defined *in one or the other way*, a subject knowing *in the same way* is immediately postulated as well. Hence, it amounts to the same thing whether I say: 'The objects have such and such inherent, specific properties'; or: 'The subject knows in such and such ways'; and also whether I say: 'The objects are to be divided into the following classes'; or: 'The subject possesses the following faculties of knowing'". *[80]

In the original version Schopenhauer continues with a sentence which is interesting because it gives the reader a glimpse of Schopenhauer's (at that time, as we saw, still tactfully concealed) true opinion of his philosophical contemporaries: "Should the identity of the subjective and the objective, claimed by the philosophy

---

later to claim "intellectual intuition" (*intellektuale Anschauung* or *Vernunft-Anschauung*) to be a faculty of the human mind as well; this pretension used to be scorned by Schopenhauer as the "sixth sense of the bats" (e.g., in his 'Critique of Kantian Philosophy').

* "Wie mit dem Subjekt sofort auch das Objekt gesetzt ist (da sogar das Wort sonst ohne Bedeutung ist) und auf gleiche Weise mit dem Objekt das Subjekt, und also Subjektseyn grade so viel bedeutet als ein Objekt haben und Objektseyn so viel, als vom Subjekt erkannt werden: genau eben so nun ist auch mit einem *auf irgend eine Weise bestimmten* Objekt sofort auch das Subjekt als *auf eben solche Weise erkennend* gesetzt. Demnach ist es ganz einerlei, ob ich sage: Die Objekte haben solche und solche ihnen anhängende und eigenthümliche Bestimmungen; oder: das Subjekt erkennt auf solche und solche Weisen; einerlei ob ich sage: die Objekte sind in solche Klassen zu theilen; oder: dem Subjekt sind solche unterschiedne Erkenntnißkräfte eigen."

of nature, mean nothing else than such being-of-the-same-kind, then I wholly agree with it. I doubt, however, whether it means nothing else, since, in order to arrive at this, no intellectual intuition, but only reflection, is required. If, therefore, one wishes to call two things one because the one thing cannot be conceived without the other, then I shall not contest this (for *re intellecta in verbis simus faciles* [i.e., 'once the matter has been understood, let us take it easy with the words']), even though I recognize here nothing but a necessary relationship — the only predicate of the two things thus related. Exactly the same relationship, however, obtains between other things, such as cause and effect, father and son, brother and sister. Neither of these can *as such* be thought of otherwise than together, since one term of each couple derives its meaning only from the other. Yet they are never for that reason called one, but always two, although it must be noted that, *in concreto*, they always have other predicates as well".*[81] Though tactfully worded (the passage was characteristically dropped in the second version, for which such courteous reserve towards his philosophical rivals had become superfluous), the point is clear: For the "philosophy of nature" (i.e., German *Naturphilosophie*, with Schelling as its most prominent representative) subject and object were *identical*, whereas for Schopenhauer they are only *correlative*.

This correlativity will never cease to be a recurrent topic in Schopenhauer's later work. It is at the root of his ongoing polemics against philosophical materialism as well as against philosophical idealism in Fichte's vein. In Schopenhauer's view, the former claims to be able to think of an object without a subject, the latter of a subject without an object. To the former the subject is therefore created by the object; to the latter it is the other way round. In fact, however, there can be no causal relationship at all between (knowing) subject and object, precisely because causality is a category of Understanding and is valid as such only *within* the realm of our representations, i.e., between *objects* (even though, as we shall see, by way of *motivation* — the fourth guise of the principle of sufficient reason — it appears as a relationship between the willing *subject* and its objects; but this *willing* subject is itself an *object* to the *knowing* subject). For

* "Sollte die von der Naturphilosophie behauptete Identität des Subjektiven und Objektiven nichts als diese Einerleiheit bedeuten, so stimme ich ihr vollkommen bei, zweifle jedoch, daß sie nur dieses meyne, weil, um zu diesem zu gelangen, keine intellektuelle Anschauung, sondern nur ein bloßes Besinnen nöthig ist. Will man also Zwei, davon Eines ohne das andre gar nicht gedacht werden kann, deswegen Eins nennen; so werde ich nicht dagegen streiten, denn *re intellecta in verbis simus faciles*, obhwohl ich nichts als eine nothwendige Beziehung anerkenne, welche jedoch zugleich das einzige Prädikat der beiden Bezogenen ist. In eben solchem Verhältniß stehn aber auch andre Dinge, z.B. Ursach und Wirkung, Vater und Sohn, Bruder und Schwester, die, *als solche*, nur zusammengedacht werden können, eines nur durch das andre Bedeutung haben und außerdem keine, die man aber nicht deshalb Eins, sondern Zwei nennt, wiewohl anzumerken ist, daß diese, wenn in concreto gegeben, immer auch andre Prädikate haben."

similar reasons, an idealism in the vein of Schelling and Hegel * is rejected as well, because "correlativity" implies not only inseparability but also distinctness.

This wholesale rejection of the various guises taken by post-Kantian metaphysics did not prevent Schopenhauer from coming to the fore, within five years of his dissertation, with a metaphysical system of his own. It is all the more striking to note that, in this first version of the *Fourfold Root*, Schopenhauer still rejects all metaphysics of whatever kind: "This much becomes manifest, however, that our investigation does not end in the rigidity of a Thing-in-itself".[†82] It is precisely because Schopenhauer was soon to change his mind in this respect (even though, in a sense, he was always to remain true to his original stance, as we shall see) that this juvenile work of his is so interesting.

Schopenhauer's later development is dimly foreshadowed once again at the end of the same section which contains the statement we have just quoted. Here he dismisses the question 'Why does consciousness appear as subject and object?' as senseless, giving as his argument that the question 'Why?' can be asked only under the authority of the principle of sufficient reason, which *presupposes* subject and object with their different forms and laws. But then he adds: "But I have a dim sense of anticipation that a wholly different part of philosophy from the one to which the present treatise belongs might some day provide us, not so much with an answer to this question as, rather, with something that makes the question superfluous, and that comes to terms with it in a wholly different way".[‡83]

We now turn to the *second* pertinent aspect of Schopenhauer's first-born in philosophy. We reiterate that he had not yet conceived a metaphysical system of his own at the time of writing the first version of his dissertation — the decisive break-through was to take place in the course of the next five years. Still, the dissertation already contains some hints about the general direction into which that "something that makes the question [of why consciousness appears as subject and object] superfluous" is going to be sought for; also what that vaguely promised,

---

* Indeed, "Schelling *and* Hegel", insofar as, for all their differences (already mentioned briefly in Chapter I, and to be addressed again in Chapter V), these two philosophers can be regarded as representatives of a way of thinking in which Thought and Being are identical. It should be remembered, however, that at the time when Schopenhauer wrote the first version of his dissertation (1812–1813) Hegel's hour had not yet struck. (It was not to strike until Hegel's 1818 appointment in Berlin; ironically, this was the very year when Schopenhauer completed his main work *The World as Will and Representation*!). Hence, we may assume that the passage in question aimed only at Schelling (on whose usage of the term "intellectual intuition", cf. our pertinent footnote on p. 43).

† "Soviel aber wird offenbar, daß unsre Untersuchung in keinem Ding an sich erstarrt."

‡ "Doch will mir ahnden, daß aus einem ganz anderen Theil der Philosophie, als der, zu welchem gegenwärtige Abhandlung gehört, nicht sowohl eine Antwort auf diese Frage, als vielmehr etwas, das die Frage überflüssig macht und auf eine ganz andre Weise beschwichtigt, uns kommen könnte."

larger work, "the contents of which would be related to the present one, as being awake is related to dreaming", will look like.

Take, to begin with, Schopenhauer's invocation, in the opening sentence of the book, not only of Kant, but also of Plato (the two thinkers especially recommended to the young student by his first philosophical teacher, Schulze-Aenesidemus, as we remember). This already suggests that, in the long run, the author was not to remain content with a Kantian critique of traditional metaphysics, but would seek for a way to satisfy his own metaphysical needs. What direction the quest was going to take is suggested in its turn by the very fact of Schopenhauer's introducing two new guises of the principle of sufficient reason besides the two older ones. We take up first his *principium rationis sufficientis agendi* and what distinguishes it from the *principium rationis sufficientis fiendi.*

Schopenhauer grants that we feel entitled to ask 'Why?' as soon as we are confronted with a decision, whether made by others or by ourselves. This implies our presupposing that the decision is the consequence of something else which has necessarily preceded it, so that, up to this point, volition obeys the law of causality. Yet he adds at once that, when acting, we are always aware that a decision, once made, does *not* necessarily follow from a representational situation belonging to one of the three other classes, but is directly dependent upon the willing subject alone; of this willing subject we perceive only the volition itself, not the situation preceding the volition. He concludes: "Thus we see here that the law of causality does not apply to Will, since according to the former every situation always follows necessarily from a preceding one; but that, since nevertheless we necessarily presuppose a sufficient reason for every decision of ourselves as well as of others, this must be the realm of a specific guise of the principle of sufficient reason, which I call the principle of sufficient reason of acting, *principium rationis sufficientis agendi*, or, for short, the law of motivation, while I call the reason presupposed by it the *motive.*"*[84]

Thus, there is something special about Will, though its metaphysical status is not yet overtly proclaimed in the first version. In the second version such a status is taken for granted, and the passage is therefore altered in a significant way. We read now that "motivation is causality seen from within". For we observe how, in the outer world, mechanical, physical, and chemical effects follow promptly upon their respective causes, yet how it is *possible* in the first place that *this* cause necessarily produces *that* effect remains a mystery. Equally mysterious would be the movements and actions of animals and of human beings (provoked as these are by causes here called "motives") if we did not know, from our own inner

---

* "Wir sehn hier also, daß für den Willen das Gesetz der Kausalität nicht gilt, da nach diesem jeder Zustand nothwendig und jederzeit aus einem vorhergehenden erfolgt; daß aber, weil wir dennoch von jedem eignen sowohl als fremden Entschluß einen zureichenden Grund nothwendig voraussetzen, hier eine eigne Gestaltung des Satzes vom zureichenden Grund herrschen müsse, den ich den *Satz vom zureichenden Grunde des Handelns*, principium rationis sufficientis agendi, kürzer, *Gesetz der Motivation*, und den durch dasselbe vorausgesetzten Grund das *Motiv* nenne."

experience, that we are concerned here with a *volition* provoked by its motive. The way the motive works is therefore known by us, not only from without (i.e., indirectly), as is the case with all other types of causes, but at the same time from within (i.e., directly); this is why we have here in our hands nothing less than the key to the riddle of the world (on this crucial point, see our exposition in Chapter VIII).[85]

Schopenhauer's later investigations, then, *did* "end in the rigidity of a thing-in-itself", and it is highly characteristic that he dropped the phrase in the second edition of the *Fourfold Root*. Still, as we have noted, there is a sense in which, this change of mind notwithstanding, Schopenhauer always remained true to his original position. For his thing-in-itself is neither Pure Intellect (i.e., the hypostasized knowing subject) nor Pure Matter (i.e., the hypostasized object) nor the Absolute in which the two are identical. That is to say, although Schopenhauer in the end arrived at a metaphysical system (thus transgressing the very border fixed by Kant), this was a metaphysics of sorts, which bore little resemblance to the metaphysical systems of German Idealism. Not the rational Idea, but the irrational Will is to stand at its centre; and his relatively low opinion of Reason is already clearly present in the first version of the dissertation.

Recall the passage in which Schopenhauer warns the reader that the way Reason is described and defined by him will not please those who hold with Kant that Reason, in the guise of Practical Reason, is the source of virtue and saintliness; and still less those who, with German Idealism, divinize Reason by calling it a faculty for gaining Absolute Knowledge. Abstract concepts (*Begriffe*) are useful, to be sure, since of the many representations (*Vorstellungen*) of which they are themselves the representations, they contain only that part which is needed at the moment, so that one is relieved of carrying with one a burden of unessential elements which would only confuse; hence, abstract concepts are indispensable in science, the aim of which in the final analysis is to reduce the particular to the general.[86] Furthermore, Reason has the even greater merit of enabling us to conduct our actions by means of concepts rather than of individual representations alone, as is the case with the animals. This is the condition of *freedom*, whereas an animal, lacking Reason, is left entirely to its immediate desire or affect. Finally, since Reason through its concepts makes possible an overall view of life and its events, it forms the condition for that equanimity and firmness which some human beings display in facing the accidents and miseries of life.[*87] Reason may, in view of all this, be called eminently *practical*; but not in the sense meant by Kant. For all this has nothing to do with morality; Reason is only an *instrument*, the

---

* In the second version, the remark on "equanimity and firmness" ("Gleichmuth und Festigkeit") is dropped, whereas the remark on the "animal", while maintained, has changed place: it appears now in the same section 'On the uses of concepts' ('Nutzen der Begriffe') in which the remark on science we have just quoted in the main text is equally to be found, whereas in the first version it occurs in the section 'Empirical truth' ('Empirische Wahrheit').

possession of which is indeed a *condition* for virtue or saintliness, but by no means its *source*, in that it is also a condition for becoming a perfect rascal!*[88]

Schopenhauer never abandoned this purely instrumental view of Reason; in this respect, the sole difference between the two versions of the *Fourfold Root* is that the first lacks a special section on 'Reason' such as he added to the second.† The new section on 'Reason' provided its author with an opportunity by means of waggon-loads of invectives to settle accounts with his philosophical rivals who happened to be precisely the prophets of Absolute Reason. But the downgrading of *Vernunft*, and the consequent upgrading of *Verstand* and *Anschauung*, had been there from the start, and it is here, in all likelihood, that the root is to be sought of one more original idea in Schopenhauer's doctor's thesis — the introduction of a *principium rationis sufficientis essendi*, to be distinguished from the *principium rationis sufficientis cognoscendi.*

A conviction he shared with Kant,‡[89] to wit, that geometry and arithmetic rest on pure intuition (*Anschauung*), made Schopenhauer impatient with the habit of making the *principium rationis sufficientis cognoscendi* enter this realm. As he sought to demonstrate in Section 40 (no. 39 in the second version) devoted to geometry, it would be possible to give such alternative proofs for geometrical

* This remark appears in Section 58, which bears the title 'Apology on imagination and reason' ('Apologie über Phantasie und Vernunft') and is to be found in the final chapter 'General remarks and results' ('Allgemeine Bemerkungen und Resultate'). This section has disappeared in the second version; in exchange for it, as we discuss at greater length in our main text, the new version contains a highly polemical Section 34 on 'Reason' ('Die Vernunft'). It is placed in Chapter 5, which deals with the second class of objects, and in it we find, not only barrages of abuse heaped upon Hegel *et al.*, but also a rejection of Kant's 'Ideas of Reason', the validity of which Schopenhauer still accepted in his first version. The same rejection is to be found in the supplement 'Critique of Kantian Philosophy' to *The World as Will and Representation*; we deal with it in some detail in Chapter VII.

† As mentioned in the previous footnote, this new section appears as Section 34 ('Die Vernunft') in the second version. Meanwhile, the first section of the chapter entitled 'On the second class of objects for the subject and on the specific guise of the principle of sufficient reason that prevails in it' ('Über die zweite Klasse der Objekte für das Subjekt und die in ihr herrschende Gestaltung des Satzes vom zureichenden Grunde'), which appeared in the first version as Section 27 under the heading 'Explanation of this class of objects. Reason' ('Erklärung dieser Klasse von Objekten. Die Vernunft'), appears now as Section 26 in a correspondingly somewhat abridged form under the equally abridged heading 'Explanation of this class of objects' (the addition 'Reason' thus having been dropped).

‡ It is true that, in the 'Transcendental Aesthetics', Kant held space and time to be, not discursive concepts, but pure (i.e., *a priori*) intuitions. But this did not prevent him from asserting in the 'Transcendental Analytic' (in the chapter on the 'system of all principles of pure understanding' ('System aller Grundsätze des reinen Verstandes'), section on the 'axioms of intuition' ('Axiome der Anschauung')) that space and time *are* discursive magnitudes insofar as everything that is extended in space and time can be *comprehended* only by our using the category of quantity.

propositions as would appeal to our immediate intuition of the *a priori* necessary properties of space, rather than to logical deduction. The latter is aimed more at persuading than at giving a true insight into the ground of Being; this might be one reason why many, otherwise excellent, brains show a dislike of mathematics.[90] One may perhaps suppose that it must have been this exposition especially which led to Goethe's original approval of Schopenhauer's way of thinking, in view of his own strained relationship with mathematics, even though it must not be forgotten that Goethe, while rejecting the mathematical treatment of physics, always spoke with respect about pure mathematics.

We finally address a *third* way in which an analysis of the *Fourfold Root* has a bearing upon our basic problem. Whereas Schopenhauer agreed with Kant about the *a priori* nature of the law of causality, he firmly rejected Kant's demonstration of the point as given in the *Critique of Pure Reason.* (In this case his criticism of the Kant of 'pure reason' was, in the first version, already sufficiently explicit not to require any change in the second). According to Kant, the primary synthesis of the sensual manifold carried out by the faculty of the imagination presents us with a certain succession of our perceptions, while leaving undecided the question as to whether, in a given case, this situation, which in our *imagination* precedes the other, is also the preceding one in the *object.* The *objective* nature of the temporal succession of our perceptions is warranted only by the faculty of understanding, with its pure concept of cause and effect. Schopenhauer argues against this that empirical phenomena may very well succeed one another in an objective manner without the first thereby becoming the cause and the latter the effect. For example, when I walk out of my house and the next instant a falling roof-tile hits me, there is no causal relationship between these two events; yet the succession is objectively determined in my perception, not subjectively by my own discretion (for in that case the succession would no doubt have been the other way round). Or take the succession of the notes in a musical piece, which is certainly not determined by the listener; yet no one will maintain that it is determined by the law of causality. In the same way the most elementary succession, namely, the rotation of the earth is without doubt known to us objectively, yet certainly not by means of the concept of causality, since its cause is wholly unknown to us.[91] According to Schopenhauer, Kant made the same mistake as Hume, though in the opposite direction. Whereas Hume had reduced all following-*from* to a simple following-*upon*, Kant interpreted every kind of constant *post hoc* as a *propter hoc*, which is just as wrong.[92]

Leaving aside whether Schopenhauer here offers a wholly correct rendition of Kant's argument, we have naturally become curious about his own mode of demonstrating the point. A first hint is supplied by the way Schopenhauer deals with the example Kant had used to illustrate his thesis that the objectivity of the succession of our perceptions is warranted only by the law of causality. Kant had compared our perception of the various parts of a house with how we perceive a ship sailing downstream. In the first case, Kant claimed, the succession in our perception is wholly subjective since we may regard the house from top to bottom or from bottom to top, whatever we prefer, whereas in the second case the succession in our perception of the local situations of the ship is objective since we cannot change it. But here he forgets, according to Schopenhauer, that in both cases we are concerned with a change in the local situation of two objects with respect to each other. In the first case, these objects are the body of the perceiver (or, more precisely, one of its parts: the eye) and the house (with regard to the parts of which the local situation of the eye successively changes); in the second case, the ship changes its local situation with respect to the stream. If only the perceiver had the power to draw the ship upstream in the same way as he has the power to move his eye in an opposite direction, then the succession might be inverted in the second case, too. We are concerned with *events* in both cases, not just in the second, as Kant thought; for our body is an object among others, and the succession of its representations (*Vorstellungen*), insofar as these are not mere phantasms, depends on the actions of other objects upon it. The succession, then, is objective, in that it is independent of the arbitrariness of the subject.[*93] But — and this is important, since in one sense it is the key point in Schopenhauer's entire philosophy — somewhat later on he remarks that our own body differs from other objects in space insofar as it is given to us *immediately*. (In a similar way, we might add — unlike Schopenhauer himself, who, in the first version, fails as yet to make the connection, however essential it is to become in his developed system — 'motives' are distinguished from 'causes' in that the former are *immediately* dependent on the subject, which, in this case, is the 'willing' subject).

The fundamental question to be asked now is how from *this* representation (*Vorstellung*) — i.e., from the representation of our own body — we can arrive at the representation of other objects in space. According to Schopenhauer, this is possible only by applying the law of causality, and it is in this direction that a better demonstration of the *a priori* nature of the law of causality may be sought than had been given by Kant.[94] Significantly, in the first version of the *Fourfold Root* this possible demonstration is given only in a brief outline (almost in passing) in the same Section 24 which has as its main object the refutation of Kant's demonstration; in the second version, however, an entirely new section has been inserted in which we find a much more elaborate account.[95] This remarkable circumstance deserves some closer attention, which it is to receive in due course,

---

[*] One might object that the second case still differs from the first in that, in the second, we are dealing with a relation between *three* objects (the perceiver's body being present here, too). Schopenhauer would probably retort that this may be true, but that the latter object remains passive in the second case. No doubt; but that does not alter the fact that we are concerned here with a *twofold* causal relationship: first, that between ship and river, and, second, that between the event as a whole and the observer, whose representation of the event is in itself a new event. In the first case there is only the causal relationship between the perceiver's body (which now takes an active part) and the house. On the problems inevitably raised by Schopenhauer's theory of causality we shall have more to say in Chapter IV, section b.

once we have provided a summing up of our findings so far, and also once we have further pursued our young philosopher's adventures on leaving Rudolstadt, the location of his *Fourfold Root*.

Apart from differences in tone and style, our comparison between the two versions of the *Fourfold Root* has brought to light two essential differences — the absence, from the first version, of an explicitly metaphysical background to the epistemology presented there; also, within that epistemology, a much less clearly elaborated account of Schopenhauer's own arguments for the *a priori* nature of the law of causality. It is well known that the former gap was to be filled in an impressive manner five years later, with the completion of *The World as Will and Representation*. It is less well known that the second gap was to be filled already *three* years later (i.e., long before the appearance of the second version of the *Fourfold Root*) with the publication of a booklet entitled '*On Vision and Colours*' (*Über das Sehn und die Farben*). This rather neglected piece of writing was the literary fruit of a memorable event in Schopenhauer's life — his personal encounter with Goethe in the autumn of 1813. The encounter would result in a short period of collaboration in the field of optics. During the same period Schopenhauer made the acquaintance of the Orientalist Friedrich Majer, who drew his attention to the *Upanishads*, which he studied for the first time in the summer of 1814, and which were to give a decisive turn to his metaphysical quest. All this is why the five years between the appearance of the first version of the dissertation and the publication of the main work were of decisive importance in Schopenhauer's development — more than that, it is the only period for which one can speak of any real development at all.[96] For after completing his philosophical system at the age of thirty, Schopenhauer certainly still wrote some important treatises, notably in the field of ethics, as we shall see. Yet in the last resort even these are no more than elaborations of thoughts already present in the main work. In order to gain a better understanding of the crucial significance of this brief period, we shall first turn our attention to Schopenhauer's optical studies and to their epistemological implications.

III

Colours

a. Goethe

In the preceding chapter we spoke of Schopenhauer's personal encounter with Goethe in the autumn of 1813, yet this was not quite the first time the two had met, or at least had been present in the same room at the same time. Here is how they had begun to make one another's acquaintance. On gaining permission from his mother to drop a commercial career, Arthur had entered the gymnasium at Gotha in the spring of 1807, in order to acquire the command of Greek and Latin required for entrance to the academy; but his stay there was not to be a very long one. A verse mocking one of his teachers led to a conflict with the rector and, rather than apologizing, young Arthur chose to leave the school in December of the same year and to complete his classical studies in private. This he did with such quick progress that (so we recall) he could enroll at the university of Göttingen in the autumn of 1809. These almost two years of private study, then, he spent in Weimar, though not in his mother's house. Well aware of her son's somewhat difficult temperament, she preferred not to live with him under one roof, and strictly rationed him as to the number of visits he was allowed to pay her. Yet Johanna Schopenhauer, née Trosiener, did admit him to the literary tea-parties she used to give in her drawing-room (just like every other self-respecting and cultivated lady at the time, and the more so since she was herself a successful author of society novels). At such evenings, then, Goethe was a regular guest, owing, no doubt, in good part to the circumstance that his hostess, unlike the majority of Weimar polite society, accepted Christiane [*97] under the liberal motto: "if Goethe gives her his name, I surely can give her a cup of tea". Probably the presence of the admired Goethe was the only reason for Arthur's attendance at these parties; but of course the great man did not take any notice of this youngster.

[*] Goethe had proceeded to formally marrying Christiane Vulpius out of gratitude towards her for saving him, by her resolute behaviour, from marauding French soldiers who had entered the house at the Frauenplan during the sack of Weimar in October 1806. The town was given free for plunder as a punishment for the little duchy's siding with Prussia — about which minister von Goethe had had his doubts — during the war which for Prussia had ended so disastrously at Jena (where the decisive battle took place, followed by the humiliating peace-treaty of Tilsit).

This, however, was to change in the course of Arthur's second stay at Weimar (November 1813 to May 1814). Not only did many violent quarrels between mother and son take place, which resulted in a definitive breakup,* but also a rather intense contact ensued with Goethe, who quite soon upon Arthur's settling down in Weimar invited him to a visit in his house at the Frauenplan. The visit passed off to mutual satisfaction, as is testified by a report from Goethe of 24 November and one from Schopenhauer of the same day,[98] all the more so since it was followed by regular visits. The motive underlying Goethe's hospitality was his hope of gaining Schopenhauer's collaboration in a matter which preoccupied him at the time — the theory of colours.

Goethe's bulky book '*On the Theory of Colours*' (*Zur Farbenlehre*) had appeared in 1810; but to his dismay it had fallen flat in the scientific world. This was all the more disappointing as Goethe himself considered this book, on which he had worked during two decades, as his most important work. As he was later to tell Eckermann, "I do not presume in the least on all my achievements as a poet. Excellent poets have lived along with me, still more excellent ones lived before me, and so there will after me. However, the fact that in my century I am the only one with a correct knowledge of the difficult science of colours, that is something in which I take some pride, and from this I derive a sense of superiority over many others."†[99] The polemical slant is unmistakeable: Goethe cannot deny that something has been done by others in this field of science during and before his life-time, yet he is the only one who has the "correct" knowledge. That is to say, with this work he quite consciously put himself in opposition to the scientific establishment, and when the approval he had somewhat naively hoped to come his way from this side failed to materialize, he took this as proof that the establishment had entered into some kind of conspiracy which rallied under the banner of Newton. After all, the *Theory of Colours* contained not only a historical section 'Materialien zur Geschichte der Farbenlehre' ('materials for the history of the theory of colours') and a didactic section, but also a polemical section 'Enthüllung der Theorie Newtons' ('unveiling of Newton's theory'), the title of which speaks for itself. It was this attack upon Newton which, indeed, was responsible for the

---

* One of these quarrels culminated in a dialogue too funny to withhold from the reader. Angered by an irreverent remark by his mother on the title of the dissertation he had brought home ("this is probably meant for pharmacists?"), Arthur exploded into the prophecy that his name would still be immortal at a time when not a single copy of his mother's novels would still be found anywhere in the world; upon which Johanna — having a ready tongue, too — retorted: "And of your book the whole first edition will still be there to be had".

† "Auf Alles was ich als Poet geleistet habe [pflegte er wiederholt zu sagen], bilde ich mir gar nichts ein. Es haben treffliche Dichter mit mir gelebt, es lebten noch Trefflichere vor mir, und es werden ihrer nach mir seyn. Daß ich aber in meinem Jahrhundert in der schwierigen Wissenschaft der Farbenlehre der Einzige bin, der das Rechte weiß, darauf thue ich mir etwas zu gute, und ich habe daher ein Bewußtseyn der Superiorität über viele."

(on the whole) negative response from the professionals and for the anti-science reputation it has earned for its author up to the present day.

Colour phenomena had already begun to interest young Goethe under the inspiration, once again, of the circle around Suzanna von Klettenberg;[100] a further, powerful stimulus came from his stay in Italy in 1787. In Rome, Goethe had regular contact with a colony of German artists, and he was eager to qualify himself in drawing and in painting. In this he had some success, in that he brought home about two thousand sketches and watercolours; yet he became aware (not without regret) that he lacked true genius in the plastic arts. What had meanwhile struck him in Rome was his discovery that those artists he had encountered, while making use of scientific knowledge in matters of perspective and anatomy, did not bother about objective explanations of their choices of colours, apparently allowing themselves to be guided in this respect by mere subjective preference.[*101] This aroused his curiosity, and, back in Germany, he began to study optical phenomena. To that purpose he borrowed prisms and other instruments from a certain *Hofrat* Büttner, a professor at the university of nearby Jena; however, other, more urgent tasks kept him from seriously engaging in experiments with these pieces of equipment. When their owner, who had grown somewhat impatient, in the end rather urgently requested their restitution, Goethe complied with the demand, but decided to have at least one quick glance through the prism before wrapping it up. Thus looking through the prism at a white-washed wall in one of his rooms, he found to his surprise that no colour spectrum appeared. This, according to his own account, gave him a sudden flash of insight that Newton must have been wrong in supposing that white light is composed of rays of different refrangibility.[†102]

It is therefore wrong to suggest (as has often been done) that Goethe was *a priori* convinced that white light, since it appears as simple and pure to the eye, cannot be composite. In point of fact he had up to then joined most other people in taking for granted the prevailing opinion.[103] But now he had found coloured fringes appearing only at some places where bright areas in the room met dark areas; in other words, he found that colour phenomena appear only at the *boundary* between light and darkness, and this made it difficult for him to continue believing that colours are already contained in white light. He now began seriously

---

[*] This in spite of the fact that Leonardo had already written a treatise on painting (posthumously published in 1651), in which he had tried to lay down the laws, not only of perspective but also those that govern the phenomenon of colour. Goethe, too, seems not to have known it, for he does not mention it in the remarks about Leonardo in the chapter on the 'history of coloration since the restoration of art' ('Geschichte des Kolorits seit Wiederherstellung der Kunst'; a title which, let it be said in passing, is characteristic of the author's classicist contempt of medieval art) from the historical part of the *Theory of Colours*.

[†] "Es bedurfte keiner langen Überlegung, so erkannte ich, daß eine Gränze nothwendig sei, um Farbe hervorzubringen, und ich sprach wie durch einen Instinct sogleich vor mich laut aus, daß die Newtonsche Farbenlehre falsch sei."

to study this field of science. By 1791 this had already resulted in a smaller work '*Contributions to optics*' (*Beiträge zur Optik*), to culminate in 1810 in the appearance of *On the Theory of Colours* in two volumes. (Volume 1 contains the didactic and the polemical sections, the first one of which had already been published separately in 1808, and volume 2 contains the historical section; the whole work is supplemented with an atlas in quarto and some explanatory text, entitled 'Sechzehn Tafeln nebst Erklärung zur Goethes *Farbenlehre*', i.e., 'Sixteen tables with explanation to Goethe's *Theory of Colours*'). This work, then, as much as, or even more than *Faust*, he considered to be his true masterpiece. So much the greater was his disappointment to find that it met with so little approval in the scientific world. Against his professed disbelief that colours are already contained in light as such, it was argued that, strictly speaking, Newton had not said that — he had asserted only that in what we perceive as white light (the light of the sun), certain physical realities (i.e., rays of different refrangibility) are present which have the power to provoke in us the sensation of different colours.[*104] Here Goethe could, and did, retort that Newton, in the very title of his first proposition, *did* speak about "Lights which differ in Colour [and that] differ also in degrees of Refrangibility",[105] an expression which in the polemical section of his book had provoked Goethe to the ironical question: "What kinds of light, then?" ("Was denn für Lichter?").[106] In fact here, too, Newton had really meant physical rays, for the characterization of which, however, he saw no other possibility than the use of the colour sensations connected with them. No alternative had indeed been open to him,[†107] as it was not until 1803 that Thomas Young was able to define homogeneous rays by numbers — namely, those of different wavelengths — which dispense with the need to introduce a qualitative terminology into an otherwise quantitative treatment of light.

Another question is whether Newton's preoccupation with the rays that compose light sources made him perhaps neglect the possibility that external factors (e.g., the properties of the refracting medium) might contribute in an essential way to the appearance of colours. In this context Goethe had a predilection for pointing to Newton's assumption that all refracting materials (in this case, flint-glass and crown-glass) disperse colours in a constant proportion and for holding this responsible for Newton's disbelief in the possibility of an

---

[*] "For the Rays to speak properly are not coloured. In them there is nothing else than a certain Power and Disposition to stir up a Sensation of this or that Colour."

[†] It would be somewhat otherwise if it were true that Newton had *a priori* denied the possibility of the existence of such *invisible* radiation as was actually to be discovered about a century later (the infrared part of the spectrum by Sir William, originally Friedrich Wilhelm, Herschel in 1800; the ultraviolet part shortly afterwards by Johann Wilhelm Ritter). Rudolf Laemmel, in his biography of Newton, mentions a letter by Newton to a certain Dr Maddock (dated 7 February 1679), who had suggested this possibility, only to see it rejected by Newton. Richard Westfall mentions this letter in his biography of Newton, but expresses his doubts about its authenticity, because "its tone of light irony resembles Newton's style about as much as a gazelle resembles a tiger".

achromatic telescope. The actual construction of such a telescope by Hall, and later by Dollond, served him therefore as a striking proof that Newton's mind had been prejudiced. Goethe was of course aware of the counter-argument that this did not imply that Newton's entire theory was invalid since one ought to distinguish between 'refraction' and 'dispersion'. But he considered this argument to be a mere verbal trick, designed only to put Newton in the right after all.*[108]

Goethe, for his part, thought nothing of heterogeneous light-rays, or, as a matter of fact, of any 'rays' at all. The spectrum is produced by the image of the sun as a whole, not by an isolated ray, far less could such a ray be split up further. To Goethe, *Light* was the Whole, infinite and indivisible; *colour* was finite and individual. This led him in his turn to stress in an at least equally one-sided manner the overall significance of external factors for the appearance of the colour phenomenon. Referring with approval to the 17th-century Jesuit Athanasius Kircher, who had described colour as *lumen opacatum* (i.e., shaded light), he defined colour as a σκιερον *(skieron)*, a kind of shadow, resulting from an interaction of light and darkness;[109] with that he placed himself in a long tradition going back to Aristotle, but also to the hermetic tradition, for which the polarity of Light and Darkness and the idea of polarity in general was so essential. Goethe was aware, of course, that in a common-sense view of things a mixture of light and darkness yields only grey. But he insisted that with the aid of a "turbid medium" (*trübes Medium*), such as vapour, dust clouds, mist, smoke or clouds (or, for that matter, certain liquid mediums — e.g., water — or solid mediums — e.g., glass — made turbid through the admixture of some chemical substance) a combination of light and darkness can be effected that gives rise to the phenomenon of colour. Looking at darkness through a turbid medium gives the sensation of *blue* (which is why the sky or mountains, seen from afar, appear blue), whereas looking at light which comes through a turbid medium from a light source behind it gives the sensation of *yellow*.[110] In the first case there is darkness behind the medium, and the only light comes from a source in front or to one side of it; therefore the light is not transmitted through the medium, but reflected by it. From these two 'polar opposites' (called by Goethe the "passive" or "minus" side, and the "active" or "plus" side, respectively) all other colours can be derived through the intensification (*Steigerung*) of one, or through the union of two of them.†[111] Since

* The matter is biographically interesting since as a matter of fact efforts spent by Newton to improve the already existing refracting telescope by grinding differently shaped lenses (in order to get rid of the chromatic interferences that used to appear at the edges of the projected image) had originally stimulated his interest in optical studies. His *experimentum crucis*, however, convinced him (wrongly) that this was impossible; he therefore gave up the attempt and turned instead to the construction of a novel kind of telescope (the reflecting telescope). That it was a modern technological problem which aroused Newton's interest in the phenomenon of colour whereas it was an aesthetic one in the case of Goethe may well be regarded as symbolic of their differing frames of mind!

† This pair of concepts (*Polarität* and *Steigerung*), which did not yet appear in the well-known fragment by Georg Christoph Tobler (composed by the latter during a visit to

any colour is *eo ipso* darker than white, its intensification must of necessity yield a still darker one. Thus intensified yellow (the brightest of colours) gradually passes into darker, more reddish, hues (which is why the sun at midday height appears as brightly yellow, but at dawn and twilight appears red due to the thicker mist), and intensified blue passes into the darkest of all colours, i.e., reddish blue (*Rotblau*). Since, therefore, the latter contains an element of redness, too, it may be said that *both* polar opposites intensify to a more or less reddish tint; a union of the intensified forms of the primary colours finally produces the 'true red', which because of its high dignity is also called '*purple*' by Goethe. This is important: what is normally called 'red' (the red component of the spectrum) is, according to Goethe, still a somewhat yellowish red. His 'purple' (or 'true red') is not exactly what is usually called 'purple' either, but rather 'crimson' or what contemporary theories of colour used to call 'magenta'. *Green*, on the other hand, results from a union not of the intensified but of the original forms of the two primary colours.[112]

Altogether we may summarize Goethe's theory thus. It is in fact a two-colour theory since (guided by the fundamental idea of *polarity*) it starts from the opposition of two fundamental colours, as can be illustrated by a linear colour-schema:

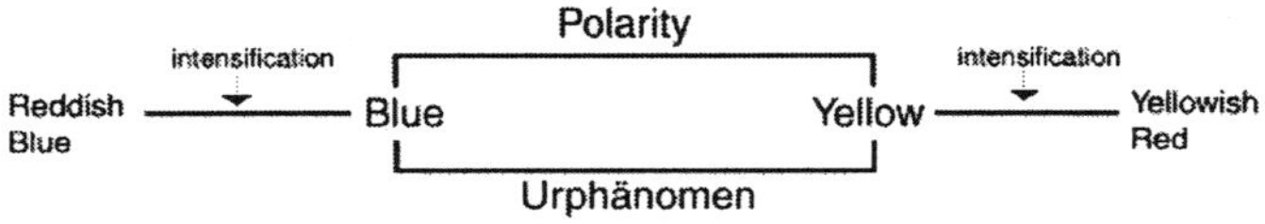

At the same time (and here it is guided by the other fundamental idea of *intensification*) the theory results in a six-colour circle, in which, otherwise than in Newton's seven-colour circle (*Opticks*, Book I, Part II, Fig. 11; here reproduced to the right) contrast colours are placed opposite one another:

Goethe in 1781 under the title 'Die Natur' and anonymously published two years later in the *Tiefurter Journal*, 32. Stück, 1783), was adopted by Goethe from Schelling, in whose philosophy of nature it plays a central part (cf. our remarks on the relation between Goethe and Romantic *Naturphilosophie* in Chapter I, p. 34). That Tobler was the true author of the fragment was already revealed by Frau von Stein in a letter (28 March 1783) to Karl Ludwig von Knebel, who took Goethe to be the author, in spite of a confirmation to the contrary by Goethe himself in a letter (3 March 1783) to the same adress. Much later however, in a letter to Chancellor von Müller (24 May 1828), Goethe writes that, although he does not remember it, he thinks it probable that he himself is the author since it is in accordance with his views at that time, and since the manuscript (found shortly before in the posthumously published correspondence of Duchess Anna Amalia of Sachsen-Weimar) betrays the handwriting of his secretary at that time.

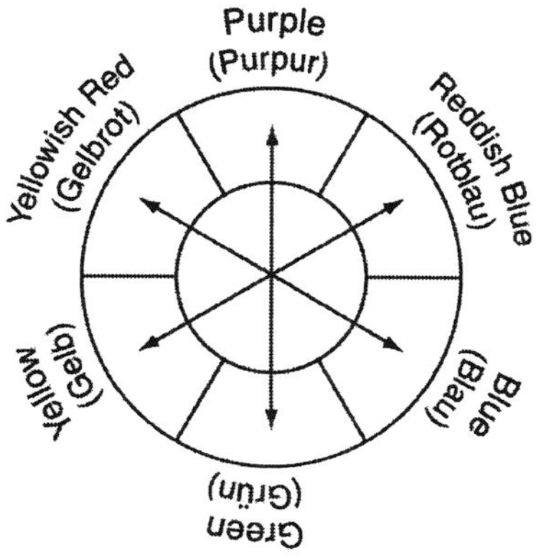

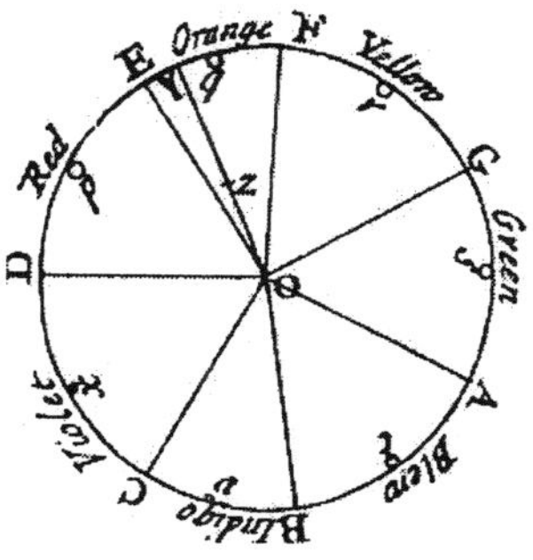

It is only fair to note that present-day theories of colours, in spite of their wholly different theoretical starting point, are here on the side of Goethe rather than Newton!* Summing up once again, Goethe's entire theory can be represented thus:

---

* According to modern theories of colours there are three kinds of colour receptors (cones) in the eye: those that are receptive of short-wave, medium-wave and long-wave light, respectively; thus there are three primary light-colours (ultramarine blue, green and red). The additive combination (cf. footnote on p. 61) of all three together naturally produces white, the additive combination of only two of them produces the three primary pigment-colours, i.e., cyanide blue (from ultramarine blue and green), magenta (from ultramarine blue and red) and yellow (from green and red). The additive combination of these with the primary light-colour left out in each case (i.e., cyanide blue with red, magenta with green and yellow with ultramarine blue) naturally produces white again; the three possible subtractive combinations of the primary pigment-colours produce the three primary light-colours again (ultramarine blue from magenta and cyanide blue, green from cyanide blue and yellow, red from yellow and magenta). Thus in the cases of magenta vs. green, cyanide blue vs. red, and yellow vs. ultramarine blue we are indeed concerned with the three pairs of complementary contrast colours we find in Goethe's colour ring (with the important proviso that their *additive* union *does* produce white; and with the terminological proviso that Goethe calls 'purple' or 'true red' what we call now 'magenta', and 'blue' what we call now 'cyanide blue' or simply 'cyanide', whereas at the same time he does not sharply differentiate between (spectral) red and orange (calling them *Gelbrot*, i.e. 'yellowish red' and *Rotgelb*, i.e. 'reddish yellow', respectively, but sometimes using 'orange' for both of them and moreover sometimes confusing them) nor between ultramarine blue and violet (calling them *Rotblau*, i.e., 'reddish blue' and *Blaurot*, i.e., 'bluish red', respectively, but sometimes using 'violet' for both of them and again sometimes confusing them). As we shall see, Schopenhauer (who rightly remarked that Goethe was inconsistent insofar as he originally called *blue* [i.e., cyanide] — instead of *violet* [today we would rather say ultramarine blue] — and yellow polar opposites) drew indeed from this the almost inevitable (but from Goethe's point of view unacceptable) conclusion that each of these complementary contrast pairs must produce white (as lights, not as pigments, to be sure).

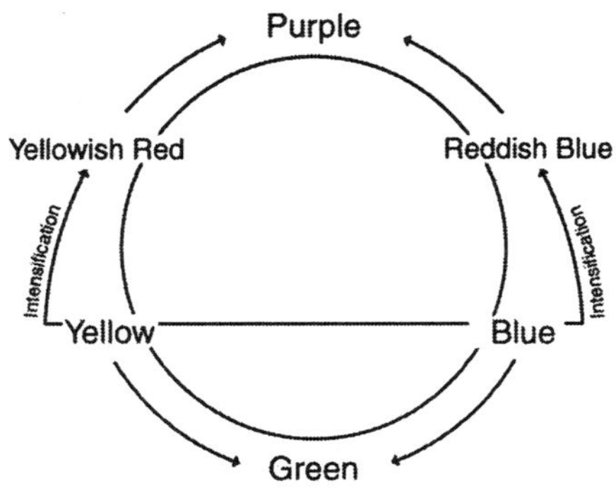

All this is theory, to be sure, but according to Goethe it can easily be demonstrated for colour phenomena produced by the eye itself like coloured shades and after-images. These he calls "physiological colours", and they are the subject matter of the first chapter of the didactical part of his *Theory of Colours*. But it can also be demonstrated by colour phenomena produced by transparent, refracting media like prisms; that is, for "physical colours" as he called them, which are the subject matter of the second chapter. These are produced in our eye by external factors; to that extent they are of an objective nature, which distinguishes them from the purely subjective 'physiological colours'. They are also transient and fugitive, however, so they must be labelled "subjective–objective". Unlike these, "chemical colours" — described in the third chapter — are fixed, because they are part and parcel of a material substratum; they are therefore labelled "objective". Here it is less easy to demonstrate the σκιερον (*skieron)*, but Goethe believes that the chemical polarity of acids and alkalis reflects the chromatic polarity of active and passive colours, in that the former are connected with the acids and the latter with the alkalis.[113] There remain the extremes of white and black. Transparent bodies constitute the highest stage of inorganic matter, to which "pure turbidity" corresponds, i.e., white. Therefore water, crystallized to snow, appears white; for the same reason dehydrated salt crystals or pulverized glass appear as white powder. "Earthes", on the other hand, are all white in their natural condition and become transparent by crystallization. Black does not appear to us as quite as elementary (*uranfänglich*) as white; but in the vegetable realm we observe it in the case of semi-burning, as with coal; metals, too, often become black through semi-oxidation. Meanwhile de-oxidation ('reduction' as we would say now) seems to produce a black colour as well (e.g., ink through the admixture of gall-nut).[114]

However, all bodies are susceptible to colour; this can either be created upon it, or (if already present) intensified in various stages or even brought to culmination point. Here chemical polarity comes in once again. Goethe sought to illustrate this in particular by pointing to what happens when metals are made to react with acids, which, according to him, produces the most important phenomena of chemical colours. Although all this means that colour, even in its material guise, is transformable, under certain circumstances it remains fixed. Thus arise the natural pigments (e.g., madder or indigo). The art of painting consists at bottom of mixing such pigments in an infinite range of possible combinations. It should be kept in mind, however, Goethe warns, that all colours mixed together, although they lose their individual natures, still hold on to their general nature as σκιερον (*skieron)*; and, since they are now no longer seen side by side, the result this time is not a harmonious composition but only *grey*. "That all colours, mixed up together, produce white, is an absurdity which, together with other absurdities, people have over a century accustomed themselves to repeat credulously and against all visual evidence".*[115] Newton could put that in his pipe and smoke it! However, it is only fair to Goethe to remark that neither Newton nor his followers at the time were able to give a satisfactory explanation of pigments and their mixing; it was not until Helmholtz' epochal investigations in the mid-19th century that the difference between what has ever since been called 'additive' and 'subtractive' colour mixing has come to be properly understood.†

As already mentioned (and not to be wondered at), most professional physicists failed to be converted to Goethe's anti-Newtonian creed. An exception was Dr Thomas Seebeck, who was to remain Goethe's chief adviser in matters of optics. A bigger response (which is not to be wondered at, either) was found with the representatives of *Naturphilosophie* like Schelling, but also with Hegel. The latter took an active part in Seebeck's experiments on what they called "entoptic" colours, i.e., the colour phenomena produced by polarized light in dual refracting crystals (more on this in Chapter V, p. 119). Goethe adopted the term, and as late as 1820 he wrote himself a little treatise on the subject (meant as a supplement to the main work). The explanation of these phenomena he sought once more only in the turbid medium (provided by the transparent material of the crystalline body) and not in (polarized) light itself; hence the scornful remarks he directed at Etienne Malus (who, as is well known, in 1808 had discovered the polarization of light, a

---

* "Daß alle Farben zusammengemischt Weiß machen, ist eine Absurdität, die man nebst andern Absurditäten schon ein Jahrhundert gläubig und dem Augenschein entgegen zu wiederholen gewohnt ist."

† In additive colour mixing the eye receives the sum of the light energies that gather in one place, e.g. when differently coloured lights are superimposed on a projection screen; since there is in this case a greater activation of one or more of the three kinds of colour receptors in the retina than the activation produced by one single colour the resulting colour is *brighter* than each of its components. In subtractive mixing there is, on the contrary, a lesser activation of one or more of the three kinds of colour receptors in the retina than the activation produced by one single colour, since by mixing pigments or by laying coloured transparencies over each other, the absorption process (and thus the diminishing of the total light energy that reaches the eye) is reinforced, and the resulting colour is *darker* than each of its components. This effect is produced by the fact that the colours of pigments and the local colours of objects result from the light they reflect after their surfaces have absorbed their specific part of the wavelength area; e.g. a red surface absorbs everything but the wavelengths corresponding to red.

phenomenon of which the precise meaning became clear only in combination with the somewhat earlier work of Thomas Young and the later work of Augustin Fresnel in the years 1818–1821). Here is what Goethe had to say about Malus: "That Newton with his prismatic experiments made the opening as small as possible in order to symbolize easily a line as a light ray has brought irreparable confusion on the world, from which perhaps centuries will still suffer. Through this small hole Malus was driven to an adventurous theory ...".*[116] Or, in his posthumous manuscripts: "The unholy little hole through which Newton has introduced his captious experiments still blinds the experimenters. Malus, too, accommodated his reflection apparatus to such an opening and Seebeck could not arrive, himself, at the true view as long as he followed this prescription."†[117] Or in the following satyrical poem, in the fourth line of which Malus is not mentioned explicitly, but implicitly since it refers to the polarization theory developed by him:[118]

Möget ihr das Licht zerstückeln,	"You may cut light to pieces
Farb' um Farbe draus entwickeln,	to develop colour after colour from it
Oder andre Schwänke führen,	or make other farces
Kügelchen polarisiren,	like polarizing little balls
Daß der Hörer ganz erschrocken,	such that the frightened hearer
Fühlet Sinn und Sinne stocken:	feels himself wholly upset;
Nein! Es soll euch nicht gelingen,	no, you will not succeed
Sollt uns nicht beiseite bringen;	in setting us aside;
Kräftig wie wir's angefangen,	with the same vigour as at the start
Wollen wir zum Ziel gelangen.	we will attain our end."

Indeed, as the reader will notice from this mocking expression "Kügelchen polarisiren", Malus was still an adherent of Newton's corpuscular theory; however, with the later explanation of 'entoptic colours' from interference of aether waves Goethe would not have anything to do, either.‡[119]

---

* "Daß Newton bei seinen prismatischen Versuchen die Öffnung so klein als möglich nahm, um eine Linie zum Lichtstrahl bequem zu symbolisieren, hat eine unheilbare Verirrung über die Welt gebracht, an der vielleicht noch Jahrhunderte leiden. Durch dieses kleine Löchlein ward Malus zu einer abenteuerlichen Theorie getrieben ...".

† "Das unselige kleine Löchlein, wodurch Newton seine captiösen Versuche eingeführt, verblindet noch immer die Experimentierenden. Malus bezog seinen Spiegelungs-Apparat gleichfalls auf eine solche Öffnung und Seebeck, so lange er dieser Angabe folgte, konnte selbst nicht zu der wahren Anschauung gelangen."

‡ Malus on his part had on 31 July 1811 written a review of the *Theory of Colours* which passes over the fourth and fifth chapter of the didactic part because, according to the reviewer, in these final chapters facts and metaphysical arguments are mixed together in a way which makes a clear analysis of them impossible ("dans ces derniers chapitres les faits et les raisonnements métaphysiques sont tellement confondus ensemble, qu'il serait impossible d'en faire analyse"), and which arrives at a general conclusion about this part in which it is acknowledged that Goethe (if you leave his vague metaphysical explanations aside) has given a rather complete description of the most important optical phenomena

In his critique of Newton, Goethe always stressed the need for patient empirical investigation uninfluenced by premature hypotheses, such as Newton's theory of refrangibility. He therefore severely criticized Newton's mode of experimentation, as we saw in our Chapter I, p. 23. We saw there, too, that Goethe was not a simple-minded empiricist either; his persistent quest for the *Urphänomen* (a concept central not only to his theory of colours but to all of his scientific work) is proof to the contrary. His unhesitating acknowledgment that a true scientific outlook must often break with common sense and visual evidence is clearly demonstrated by comments like the following: "The Copernican system is based on an idea that was difficult to grasp and that, moreover, contradicts our senses every day. We just repeat what we neither know nor understand. The metamorphosis of plants contradicts our senses, too."*[120] And of course his own explanation of colour phenomena (i.e., the marriage of the polar opposites of Light and Darkness under the auspices of the turbid medium — an *Urphänomen* if ever there was one!) transcends pure experience no less than Newton's rays. The essential difference is that Goethe's type of explanation is of a qualitative rather than of a quantitative nature. This is in agreement with his general disinclination towards mathematics and with his predilection for a kind of science which appeals to the whole person, that is, not only to the analytical and abstracting but also to the intuitive and poetic propensities. The spirit of the age, he felt, still stood in the way of a breakthrough

---

which betrays his knowledge of all experiments and his having repeated them himself, but which makes clear, too, that he has never measured any phenomenon; hence the vagueness of his theory ("Si l'on considère cette première partie de l'ouvrage de M.Goethe, indépendamment de la métaphysique et des explications vagues qu'elle renferme, on y trouve une description assez complète des principaux phénomènes de l'optique. On voit que l'auteur a connu toutes les expériences, et les a répetées lui-même; mais on s'apperçoit en même temps qu'il n'a mesuré aucun phénomène, ce qui est la cause du vague qui règne dans sa théorie"). He likewise passed over the historical part; and as to the polemical part: after having mentioned some concrete points of Goethe's critique of Newton he proceeds to the way in which Goethe speaks about Newton and his partisans, namely, as if they belong in a lunatic asylum and that their skulls ought to be inspected by Dr Gall [the famous phrenologist, PL], and expresses his amazement about Goethe's using such arguments in a work on physics which, according to the reviewer, makes clear that the author does not possess that frame of mind which belongs to a man who sincerely seeks truth, so that one may safely assume that he will not get many followers and that it is not in this book that one will track down the real errors that Newton may have committed ("On est surpris de voir M.Goethe employer de semblables arguments dans un ouvrage de physique, et l'on s'apperçoit trop souvent qu'il n'est pas dans la disposition d'esprit qui convient à ceux qui cherchent franchement la vérité. Il est donc probable, qu'il fera peu de prosélites, malgré son excessive intolérance. Comme il condamne indistinctement toutes les opinions du Traité d'optique, ce n'est pas dans son livre qu'on osera rechercher les erreurs que Newton peut avoir commises").

* "Das Copernikanische System beruht auf einer Idee, die schwer zu fassen war und doch täglich unseren Sinnen widerspricht. Wir sagen nur nach, was wir nicht erkennen, noch begreifen. Die Metamorphose der Pflanzen widerspricht gleichfalls unseren Sinnen."

toward such a new scientific outlook, but this being so it was all the more urgent to seek potential disciples to whose hands the Goethean legacy could be entrusted and who might in future bring about its inevitable (of this he was convinced) triumph.

b. Goethe and Schopenhauer

Could young Schopenhauer be such a disciple? He, after all, had brought from Rudolstadt a philosophical dissertation in which (as we discussed in Chapter I, p. 50) he had explained the dislike many sharp minds feel for mathematics through the purely logical–deductive manner in which this science is usually presented. Also, and of more general significance, he had defended there the priority of *Anschauung* (intuition) over abstract *Begriffe* (concepts). Small wonder that such a philosophical attitude met with Goethe's approval; in fact, the reading of the dissertation (a copy of which the proud author dedicated and sent to Goethe) made the customarily rather reserved poet at long last address the young philosopher at one of Johanna's tea-parties (although courtesy towards his hostess, who happened to be the philosopher's mother, may also have played a part). Yet within no more than a few weeks of discussion and joint experimentation, Goethe vented his annoyance over his opinionated disciple in the following epigram: "Trüge gern noch länger des Lehrers Bürde / wenn Schüler nur nicht gleich Lehrer würden" ("I would gladly go on carrying the teacher's burden for a while / if only pupils would not at once turn into teachers themselves").[121] And when, one year after his departure from Weimar to Dresden,* Schopenhauer put down in writing his own theory (one in which, for all the inspiration gratefully received from Goethe, he claimed nothing less than to give for the first time a real *theory* of the origin of colours to which Goethe had only contributed important materials!) and sent Goethe the manuscript with a request to give it his blessing by providing it with a preface, the latter, as one might expect, kept aloof, so that in the end there was nothing for it but to have it printed, in the spring of 1816, without the hoped-for Goethean sanction.

Much later, Schopenhauer himself (with a, for him, quite remarkable *sans rancune* due to an admiration for Goethe left unbroken by the disappointment the latter had caused him) was to give the following explanation of the regretted estrangement: "The amazing *objectivity* of his mind, which always impresses his poetry with the stamp of genius, is precisely what prevented him from returning to the *subject*, in this case to the seeing eye itself, in order to pick up there the ultimate threads on which the whole phenomenon of the world of colours depends; I, on the other hand, coming as I did from the school of Kant, was optimally prepared to satisfy this demand, so I was able, one year after being withdrawn from Goethe's personal influence, to find the true, fundamental, and irrefutable

* Dresden is where he completed his *magnum opus*, *Die Welt als Wille und Vorstellung.*

theory of colours."*[122] *Excusez du peu!* Yet this modest statement (a modesty which, as we shall see, he likewise had no compunction whatsoever in displaying in his personal correspondence with Goethe) does make the essential point: Schopenhauer's epistemological opinions (rightly or wrongly linked here by him to Kant — this will be the subject of our next chapter) caused him to give a purely physiological interpretation of the phenomenon of colour, which Goethe could not accept. Yet, as we have seen, it was his epistemological study *On the Fourfold Root of the Principle of Sufficient Reason* that had originally drawn Goethe's sympathetic attention towards him; and Schopenhauer, on his part, never ceased to share Goethe's objections to Newton! In order to gain a grasp of this comedy of errors we shall now take a closer look at *On Vision and Colours,* in which Schopenhauer, as noted at the end of the preceding chapter, for the first time gave an elaborate account of the essentials of his physiology-based theory of causality, in close association with his no less physiology-based theory of colours.

c. Schopenhauer

"Alle Anschauung ist eine intellektuale" ("All intuition † is of an intellectual nature") — this is the lapidary statement with which the first chapter 'On Vision' ('Vom Sehn') of *On Vision and Colours* opens.‡[123] This is not of course to be understood in the Schelling-like sense of a higher form of cognition, for as we saw in our Chapter II, already in his dissertation Schopenhauer spoke with polite irony of this would-be faculty of the human mind, just as in his later work he would repeatedly speak with scorn about that "sixth sense of the bats" which he mockingly confessed not to possess.[124] Quite to the contrary, we are concerned

---

* "Gerade die erstaunliche *Objektivität* seines Geistes, welche seinen Dichtungen überall den Stämpel des Genies aufdrückt, stand ihm im Wege, wo es galt, auf das *Subjekt*, hier das sehende Auge selbst, zurückzugehen, um daselbst die letzten Fäden, an denen die ganze Erscheinung der Farbenwelt hängt, zu erfassen; während hingegen ich, aus Kant's Schule kommend, dieser Anforderung zu genügen auf's Beste vorbereitet war: daher konnte ich, ein Jahr nachdem ich Goethes persönlichem Einfluß entzogen war, die wahre, fundamentale und unumstößliche Theorie der Farbe herausfinden."

† On the usage of the terms "intuition" and "perception" in the present book, see footnote on p. 43.

‡ The reader should be aware that, just as with the *Fourfold Root*, there exist two versions of *On Vision and Colours* with a considerable time lag between them. Other than with the second version of the *Fourfold Root*, the principal thesis of *On Vision and Colours* displays no further essential development but (apart from a new preface in which Schopenhauer gives a more detailed account of his relation with Goethe and of the differences between them than naturally was possible in the first version) only some (if sometimes considerable) enlargements, some updating (especially on the wave-theory), and, of course, some freshly added mud-slinging. Therefore we feel entitled to refrain in this case from a systematic comparison between the two versions. Only when we refer to an argument which as such appears solely in the second version, we mention its absence from the first version.

here with cognition in the primitive, still unreflected guise that humans share with the animals and which underlies all intuition. For 'intuition' (*Anschauung*), according to Schopenhauer, should be clearly distinguished from mere 'sensation' (*Empfindung*). Intuition involves the recognition of objects, which is possible only because our Understanding (*Verstand*) relates any impression the body receives to its *cause*, which it then localizes in *space*. This immediate inference from an effect to its cause is a cognition from pure Understanding, not a conclusion of Reason (*Vernunft*). Certainly the law of causality in its articulated guise is a reflection, and to that extent it belongs to the domain of Reason; but its *unconscious*, immediate application is the work of Understanding, which always precedes reflection. And it likewise precedes intuition, since it is the very thing which makes intuition possible. The senses alone will not do. They are no more than spots on the body which are to a heightened degree susceptible to the effects of other bodies. Each of the senses is susceptible to one specific kind of effect, to which the others are barely, if at all, susceptible. So far Schopenhauer is in accord with Johannes Müller's famous law of specific sense-energies (which as such had not yet been proclaimed at the time when Schopenhauer wrote the first version of *On Vision and Colours*); in the second version (which is from 1854, when this law had been known for a long time) he departs from it insofar as he maintains next that the difference in sensation between the five senses has nothing to do with the nervous system itself, but only with the manner in which it is affected.[125] It is somewhat astonishing, therefore, to note that he does not refer to Müller's theory in this revised edition, not even to dispute it. Maybe Wilhelm Ostwald was right in ascribing this omission to the circumstance that Schopenhauer used to take more notice of the discoveries of foreign than of German investigators, whom he was *a priori* inclined to consider as corrupted by university philosophy.[*126]

However this may be, what interests us here is that Schopenhauer's theory of intuition enabled him to give a (so he thought) better demonstration of the *a priori* nature of causality than had been given by Kant (this is the topic on which we ended our previous chapter, as the reader will remember). "The child", Schopenhauer asserts in an interesting passage which reads as an anticipation of modern developmental psychology and which we quote at some length as it will prove of crucial importance for both this and our next chapter, "the child, during the first weeks of its life, feels with all its senses; but it does not intuit, it does not apprehend; therefore it gazes stupidly into the world. Meanwhile it soon begins to learn to use Understanding, to apply the law of causality — of which it is aware prior to any experience — and to connect this law with the likewise *a priori* forms of all knowledge: space and time. Thus it proceeds from sensation to intuition, to apprehension; and from now on it looks with sensible, intelligent eyes into the world. Since, however, each object affects each of the five senses in a different manner, the effect nonetheless pointing to one and the same cause — which, for

---

[*] It must be noted, however, that Schopenhauer does mention Müller with approval in another context, which proves an at least partial acquaintance with the latter's work.

this very reason, presents itself as an object — the child learns to intuit by comparing the diverse impressions it receives from one and the same object. It handles what it sees, looks at what it handles, traces back the sound to its cause, invokes the help of smell and taste, and finally it takes into account the effects upon the eye of distance and of an object's exposure to light. It gets acquainted with the effects of light and shadow and finally, with many an effort, with perspective, too, the knowledge of the latter resulting from the union of the laws of space and of causality, which are both *a priori* present in consciousness and are only waiting to be applied. Here even the modifications which, in looking at various distances, are undergone by the inner conformation of each eye as well as by the position of both eyes in respect of one another, must be taken into account. On the level of Understanding, all these combinations are already being made by the child; on the level of Reason, i.e. *in abstracto*, only by the optical expert. This is how the child turns the manifold sense data into intuition, in accordance with those laws of *Understanding* (of which it has *a priori* knowledge) by virtue of which the world presents itself as an object to the child. Only much later it learns to use *Reason*; only then it begins to understand speech, to speak itself and, strictly speaking, to *think*".*[127]

Schopenhauer illustrates all this by referring to some well-known optical phenomena. Thus he points to the circumstance that we perceive each object as single, although our two eyes are affected by it independently, so that we receive a dual impression; and in the second version of the treatise he also mentions our

* "Das Kind, in den ersten Wochen seines Lebens, empfindet mit allen Sinnen: aber es schaut nicht an, es apprehendirt nicht: daher starrt es dumm in die Welt hinein. Bald indessen fängt es an den Verstand gebrauchen zu lernen, das ihm vor aller Erfahrung bewußte Gesetz der Kausalität anzuwenden und es mit den ebenso *a priori* gegebenen Formen aller Erkenntniß, Zeit und Raum, zu verbinden: so gelangt es von der Empfindung zur Anschauung, zur Apprehension: und nunmehr blickt es mit klugen, intelligenten Augen in die Welt. Da aber jedes Objekt auf alle fünf Sinne verschieden wirkt, diese Wirkungen dennoch auf eine und die nämliche Ursache zurückleiten, welche sich eben dadurch als Objekt darstellt; so vergleicht das die Anschauung lernende Kind die verschiedenartigen Eindrücke, welche es vom nämlichen Objekt erhält: es betastet was es sieht, besieht was es betastet, geht dem Klange nach zu dessen Ursache, nimmt Geruch und Geschmack zu Hülfe, bringt endlich auch für das Auge die Entfernung und Beleuchtung in Anschlag, lernt die Wirkung des Lichts und des Schattens kennen und endlich, mit vieler Mühe, auch die Perspektive, deren Kenntniß zu Stande kommt, durch Vereinigung der Gesetze des Raums mit dem der Kausalität, die beide *a priori* im Bewußtseyn liegen und nur der Anwendung bedürfen, wobei nun sogar die Veränderungen, welche, beim Sehn in verschiedene Entfernungen, theils die innere Konformation der Augen, theils die Lage beider Augen gegen einander erleidet, in Anschlag gebracht werden müssen: und alle diese Kombinationen macht für den Verstand schon das Kind, für die Vernunft, d.h. *in abstracto*, erst der Optiker. Dergestalt also verarbeitet das Kind die mannigfaltigen Data der Sinnlichkeit, nach den ihm *a priori* bewußten Gesetzen des *Verstandes*, zur *Anschauung*, mit welcher allererst die Welt als Objekt für dasselbe daist. Viel später lernt es die *Vernunft* gebrauchen: dann fängt es an die Rede zu verstehn, zu sprechen und eigentlich zu *denken*."

perceiving objects upright, even though (due to the crossing of the light rays when these pass through the pupil) their impression on the retina is upside down. In all such cases Understanding immediately, and therefore unconsciously, corrects the false impression, tracing back the effect to its cause by following the direction from which the sensation came. It would not be able to do this if it had no knowledge of an object, which (as he adds in the second version) proves the absurdity of deriving the law of causality from experience as not only Englishmen and Frenchmen still do (which is forgiveable since Kantian philosophy has not yet penetrated their countries), but even German *Philosophaster* who in this way demonstrate their stupidity (we return to this subject in Chapter IV, p. 100).[128]

Thus the senses provide only the starting point for our perception of the outer world. Their sensations are the data, out of which alone conscious intuition arises through Understanding, that is, physiologically speaking, in the *brain*. A very important place among those sensations belongs to the impression of light upon the eye, which provokes an activity of the retina of which colour vision is a specific modification. Light and colour are already in the eye as sensations before Understanding transforms sensation into intuition. Certainly, after this transformation has taken place we come to regard light and colour as *qualities* (i.e., as *effects*) of the affecting object, which is why from then on we say 'this object is red'. But this does not alter the fact that the colour in itself is an affection of the eye and, as such, independent of the object, the latter existing only for Understanding (i.e., in the brain). Hence, when we talk about colours we are no longer concerned with the function of the brain, but only with the function of a small piece of nervous tissue at the back of the eyeball: the retina.[129]

Colour, then, not as "actions and sufferings of light", as Goethe had said in the Preface to the didactic part of his *Theory of Colours*,*[130] but as actions and sufferings of the retina; this is the core of the difference between Schopenhauer and Goethe, whom he meanwhile did join with in insisting, against Newton, upon the homogeneity of white light. In the second chapter 'On colours' ('Von den Farben') of *On Vision and Colours* Schopenhauer proceeds to work out his theory in detail. According to him the fundamental error of all previous theories (including that of Goethe!) lies in their premature quest for the cause of the sensation of colour without a prior thorough examination of the effect, i.e., of colour as a physiological phenomenon. Sensation as such is never something purely passive; it is the reaction to a received stimulus. In the case of the eye the stimulus is light; the reaction is an activity of the retina. If the latter receives the full effect of light, it develops a full activity; if light is absent, i.e., if darkness reigns, the retina remains passive. Bodies which, under the influence of light upon them, act in their turn upon the eye with the same strength as light itself would do, are called 'shining' or 'mirrors' if they do so in concentrated form, and 'white' if they do so in diffuse form. Bodies which, under the influence of light upon them,

* "Die Farben sind Thaten des Lichts, Thaten und Leiden."

do not in their turn act upon the eye, are called 'black'. Therefore, 'white' has essentially the same effect upon the eye as 'light' and this is why, under the influence of either of them, the retina is fully active, whereas in the absence of both, i.e., in the case of either darkness or black bodies, the retina is passive.

Now the influence of 'light' and 'white' upon the retina and its resulting activity can occur to different degrees. Between the extremes of full activity and full passivity, i.e. between light and darkness and between white and black, lie the domains of dimmed light (*Halbschatten*) and of 'grey', respectively. We may speak in this case of a quantitatively divided activity of the retina, which can itself be differentiated into intensively and extensively divided activity. The latter appears when the different parts of the surface of the retina are active to different degrees. This explains a phenomenon described by Goethe in the first chapter of the didactic section of his *Theory of Colours*, namely, that a black cross on a white background, if we gaze at it for some time and then turn our eyes away and watch a grey or a dim surface, produces the opposite phenomenon, i.e., a white cross on a black background. Those spots on the retina that were affected by the white background are so exhausted by this stimulus that they cannot at once be activated by the much weaker stimulus of the grey surface. This, in its turn, now acts with its full vigour upon those other parts of the retina that had previously been affected by the black cross and consequently were at rest, thus provoking the full activity of the retina.[131]

But there is still another possibility, namely, a qualitatively divided activity of the retina, and this is the point at which colour proper comes in. If you look intensely at a white circle on a black background for twenty to thirty seconds and then turn your eyes away towards a lightly grey or a dim surface, you see a black circle on a white background for the reason just given. But if you put a yellow circle in place of the white one and then look at the grey surface, you will see, there, not a black circle but a violet one. An orange circle will likewise produce a blue 'spectrum';* a red circle ("red" in Goethe's sense of "true red", i.e., magenta, PL) will produce a green spectrum, and so on. Schopenhauer explains all this as follows. Yellow, being the brightest colour, is somewhat darker than white; violet, being the darkest colour, is somewhat brighter than black, such that violet is as much brighter than black as yellow is darker than white. In consideration of the view that white corresponds to full activity of the retina and black to its full passivity, one may say, according to Schopenhauer, that yellow corresponds to 3/4 of the full activity and violet to 1/4. Since the retina naturally aims at full activity, the impression of yellow produces a violet spectrum, which makes up for the missing quarter. Both colours are therefore complementary and together form a

* The reader should be warned that the word "spectrum" is used by Schopenhauer as well as by Goethe in the unusual sense of 'after-image'.

pair.*[132] Orange, being darker than yellow, is therefore further removed from full activity. It is as much darker than white as blue is brighter than black; both colours are in the proportion 2/3 to 1/3. They, too, form a pair of colours — a pair that likewise represents two unequal halves of the full activity. But red, being the darkest of bright colours, and green, being the brightest of dark colours, represent the two wholly equal halves of the full activity; they are therefore in the proportion of 1/2 to 1/2. In this pair of colours polarity is at its most evident, which is why red and green are generally considered the most harmonious colours. Thus we get the following schema (to be found as such only in the second version):

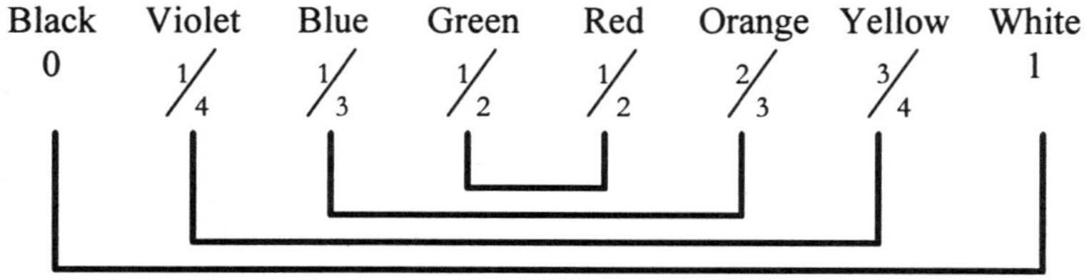

It might be objected, Schopenhauer says, that these fractions introduce a quantitative element into a theory predicated on the characterization of the process of observing colour as a qualitatively divided activity of the retina. But this cannot be avoided, as *every* qualitative division is at the same time a quantitative one, even though in a subordinated way. For example, the chemical division of matter is qualitative (in contrast to the purely quantitative mechanical division); yet it is at the same time quantitative insofar as it divides mass as such. Furthermore, it might seem arbitrary to select these particular six colours (or rather, in view of the essentially polar nature of the phenomenon of colour, these three pairs of colours), since the colour circle has no internal boundaries, but is a coherent and continuous entity in which all colours shade into one another by imperceptible nuances, so that, strictly speaking, there is an infinite number of colours. However, it is only these six colours that, while rarely appearing in nature as pure, have received fixed, generally agreed-upon names with all peoples and at all times, whereas all other colours have always been named after that one among the six to which each of them appears closest. And this suggests that they were considered as deviations, in the one or in the other direction, of the pure colour. For example, we may speak of greenish or reddish yellow, whereas no one will call green yellowish blue. And this precisely because these three pairs of colours have an arithmetic relationship to one another that is expressible in simple whole numbers, which provides the justification of Schopenhauer's fractions, although he admits that, strictly speaking, they cannot be demonstrated.[133]

---

* As already mentioned in a footnote on p. 57, and as we shall see further on, Schopenhauer differs here from Goethe insofar as he considers not yellow and blue but yellow and violet to be the original polar opposites.

The essential point for Schopenhauer in all this is that he believes, by means of his theory, for the first time to have given a real deduction of Goethe's σκιερον (*skieron)*. For the qualitatively divided activity of the retina implies its partial inactivity in the process of seeing colour; and since inactivity of the retina means *darkness*, each colour necessarily contains an element of darkness. Colour is nothing but the visibility of a marriage of Light and Darkness that results necessarily from the partial activity of the retina; therefore, Goethe had been wholly correct in saying that colour is of an essentially shadow-like nature. But one must keep in mind that an activity of the retina that is partial only insofar as the *intensity* is concerned yields no specific effect, since it is no more than an accidental and gradual diminishing of the full activity. It is only the *qualitatively* divided activity which produces the cheerful, stimulating impression of colour, as distinct from the equally bright, yet sad, grey: and this precisely because of the polar separation by virtue of which the lively activity of one half has all of the other half for its support. Metaphorically speaking, one might therefore call grey a simple, *mechanical mixture* of Light and Darkness, as opposed to which colour represents a *chemical union* and an intense interpenetration of Light and Darkness. For in the latter case the two neutralize one another, so to speak, each renouncing its own nature in yielding a new product which, though still faintly resembling both, still has a pronounced character of its own.[134]

All this meant that, although Goethe's notion of the σκιερον (*skieron)* was deemed basically sound by Schopenhauer, he was forced to admit that there are also a few grains of truth in Newton's heresy. First, the phenomenon of colour *is* indeed the result of a divisional process, as in Newton's theory, only it is to be sought in the eye rather than in light itself. Also, more importantly, it follows from Schopenhauer's theory that under certain circumstances white light *can* be produced from a number of coloured lights, only the number is two, not seven. For if, as Schopenhauer's theory asserts, two opposite colours are complementary with regard to the full activity of the retina to whose division they owe their existence, then there can be no doubt that the union of two such colours will produce that full activity and consequently the impression of pure light or white. And indeed, when we gaze at red (in the Goethean–Schopenhauerian sense of "true" red, i.e., magenta) and immediately thereupon at green, no after-image will appear.

It would be more convincing, Schopenhauer thought, if this complementarity could also be demonstrated for physical or even for chemical colours. To that end it would be necessary for the two external causes of the two complementary colours to be made to act simultaneously yet separately upon the same spot of the retina. This effect cannot be produced by mixing two contrasting chemical pigments, for these will only act simultaneously, not separately, on the same spot of the retina. Consider, moreover, their material nature, owing to which, although after unification they cease to produce a qualitatively divided activity of the retina (i.e., to act as colours), they will produce instead an intensive, quantitatively divided activity (i.e., to act as grey). But it may be done with prismatic colours. If one covers the violet band of one prismatic spectrum with the orange band of another, the result will be 'true' red (i.e., magenta); if one then directs the green band of a third prismatic spectrum toward that spot, this will appear as white. Goethe himself had performed this experiment, but his polemic against Newton had prevented him from accepting its outcome as conclusive evidence for the possibility of producing white light out of coloured lights. Goethe's argument had been that in this case the threefold sunlight is powerful enough to make the element of darkness invisible, although it remains present. Yet this argument does not convince Schopenhauer because, if every colour contains an element of darkness, as Goethe himself had taught, then in this experiment not only the element of light but the element of darkness, too, must be threefold; and so they neutralize each other. Therefore, it is not the increased illumination but rather the opposition of the colours which produces here the impression of pure light, or of white. This can be demonstrated even more easily by laying two prismatic colour spectra across each other in such a way that the violet of the first covers the yellow of the second, and the blue of the first the orange of the second; once again, in both cases the result will be white, but now the white spot will be twice as wide as in the previous experiment since both pairs of colours lie next to each other.[135] This, indeed, was Newton's thirteenth experiment in the second part of the first book of the *Opticks*, and it was therefore this passage in particular which (according to Schopenhauer's account in the second edition of *On Vision and Colours*)[136] provoked Goethe's anger and made him regard his former disciple as an adversary. Wrongly so, Schopenhauer assures us, for in fact this experiment does not at all demonstrate that Newton was right, as it is only two colours, not seven, which in this case cover each other. Newton never understood the true, physiological nature of colour and thus attributed to light rays what really ought to be attributed to the activity of the retina. Still, Schopenhauer went on, it cannot be denied that Goethe himself went too far in rejecting (for quite the same reason, since he, too, failed to understand the true, physiological nature of colour grasped for the first time by Schopenhauer!) any possibility of producing white out of colours.[137] Thus Schopenhauer arrived at the paradoxical conclusion that Newton rightly acknowledged the possibility of producing white through a union of colours, but for the wrong reason and on behalf of a false theory, whereas Goethe wrongly denied this possibility, but in the context of an otherwise correct assembly of facts (an assembly which had then enabled Schopenhauer to be the first to formulate an adequate theory!).

We may sum up all this by saying that Newton described colours as due to properties of the rays that according to him compose light, thus concentrating his investigation on the *light sources* as such, whereas Goethe stressed the contribution of the physical *media* encountered by the light as it travels from its source to the eye of the beholder, and Schopenhauer, while concentrating on the *eye*, anticipated (though in a confused sense as we shall see) later theories on the function of the retinal response in that eye. And this difference between Goethe and Schopenhauer gives rise to the question of whether Goethe was really so wrong in failing to recognize Schopenhauer as his true heir? Let us try to answer that question by a closer scrutiny of both the differences and the agreements between the two men.

d. Schopenhauer and Goethe

First of all, Goethe surely had some personal reasons for keeping this self-appointed pretender at arm's length. What, for example, to say of a 27-year-old, still wholly obscure philosopher who has the conceit to write in a letter "When I compare your *Theory of Colours* to a pyramid, then my theory is its peak, the indivisible mathematical point down from which the whole large building extends, and which is so essential that, without it, it would no longer be a pyramid, whereas one may cut off from its bottom ever more slices without its ceasing to be a pyramid ...",*[138] and who, adding insult to injury, frankly grants that, to him, the essay on colours is only a side-issue because he is preoccupied all the time with wholly other theories (namely, the construction of his metaphysical system)?[139] Poor old Goethe, who had dabbled in the science of colours for about twenty years, had been lucky enough now to find a young disciple who in a few weeks and, so to speak with the left hand — after all, he had more important things to do! — had crowned the master's work by, for the first time, making a real *theory* out of the mass of material laboriously collected by the master ... But, setting aside such personal irritations (which, for the rest, Goethe did not allow to run away with him; the tone of his correspondence with Schopenhauer remained on the whole friendly, if in a somewhat condescending way), it will have become clear from the above that, from his point of view, Goethe also had good, objective reasons for rejecting Schopenhauer's handling of the problem of colour. It was not that Goethe did not recognize the important role of the beholding eye in the production of colour phenomena. On the contrary, for him it went without saying that we can perceive external objects only insofar as the perception is, as it were, programmed in our sense-organs; and it was therefore not for nothing that, as we have already seen, his *Theory of Colours* opened with a chapter on 'Physiological Colours' in which he called these the "foundation of our whole doctrine",†[140] and in which he discussed phenomena like complementary after-images and coloured shadows. His later acquaintance with Purkyně's *Beiträge zur Kenntnis des Sehens in subjektiver Hinsicht* (*'Contributions to the Knowledge of Subjective* Vision*'*) had further deepened his insight into the physiological implications of his own theory. Quite in the spirit of what we heard Schopenhauer arguing, Goethe writes therefore on physiological colours: "these phenomena are of the highest importance, since they turn our attention to the laws of vision and are a necessary

* "Vergleiche ich Ihre Farbenlehre einer Pyramide, so ist meine Theorie die Spitze derselben, der untheilbare mathematische Punkt, von dem aus das ganze große Gebäude sich ausbreitet, und der so wesentlich ist, daß es ohne ihn keine Pyramide mehr ist, während man von unten immer abschneiden kann ohne daß es aufhört Pyramide zu seyn."

† "... diese Farben, welche das Fundament der ganzen Lehre machen ..."

propaedeutic to a subsequent observation of the colours. *The eye actually desires totality and closes in itself the colour circle* [my italics, PL]. In the violet demanded by yellow lies purple [*Roth*] and blue; in orange, the yellow and purple to which blue corresponds; green unites blue and yellow and demands purple, and thus in all degrees of the most different mixtures".*[141] But, mark you, Goethe speaks of a necessary *propaedeutic*; it is a far cry from there to Schopenhauer's claim that the consideration of colour as a physiological phenomenon is the most essential half of the whole theory of colours to which the second half (dealing with 'physical' and 'chemical' colours) stands in a dependent and subordinate relation.[142] In other words, for Schopenhauer colours were *only* 'actions and sufferings of the retina'; for Goethe they certainly were that, too, but also and foremost "actions and sufferings of light". In fact, to his harmonizing way of thought these two things belonged undissolubly together since the eye was for him a creature of Light. We recall from Chapter I, p. 24 what he stated in the introduction to the didactic part of the *Theory of Colours*: "The eye owes its existence to Light. Out of a neutral animal auxiliary organ Light calls into being an organ that is to become its equal counterpart; thus the eye forms itself *by* Light *for* Light, in order that the inner light may meet the external light." In Schopenhauer's 'crowning' of his own doctrine Goethe must have perceived a subjectivist one-sidedness which might have a dangerous consequence for this doctrine, to wit, the possibility of some kind of division of labour which would confine his own qualitative way of treating the phenomenon of colour to the 'subjective' realm (the "inner light"), leaving the — for Schopenhauer, but not for Goethe, less important — 'objective' realm (the "outer light") to Newton's quantitative treatment. For could Schopenhauer's purely physiological interpretation of Goethe's σκιερον (*skieron)* not be reconciled in principle with Newton's physical theory in that the two theories evidently deal with different aspects of reality?†[143]

Now this was certainly not Schopenhauer's own opinion. In Section 13 of the second version of *On Vision and Colours*, which deals with the external stimuli that provoke the qualitative division of the retina's activity,[144] he lists his arguments for taking Goethe's side against Newton at the level of physical colours. (Incidentally, the fact that this exposition had to wait until the thirteenth section is of course quite in agreement with Schopenhauer's claim that it is essential first to study thoroughly the physiological effect before turning to the physical cause.) First, since the colours themselves, as well as their mutual relationships and the

---

* "Diese Phänomene sind von der größten Wichtigkeit, indem sie uns auf die Gesetze des Sehens hindeuten und zu künftiger Betrachtung der Farben eine notwendige Vorbereitung sind. Das Auge verlangt dabei ganz eigentlich Totalität und schließt in sich selbst den Farbenkreis ab. In dem vom Gelben geforderten Violetten liegt das Rothe und Blaue; im Orange das Gelbe und Rothe, dem das Blaue entspricht; das Grüne vereinigt Blau und Gelb und fordert das Rothe, und so in allen Abstufungen der verschiedensten Mischungen."

† Such a gentleman's solution (which would, we repeat, have been wholly unacceptable to Goethe) was to be suggested much later by Werner Heisenberg.

regularity of their appearance, are all in the eye itself, the external cause of the appearance of colour in the eye (i.e., of the appearance of the polarity of its retina) can never be more than a *stimulus*, comparable to the role the rubbing of a body plays in provoking the electricity which is latent in it (i.e., the polar separation of +e and –e). Therefore a combination of his own theory with Newton's (we note that Schopenhauer had at least once considered the possibility!) would only be feasible by assuming the oddest *harmonia praestabilita* ever conceived of by a human brain. For in that case certain colours, although like all other colours they arise in the eye in obeyance to the law that rules its functioning, must nevertheless have causes that are already present in light itself, namely, in its components. Furthermore, since every colour is the qualitative half of the retina's full activity, for the production of which another, complementary colour is needed, there exist only pairs of colour and no single colours, let alone an odd number of them (namely, seven).[*145] Nay, more: since the colours form a continuous circle without internal boundaries or fixed points, they are infinite in number, so it is impossible not only to assume seven really existing colours but even to assume any distinct number of them at all. For it must be repeated that the three pairs of colour (i.e., the six colours) described by Schopenhauer are distinct in his view from all other colours only by the circumstance that their polar relationship can be expressed in integer numbers and is therefore easiest to demonstrate. Well, then, since the infinity of the number of colours is a consequence of the infinitely modifiable divisibility of the activity of the retina, there must be a correspondingly infinite modifiability in the external cause which acts as stimulus. Now this demand is deemed by Schopenhauer to be met much better by Goethe's turbid medium (which can be more dense or more transparent to an infinite variety of degrees, and which can be illuminated either in front or behind in all kinds of different proportions) than by Newton's seven, rigid and homogeneous lights. It is true, Schopenhauer admits, that Newton sometimes suggests there to be in fact an infinite number of homogeneous lights; yet, although he grants that this esoteric version of Newtonianism would more or less do away with the preceding objection, it would, on the contrary, reinforce the following, namely that it follows from the Newtonian construction that all colours would be in the same proportion to white (one to seven),[†] whereas in fact even the darkest of them, which is violet,

---

[*] Schopenhauer rightly believed that Newton distinguished between seven colours only for the benefit of an analogy to the seven intervals of the gamut; according to him there were only four colours in the prismatic spectrum, since he (like Goethe) did not sharply differentiate between violet and ultramarine blue on the one hand and between (prismatic) red and orange on the other; and since, moreover, he claims that green appears only when, at a greater distance from the prism, the blue (i.e., cyanide) and yellow components overlap. As already mentioned in our footnote on p. 59, contemporary theories of colour, just like those of Goethe and Schopenhauer, accept six basic colours roughly corresponding with theirs in fact, though not in terminology.

[†] It is unclear why this should imperatively follow from Newton's theory. Moreover, Schopenhauer fails to mention what for him must have been the most important reason for

is in the proportion of one to four. In the esoteric version this would mean that (still more absurd) each colour would relate to white in the proportion of an infinitely small fractional number to one, and thus wholly disappear into darkness and become invisible. Here, too, Goethe's turbid medium provides the most satisfactory solution. According to Goethe, colour is produced either by the darkening of light or by the illumination of darkness, provided that both do not take place immediately (for that would produce only grey) but through the intervention of a turbid medium which is, so to speak, the *menstruum* of the chemical interpenetration of Light and Darkness. Now Schopenhauer's theory of the polarity of the eye, i.e., the qualitative division of its activity, evidently demands an external cause by which light is diminished in such a way that each colour contains as much light as its opposite contains darkness. And this demand is met perfectly by the turbid medium which, under opposite illumination, always gives rise to two physiologically complementary colours, varying by the degree of the density of the dividing wall between them.

On two points, however, Goethe must, according to Schopenhauer, be corrected. First, although Goethe knew that physiologically the polar opposite of yellow is violet, he nevertheless put forward yellow and blue as physical opposites, which would imply a strange incongruity between physiological and physical colours and can therefore be accepted only symbolically, making yellow and blue stand for the whole range of warm (active) and cold (passive) colours, respectively. Actually, according to Schopenhauer, the original polar opposites are yellow and violet; increasing the density of the turbid medium modifies these into orange and blue, then into red and green (note here that Schopenhauer, although accepting the Goethean principle of 'polarity', thus tacitly drops the no less fundamental one of 'intensification'!); and finally, when the density has become impenetrable, into white and black. Second (and this point is by far the most important, as has been rightly remarked by Hübscher),[146] Goethe's *Urphänomen* is not the real *Urphänomen*, in that it is not something simply given beyond any possible explanation, but only the external cause which provokes the division of the activity of the retina. In the organic capacity of the retina to divide its activity into two, qualitatively opposite, equal or unequal halves we have the real *Urphänomen*, for this capacity (like so many others in physiology) could only be explained by falling back upon some *causa finalis*, for example, by saying that, thanks to colour, we have one more tool available for discerning and recognizing the things around us.[147]

So far Schopenhauer. What is one to make of these arguments from a Goethean point of view? This above all, that Schopenhauer's effort to produce arguments for Goethe's physical theory, however well intentioned, cannot conceal the

---

opting for Goethe's theory: it is only the Goethean polarity of cold (or dark or passive) and warm (or bright or active) colours which can account for the otherwise (at least for us) rather inexplicable fact that red (magenta) and green appear as different colour sensations, although according to Schopenhauer they both represent 1/2 of the activity of the retina!

fundamental fact that his 'correction' of Goethe, although presented by him as of secondary importance, reveals a radically different frame of mind. For to Schopenhauer's contention that the polarity of Light and Darkness is not the real *Urphänomen* but only the external cause which provokes the real *Urphänomen*, which is the polar activity of the retina, Goethe would respond that precisely this *derivative* nature of the latter shows that it cannot be the *Urphänomen* itself. It can only be one of its aspects, namely, its mode of expression in the animal and human organism, i.e., in one particular form (even if the highest developed one) of the myriad forms of Nature. The organic capacity of the retina to translate the marriage of Light and Darkness into the phenomenon of colour is not therefore, according to Goethe as we imagine him here, beyond any further explanation, as Schopenhauer thinks it is, but rather a striking proof of Nature's all-pervasive unity. "Wär nicht das Auge sonnenhaft ..." (p. 24). Young Schopenhauer's failure to recognize this fundamental fact had lured him into the heresy of acknowledging that white may be produced from colours, so we hear Goethe rage. True, he had acknowledged the possibility only for coloured lights ('physical colours'), not for pigments ('chemical colours'); but this, in Goethe's eyes, would hardly diminish Schopenhauer's guilt, since in this respect the difference between the two categories of colours was irrelevant to Goethe.

It is of course otherwise from our contemporary point of view; and the modern reader may therefore notice with interest that Schopenhauer's heretical refashioning of Goethe's theory of colours (which accepted the possibility of producing white from physical — though not from chemical — colours, and which, therefore, recognized at least in principle two different kinds of colour-mixing) carried him to the threshold of the later theory of additive and subtractive colour-mixing we mentioned above. But the revival by Young, Fresnel and Arago in a modified form (transverse instead of longitudinal waves) of Huygens' wave theory, which was to form the basis of the theory of additive and subtractive colour-mixing,* was vehemently rejected and scorned by him in the second edition of *On Vision and Colours*, with Goethe's failure to say anything sensible about it being ascribed by him to the latter's growing old.†[148] According to Schopenhauer

* In fact there is a historical rather than a necessary connection here insofar as in principle it might also have been possible on the basis of a modified emission theory; yet not in any way on the basis of the Goethean–Schopenhauerian theory of the homogeneous nature of light.

† "Goethe on the other hand was too old when these phenomena [the polarization of light] were discovered, and began to ramble." ("*Goethe* wieder war zu alt, als diese Phänomene entdeckt wurden, — fieng an zu radotiren."). This remark, as well as Schopenhauer's polemic against the wave theory generally and also (in connection with it) against some contemporary physicists, is to be found in the fourteenth and final section ("Some additions to Goethe's doctrine on the origin of physical colours"; "Einige Zugaben zu Goethes Lehre von der Entstehung der physischen Farben") of the second edition — a section which is therefore a considerably enlarged version of the seventeenth section (simply headed "Conclusion"; "*Beschluß*") of the first edition. In his rejection of the wave theory

the wave theory was nothing but one more desperate attempt by French physicists to save Newton's doctrine of the *objective* existence of colours, i.e., of seeking the divisional process which gives rise to the phenomenon of colour in light itself instead of in the eye. And indeed, in respect of what was the principal bone of contention for both Goethe and Schopenhauer, to wit, the supposed heterogeneity of white light, there was no difference between the emission theory and the wave theory; Goethe's and Schopenhauer's failure to appreciate the latter's importance was therefore not so amazing as is sometimes suggested, but on the contrary from their point of view wholly consistent. Just so, Schopenhauer should be regarded only with many reservations as a precursor of the modern physiological explanation of colour vision, although the distinct functions of visual rods (responsible only for greater or smaller brightness) and cones (responsible for colours) postulated by the latter bear a certain resemblance to Schopenhauer's differentiation between a 'quantitative' and a 'qualitative' division of the activity of the retina, and even though there is also some superficial resemblance between Schopenhauer's 'qualitatively' divided activity of the retina and the three-receptor theory mentioned in our footnote on p. 59. This theory, which has the wave theory as its starting-point, was originally, still more like a guess, formulated by Young, later revised by Grassmann, experimentally corroborated by Maxwell and (under the influence of the latter two) accepted by Helmholtz, who had originally rejected it.[149]

How ambiguous Schopenhauer's position was in sticking to Goethe's theory of colours yet 'improving' it in such a way as to admit the possibility of producing white light out of a number of coloured lights, and how inevitably it predisposed him to 'falling between two stools', so to speak, appears with all desired clarity from a polemic he engaged in in the second version of *On Vision and Colours* against the Austrian Jesuit Scherffer and the French physicist Pouillet. The former had explained the phenomenon of after-images (the 'physiological spectrum', in Goethe's terminology, that was adopted, as we saw, by Schopenhauer) by means of the exhaustion that befalls the eye upon its prolonged exposure to one particular colour. It then loses its receptivity for this specific kind of light rays, so that immediately afterwards it will not see white but rather a compound of the remaining light-rays. Schopenhauer, who accepted (as we saw) the 'exhaustion' theory for what he called the '*quantitatively* divided activity of the retina', would have nothing of this, since it would mean that, e.g., a yellow after-image, since it

---

Schopenhauer points especially (and, in retrospect surely, with justice) to the "completely hypothetical, nay, mythological nature" of "a certain aether", the vibrations of which are supposed to be the cause of the colour-phenomenon; particularly amusing in this context is for him the circumstance that the quickest vibrations are attributed to the darkest and least effective of all colours, namely violet, and the slowest to the lively red which even excites animals (evidently Schopenhauer, like most people for that matter, still thought that the bull is provoked by the red colour of the swung cloth rather than by the swinging itself). Somewhat further on he attempts, therefore, an explanation of his own for the phenomenon of polarized light; we refer the interested reader to Schopenhauer's text itself.

follows upon a violet image, would be the compound of all components of the spectrum apart from violet. "Just try to concoct yellow out of these" ("Daraus Gelb zu brauen probire man!"), Schopenhauer snorted,[150] similarly dismissing Pouillet's assertion that "orange and green give yellow" ("l'orange et le vert donnent du jaune"). In both cases Schopenhauer's argument is the typical common-sense consideration that such mixtures never yield anything other than 'street-mud colour'. Both gentlemen would in all likelihood have replied that this is of course perfectly true of pigments, but that they were speaking of prismatic colours — a domain in which, *nota bene*, the possibility of 'concocting' white out of 'red' (i.e., magenta, itself first to be produced by covering the violet band of one prismatic spectrum with the orange of an other one as we remember) and green had been demonstrated by Schopenhauer himself, which is hardly less absurd, one might say, from a common-sense point of view! In other words, *as a consequence of his revision of Goethe's doctrine Schopenhauer was no longer entitled to use such common-sense arguments*; and Goethe may instinctively have felt that, in the hands of the (in Schillerian terms) 'sentimental' Schopenhauer, his scientific pet-child had lost its divine 'naiveté'![*151]

Meanwhile, the problem for Schopenhauer was that it would have been hard for him to fit the phenomenon Pouillet had pointed to into his own theory. For if Schopenhauer had bothered to use his prisms — even if only for the fun of it — to test Pouillet's assertion, he might have been led to the conclusion that the latter was right,[†] in spite of the obvious fact that 2/3 (his fraction number for orange, as we remember) + 1/2 (likewise for green) does not yield 3/4 (the fraction number

---

[*] Allusion to Schiller's famous essay 'Über naive und sentimentalische Dichtung' ('On naive and sentimental poetry'). Goethe's attitude towards his self-opinionated disciple, and towards any criticism of his theory of colours in general, is illuminatingly revealed in a conversation with Eckermann on 19 February 1829. Eckermann had allowed himself to put forward some criticism of Goethe's explanation of coloured shadows, and, consequently, provoked the master's wrath; according to Eckermann's account Goethe compared the fate of his theory of colours with that of Christianity: for a time one thinks one has faithful disciples, but they deviate and become a sect. "You are a heretic like others" ("Sie sind ein Ketzer wie die anderen auch"), poor Eckermann was told; "the others" are not mentioned by name in the text, but are marked by dots for reasons of discretion; but in the editor's postscript a letter from Eckermann to Varnhagen von Ense of 14 June 1836 is quoted, from which it appears that Schopenhauer was one of them. Eckermann seems to have been a little puzzled by Goethe's intolerant attitude to even the slightest objection to his theory of colours, since this stands in marked contrast with Goethe's liberal attitude to criticism of his literary work; he seeks to explain this difference through Goethe's sense of being misjudged as a scientist (whereas there was no reason for such feelings for Goethe as a poet).

[†] Provided that Pouillet, like Schopenhauer, and also Goethe, did not sharply differentiate between spectral red and orange (cf. our footnotes on pages 59 and 75); for, as we know, the additive combination of red and green light does indeed produce yellow light.

for yellow).* However, let us now put ourselves in the position of an imaginary reader convinced that Schopenhauer's physiological theory was 'the truth, the whole truth, and nothing but the truth'. The arguments given by Schopenhauer in favour of Goethe's *physical* theory might have convinced him of the greater plausibility of the latter as compared to Newton's theory; but he might also remark that all this would not amount to a real *proof* of Goethe's theory. And indeed, Schopenhauer himself said as much in the Latin version of *On Vision and Colours*, which dates from 1830. Here he admits explicitly that his own theory is independent of Goethe's insofar as it would still be true as a *physiological* theory even if, as physicists, Newton and Goethe had *both* been wrong.[152]

On balance, it seems that the lasting merits of Goethe's *Theory of Colours* are to be found, above all, in its historical section (which there was no need, in our context, to treat systematically, but which offers in many ways a fine piece of history of science writing) and in the first and last chapters of the didactic section. Its first chapter, which (as we have seen) deals with 'physiological colours', contains Goethe's expositions on the subjective perception of contrasting colours and on coloured as well as uncoloured complementary after-images — expositions which were to be fruitful for physiologists like Purkyně, Müller and Hering. In a footnote on p. 59 we have already pointed to the fact that his three colour pairs basically correspond with contemporary views; it is indeed not surprising that Goethe's phenomenological (to use a modern term) method, inappropriate though it was in the domain of physics, showed itself to best advantage when dealing with our direct perception and distinction of colours. Newton, for his part, confessed that he needed "an Assistant, whose Eyes for distinguishing Colours were more critical than mine" to note by right lines, drawn across the spectrum falling upon a paper, the confines of the colours; and at another place he explicitly warned: "I speak here of Colours as far as they arise from Light. For they appear sometimes by other causes: as when by the power of Phantasy we see Colours in a dream, or a Mad-man sees things before him which are not there; or when we see Fire by striking the Eye, or see Colours like the Eye of a Peacock's Feather, by pressing our Eyes in either corner whilst we look the other way".[153] These colour phenomena (or at least the two mentioned last) thus consciously left out by Newton are precisely the 'physiological colours' of the first chapter of the didactic part of Goethe's *Theory of Colours*. Its sixth and last chapter is about 'sensual and moral effects of colour' ('Sinnlich-sittliche Wirkung der Farbe'). This has inspired many painters and theorists of painting, such as Goethe's contemporary and compatriot, Philipp Otto Runge, or the famous 19th-century English painter William Turner (who had read Eastlake's translation of Goethe's *Farbenlehre*,

---

* Nor does, by the way, 1/4 (Schopenhauer's fraction number for violet) + 2/3 (likewise for orange) yield 1/2 (the fraction number for 'red', i.e. magenta); but characteristically Schopenhauer had no problem with the latter's resulting from the former two, probably because the prismatic result happens to coincide here roughly with the pigmental result.

published in 1840 in Newton's own country!), and was attracted especially by Goethe's theory of the 'turbid medium', as is not to be wondered at since in his paintings there is certainly no lack of vapours and other turbidities), and also 20th-century Expressionists like Ernst Ludwig Kirchner and Vassily Kandinsky.[154] This latter aspect is one to which Schopenhauer did not contribute anything; the more so to the 'physiological' aspect, as we have seen. This is not to be wondered at either, since it is not by chance that his book bears the title "On *Vision* and Colours" (our italics); after all, Schopenhauer's interest in Goethe's investigations had its origin not so much in an interest in the phenomenon of colour as such, but was intimately and indeed crucially linked to his own preoccupations with perception in general which were to lead to a new, physiology-based demonstration of the *a priori* nature of *causality*.

Here we cannot fail to note the numerous, quite striking similarities between Schopenhauer's theory and the expositions given by Helmholtz in his 'Über das Sehen des Menschen' ('On human vision'; originally a lecture given in Kant's native Königsberg on 27 February 1855). For example, we read there: "We never perceive immediately the objects of the external world, but we perceive only effects of these objects on our nervous organization; and this has been so from the first moment of our life. In what way, then, have we first made the transition from the world of the affections of our nerves to the world of reality? Evidently, only by making an inference: we must presuppose the presence of external objects as the *cause* of the stimulation of our nerves, for there can be no effect without a cause".[*155] In a response to a letter by Johann August Becker (one of Schopenhauer's 'apostles'), who refers to this lecture, Schopenhauer calls Helmholtz a rascal (*Lump*) and accuses him of plagiarism.[156] Helmholtz surely never heard of this correspondence, but later the same reproach was made in public by Schopenhauer's 'evangelist' Frauenstädt. Helmholtz never responded directly (which may or may not be interpreted as a tacit confession), but he did so indirectly in a much later lecture held in Berlin on 3 August 1878. Here he described the mental processes that lead to the gradual constitution of our representations of reality, like he had done before, not only in 'On human vision', but also in his fundamental *Handbuch der physiologischen Optik* (1856)). Other than in those works of the 1850s, however, Helmholtz here refrained from using the term "unconscious inferences" ("*unbewusste Schlüsse*"), in order, so he said, to avoid confusion with the "in my opinion, wholly unclear and unjustified notion labelled with this name by *Schopenhauer* and his successors".[†157] All the more

---

[*] "Wir nehmen nie die Gegenstände der Aussenwelt unmittelbar wahr, sondern wir nehmen nur Wirkungen dieser Gegenstände auf unsere Nervenapparate wahr, und das ist vom ersten Augenblick unseres Lebens an so gewesen. Auf welche Weise sind wir denn nun zuerst aus der Welt der Empfindungen unserer Nerven hinübergelangt in die Welt der Wirklichkeit? Offenbar nur durch einen Schluss: wir müssen die Gegenwart äusserer Objekte als *Ursache* unserer Nervenerregung voraussetzen, denn es kann keine Wirkung ohne Ursache sein."

[†] "Ich habe später jenen Namen der unbewussten Schlüsse vermieden, um der Verwechslung mit der, wie mir scheint, gänzlich unklaren und ungerechtfertigten

reason, then, for us to subject to a critical analysis this basic constituent of Schopenhauer's philosophy — the one embodied in Helmholtz' term "unconscious inference".

Vorstellung zu entgehen, die *Schopenhauer* und seine Nachfolger mit diesem Namen bezeichnen."

IV

THE ANTINOMY: CAN THE WORLD BE IN MY HEAD, YET MY HEAD BE IN THE WORLD?

a. Schopenhauer and Kant

We have seen in the preceding chapter that Schopenhauer's profound admiration for Goethe did not prevent him from refashioning Goethe's theory of colours in a Kantian way, or at least in a way considered by him to be Kantian. But we also recall from the second chapter that his no less profound admiration for Kant had not prevented him either, already in the first version of his dissertation, from making some critical remarks about Kant, too, particularly where the latter's proof of the *a priori* nature of the law of causality was concerned. However, Schopenhauer did not put together an elaborate, public exposition of his attitude towards Kant until the appearance in 1818 of his own philosophical system, the first volume of which contains a supplement entitled 'Critique of Kantian Philosophy'. In the main work itself, too, we find an interesting chapter in which (foreshadowing the argument of the supplement) Kant's four 'antinomies' are dismissed as "sham fights" (*Spiegelfechtereien*). These antinomies are the mutually contradictory, yet each in itself equally irrefutable, theses into which any attempt at constructing a 'rational cosmology' must necessarily run.* But according to Schopenhauer, only the successive antitheses (namely, the infinity of the world in space and time, its infinite divisibility, and the non-existence of God and of free will) are in accordance with the principle of sufficient reason (the *a priori* form of all knowledge), the chain of conditioning relationships being necessarily infinite. Against these "sham fights" Schopenhauer posits what he takes to be the one and only real, since really insoluble, antinomy — really insoluble in that it does not derive from a pseudo-science like rational cosmology, but rather from the apparently contradictory results of real science and of critical philosophy, respectively.

---

* I.e., one of the three metaphysical pseudo-sciences which result from Reason's futile, if inevitable, attempt to transcend the realm of experience (the other two being 'rational psychology' — which gives rise to 'paralogisms' as we shall see below — and 'rational theology', which presents us with 'proofs' of God's existence).

Schopenhauer phrases this antinomy as follows: "Thus we necessarily see, on the one hand, the existence of the whole world to be dependent on the first knowing creature, however imperfect the latter may be; but, on the other hand, and equally necessarily so, this first knowing animal to be fully dependent on a long chain of causes and effects which has preceded it and into which it enters itself as a small link."*[158] For, although it is perfectly true that materialism forgets that there is no object without a subject, so that it makes no sense to speak of suns and planets without an eye that beholds them and an intellect that recognizes them, it is equally true that the (*a priori* certain) law of causality compels us to admit that in the course of time every higher organized state of matter has always been preceded by a more primitive one (animals were earlier on the spot than human beings, fish earlier than land animals, plants earlier than fish, and inorganic nature earlier than organic nature as a whole), so that the original mass had to undergo a long chain of mutations †[159] before the first eye could open itself — an eye upon which, nonetheless, the existence of the whole world depends since the latter is nothing but a representation (*Vorstellung*) and cannot as such be thought of without a knowing subject as its bearer. An antinomy if ever there was one! And in Chapter 22, 'Objective View of the Intellect' ('Objektive Ansicht des Intellekts') of the second volume (which consists of a range of comments on, and elaborations of, the philosophical system expounded in the first volume) we read that there are two fundamentally different ways of contemplating the intellect: the *subjective* way takes *consciousness* as its starting-point and culminates in Kant; the *objective* way (culminating in Cabanis more than in anybody else) regards consciousness as nothing but a function of a part of the bodily organism, namely, the *brain.*[160] Clearly, the same antinomy reappears here on the ontogenetic level that, in the passage just quoted, we met on the phylogenetic level; only, this time

* "So sehn wir einerseits nothwendig das Daseyn der ganzen Welt abhängig vom ersten erkennenden Wesen, ein so unvollkommenes dieses immer auch seyn mag; andererseits eben so nothwendig dieses erste erkennende Thier völlig abhängig von einer langen ihm vorhergegangenen Kette von Ursachen und Wirkungen, in die es selbst als ein kleines Glied eintritt."

† This sounds like an anticipation of Darwinism; and indeed, there is an evolutionist streak in Schopenhauer's thought. However, one should always keep in mind that, for him, processes in *time* belong only to the phenomenal world, whereas the true philosophical spirit is interested only in that which remains identical under all apparent change (which explains among other things his relatively low opinion of history). To him the different stages of complexity in nature are nothing but objectivations of what he holds to be the Platonic Ideas, which exist from all eternity. Consequently he dismissed Lamarck as a typical representative of a materialist science that fancied itself able to explain higher phenomena ('life') from lower, mechanical, physical, and chemical phenomena. When near the end of his life (1860) he made the acquaintance of Darwinism (it was to remain a second-hand acquaintance acquired by reading a review of the *Origin of Species* which had appeared in 1859), he dismissed it as "vulgar empiricism" and as just another version of Lamarckism (cf. also footnote on p. 113).

Schopenhauer does not use the expression. For here, too, one might say that, on the one hand, consciousness cannot exist without a material substrate — the brain — whereas, on the other hand, the brain (and the bodily organism as a whole) exists only as a representation of the knowing subject. Tying up the phylogenetic and the ontogenetic versions, we arrive in fact at a *circulus vitiosus* (the expression is not used by Schopenhauer). For in Schopenhauer's vision the world is a representation (*Vorstellung*) of my intellect, which means that *the world is in my head*,[*161] since my intellect is nothing but a function of my brain; this brain, however, is the product of a long evolution of matter, so *my head is in the world*;[†162] yet the existence of this brain — and of the entire material world of which it is a part — is known to me only by my intellect and is therefore only a *Vorstellung* of the latter again ... and so on!

At this point we may well imagine Kant's rejoinder from the celestial realm: 'My dear sir, your own antinomy is a sham fight since it has no bearing upon the real problem of critical philosophy, which is not the *reality* of the visible world but the *truth* of our judgements about it. My problem was not the *real dependence* of spirit on matter or the other way round, but the *ideal interdependence* of truth-judgements. What interested me in the first place was not, therefore, the psychological genesis of knowledge but its internal structure. Not the human subject in the sense of the individual human *psyche*, but the interindividual phenomenon of science as it actually lies before us was the object of my analysis. True, I said in the Preface to the second edition of my *Critique of Pure Reason* that Reason understands only that which it produces itself after its own design; but I did so because I was so powerfully impressed by the succes of mathematical and empirical science, and therefore I set myself the task of demonstrating the conditions which make it possible. 'Possible', not in the sense of: 'What are its psychological origins?', but in the sense of: 'What are its necessary governing principles?'. It was clear to me that these cannot be deduced from the supposed external objects of science but only from the structure of scientific knowledge itself; that was my 'Copernican Revolution'. But the *a priori* nature of the

---

[*] In Schopenhauer's wording: "In having brought this thesis ["the world is my representation", PL] to consciousness and in having connected with it the problem of the relationship of the ideal to the real, i.e., of the world in the head to the world outside the head, consists — apart from the problem of moral freedom — the distinctive character of modern philosophy." ("Diesen Satz zum Bewußtsein gebracht und an ihn das Problem vom Verhältnis des Idealen zum Realen, d.h. der Welt im Kopf zur Welt außer dem Kopf, geknüpft zu haben, macht, neben dem Problem von der moralischen Freiheit, den auszeichnenden Charakter der Philosophie der Neueren aus.").

[†] In Schopenhauer's wording: "Under all *transcendental* ideality the objective world retains *empirical* reality: the object is, to be sure, not a Thing-in-itself, but as an empirical object it is real. True, space is only in my head; but empirically my head is in space." ("Bei aller *transcendentalen* Idealität behält die objektive Welt *empirische* Realitat: das Objekt ist zwar nicht Ding an sich; aber es ist als empirisches Objekt real. Zwar ist der Raum nur in meinem Kopf; aber empirisch ist mein Kopf im Raum.").

principles formulated by me should not be understood in a *temporal*, nativist sense (in that case it would be of the same order as, for example, the responsibility of the retina for our seeing the world in different colours, a phenomenon you are so interested in), but in a *logical* sense. Reason, like the tree from the Gospel, can be known only by its fruits. To use the vocabulary of a 20th-century philosopher who considered himself to be some kind of disciple of mine, a certain Popper — later advanced to Sir Karl —, the essence of my 'Copernican Revolution' was therefore the shift of attention, away indeed from 'World 1', but not towards 'World 2', but rather towards 'World 3'![163]. And indeed, had it been a shift towards 'World 2', this would, firstly, not have been a very original step in view of what Descartes had already wrought, and, secondly, I would have been clothing it in rather odd terms. For the introduction of the heliocentric system meant that, although Man's movements should indeed be included in the calculations (which is why I chose this terminology for my own enterprise), Man no longer stood in the centre of the world; viewed that way, Copernicus' deed was rather the opposite of a shift towards the subjective. It is therefore somewhat misleading to summarize, as you do in the opening sentence of your otherwise highly readable *magnum opus* (how I envy your literary skill and your ability to put forward difficult philosophical problems in such a clear and elegant manner!), the content of critical philosophy in the lapidary sentence: "The world is my representation" ("Die Welt ist meine Vorstellung"). For, to quote another 20th-century philosopher, who also regarded himself as in many ways a disciple of mine, a certain Wittgenstein: "the thinking, representing subject does not exist", or: "the subject does not belong to the world; rather, it is the boundary of the world", or again: "the I of solipsism contracts into an extensionless point and what remains is the reality coordinated with it."[*164] So in fact you might as well say that critical philosophy annihilates the I as that it annihilates the world; or, better still, that it overcomes the hoary Cartesian dualism by depriving the thinking subject of its substantial character as well as depriving the object, extended in space, of any metaphysical reality beyond the forms under which it is knowable as such ('substance' being only one of these forms). It might therefore be said that in critical philosophy the two appear as a *correlative unity*.'

But of course Schopenhauer would not yet own defeat. He would indignantly retort, first, that he never said that the world is *only* my representation (on the metaphysical level it is also *Will*); second, that he, too, even more so than Kant himself, always insisted upon the strict correlativity of subject (*knowing* subject, to be sure) and object and upon the impossibility for the former to become an object of knowledge itself; third, however, that, although this is true *in abstracto*, in concrete reality this knowing subject (Kant's 'transcendental subject') appears only as the instrument of an (animal or human) individualization of Will that, in

---

[*] "Das *denkende, vorstellende* Subjekt gibt es nicht"; "Das Subjekt gehört nicht zur Welt, sondern es ist eine Grenze der Welt"; "Das Ich des Solipsismus schrumpft zum ausdehnungslosen Punkt zusammen, und es bleibt die ihm koordinierte Realität"; "Das Subjekt gehört nicht zur Welt, sondern es ist eine Grenze der Welt."

order to serve Will's universal aims, has acquired *consciousness*, only in the *inside* of which the representation of the outer world comes into being;[*165] and that, last but not least, Kant himself originally took the same position on the latter issue, and only changed the passages concerned in the second edition of the *Critique of Pure Reason* out of fear of censorship.

The first argument, as we have imagined it, is of course valid (although Schopenhauer's identification of Kant's intrinsically unknowable 'Thing-in-itself' with 'Will' was in itself a revolutionary deviation from the master's way!); and so is the second. Indeed, we already know from our discssion (on p. 44 of Chapter II) of the first version of the dissertation that, according to Schopenhauer, the knowing subject can be known only by our studying the various classes of its objects. And in the second volume of the *magnum opus* we read in the first chapter, entitled 'On the fundamental view of idealism' ('Zur idealistischen Grundansicht'): "for quite as wrong as the thesis of the uneducated mind 'the world, the object, would still be there even if there were no subject', is this: 'the subject would still be a knowing subject even if it had no object, i.e., no representation, at all'. Consciousness without an object is no consciousness."[†166] 'No object without a subject', but also: 'no subject without an object'! Schopenhauer may indeed be awarded the credit that, long before Brentano and Husserl, he in effect (though not using the term) insisted on the *intentional* nature of consciousness (which, we repeat, is implicit in Kantianism, but was put forward by Schopenhauer more clearly than by Kant himself). Hence, Kant (remembering his own polemic against 'rational psychology' — soul as substance (*Seele als Substanz*) — in the section on the 'paralogisms' from the *Critique of Pure Reason*) would certainly not have objected to Schopenhauer's writing elsewhere of the thinking subject: "This focus of the brain's activity (in other words, the knowing subject) is, as an indivisible point, certainly simple by nature, yet not for that reason a substance (soul), but rather a mere state of things."[‡167] Thus we may

---

[*] This argument was in fact used by Schopenhauer in a letter to his faithful disciple Julius Frauenstädt, who had some difficulties with Schopenhauer's refutation of Fichteanism. Moreover, in the second volume of the main work Schopenhauer, in speaking of Kant's "transcendental unity of apperception", says that what gives unity and coherence to consciousness cannot be a representation itself, since it is the substrate of all representations (up to this point Kant would agree, PL); it must be the *prius* of consciousness and the root of the tree, which is *will*. 'Will' has produced consciousness, it gives it its unity, and it accompanies all its representations and thoughts as its *basso continuo*, so to speak.

[†] "Denn eben so falsch wie der Satz des rohen Verstandes 'die Welt, das Objekt, wäre doch da, auch wenn es kein Subjekt gäbe', ist dieser: 'das Subjekt wäre doch ein Erkennendes, wenn es auch kein Objekt, d.h. gar keine Vorstellung hätte'. Ein Bewußtsein ohne Gegenstand ist kein Bewußtseyn."

[‡] "Jener Brennpunkt der Gehirnthätigkeit (oder das Subjekt der Erkenntniß) ist, als untheilbarer Punkt, zwar einfach, deshalb aber doch keine Substanz (Seele), sondern ein bloßer Zustand."

assume that, in our imagined dispute, Kant would have given way to Schopenhauer's first two arguments, albeit with the significant proviso that it is a transgression to speak of 'Will' as Thing-in-itself. The more firmly would he have disputed the third argument, urging that it is precisely here that we are at the heart of the divergence, to wit, in Schopenhauer's thesis that "the world is my representation"! And the fourth argument would of course have been rejected by him with indignation.

Now this fourth argument greatly contributes to a better understanding of the third; it seems appropriate, therefore, to turn our attention to it first. Leaving aside the question of whether Schopenhauer's 'censorship' assumption was justified (the first editor of Kant's complete works, Karl Rosenkranz, although complying with Schopenhauer's request to reissue the first version of the *Critique*, firmly rejected the suspicion),*[168] the fact still remains that the edition of 1787 differs in many places from that of 1781. Schopenhauer mentions here as especially significant the mutilation of the first chapter of the second book of the 'Transcendental Dialectic', and the insertion, in its stead, of a 'Refutation of Idealism' in the 'Transcendental Analytic'. Let us find out, therefore, what a close look at these texts reveals.

The first chapter of the second book of the 'Transcendental Dialectic' is concerned with the 'paralogisms' (i.e., the formal defects of a syllogistic conclusion) into which any attempt at a rational psychology (see footnote on p. 83) must necessarily run. Kant distinguishes four such paralogisms, the first three of which need not concern us here since they deal with the 'soul' itself and with its alleged, substantial nature. The fourth, however, is about the soul's relationship with possible objects in space. The syllogism in question runs as follows: "Anything of which the existence can solely be inferred as a cause of given perceptions has only a doubtful existence. Now, all outer phenomena are of such a kind that their existence cannot be immediately perceived but only inferred as the cause of given perceptions. Therefore the existence of all objects of the outer sense is doubtful."†[169]

In his treatment of this paralogism Kant concedes that the existence of a 'real' object (i.e. of a thing-in-itself) is never given in perception but can only be inferred as its external cause. To that extent the philosophical position usually called "idealism" is correct. But, Kant continues, one must distinguish between *transcendental* idealism and *empirical* idealism. The former — his own position

---

* Schopenhauer wrote in response that he did not so much mean fear of censorship from above but rather some kind of inner censorship — the fear that people would regard Kant's doctrine as a new version of Berkeleyanism and thus misjudge his originality.

† "Dasjenige, auf dessen Dasein nur als einer Ursache zu gegebenen Wahrnehmungen geschlossen werden kann, hat eine nur zweifelhafte Existenz. Nun sind alle äußeren Erscheinungen von der Art: daß ihr Dasein nicht unmittelbar wahrgenommen, sondern auf sie, als die Ursache gegebener Wahrnehmungen, allein geschlossen werden kann: Also ist das Dasein aller Gegenstände äußerer Sinne zweifelhaft."

— regards all phenomena as mere representations, not as things-in-themselves; consequently, space and time are forms of our sense-intuition (*Anschauungsformen*) only, not conditions of the objects as things-in-themselves. But this is precisely why the transcendental idealist has no need to leave the realm of self-awareness in order to acknowledge the existence of material things, since these are nothing but a certain kind of representations that are called *outer* ones, because they are also perceived in *space*; as such they exist just as certainly as I myself exist as a thinking subject, the only difference between the two being that the representation of myself belongs only to the *inner* sense (which is perceived only in *time*) and the representation of extended bodies belongs also to the *outer* sense.

According to Kant, the transcendental idealist is therefore an empirical *realist*; for he grants reality to matter as a phenomenon — a reality that need not be inferred but is directly perceived. The empirical idealist, in contrast, is a *transcendental* realist because he can conceive of external phenomena (insofar as these are acknowledged to be real) only as things-in-themselves that exist independently of us.[170] In order to avoid the pitfall of materialism he is therefore compelled either to deny flatly or at least to doubt the existence of matter. In the first case we are concerned with a *dogmatic* idealist, in the second with a *sceptical* one. The latter (of which Descartes is the prototype) is prepared to believe in the existence of matter (thus becoming a *dualist*), but he finds in direct perception insufficient reason for this belief. This turns him into a partisan of human reason insofar as he urges us always to keep our eyes open and not to consider the results of common experience as our legitimate and secure possession. The dogmatic idealist, however — no name is explicitly mentioned here, but Kant evidently aims at idealists of the Berkeleyan type — is in fact quite as uncritical as his counterpart, the dogmatic materialist, since the non-existence of matter (considered as a thing-in-itself) is just as unprovable as its existence.[*171]

So far Kant in the first edition of the *Critique of Pure Reason*. The distinction between transcendental and empirical idealism was whole-heartedly adopted by Schopenhauer, especially so in the first chapter of the second volume of *The World as Will and Representation* mentioned above. Now it is true that in the second edition of the *Critique* this distinction, although certainly not abandoned, has been pushed somewhat to the background, at least in the chapter in question. The long digression on the fourth paralogism we have just summarized is reduced here to the statement that the distinction between my own existence as a thinking subject and the existence of other things outside me is just as much a mere *analytical* judgement as the three preceding paralogisms (the treatment of which has also been substantially reduced here), since 'being distinct of mine' is the very definition of 'other' things. The mere notion of 'I' is nothing but the functional condition of knowledge; and it is by no means certain that the thinking subject, regarded as a substance, has more reality, as such, than material things; indeed, it

[*] In justice to Berkeley it should be noted that he did not deny the existence of matter but only stressed its dependence on mind, with God as its warrant.

is not even certain whether, without my having representations of these (of which my body is one) consciousness of myself as a thinking subject would be possible at all.[172] Furthermore, it is equally true that (in the fourth section of the second chapter of the second book of the 'Transcendental Analytic', which deals with 'the postulates of empirical thinking in general') the second edition contains the very 'Refutation of Idealism' Schopenhauer so strongly disapproved of. In this new section the two versions of what was in the first edition called 'empirical idealism' are once more described and rejected, but now *without* contrasting them with a good, 'transcendental', brand of idealism.

So at first sight the conclusion seems obvious that Kant had meanwhile become inclined to avoid the term 'idealism' as a label for his own philosophical position because he felt the need to demarcate it more carefully than he had done in the first edition, from Berkeley's philosophical position. (Berkeley is now being mentioned explicitly as the prototype of 'dogmatic' idealism, alongside Descartes as the prototype of 'problematic' idealism; this predicate has come to replace the 'sceptical' idealism of the first edition.) This conclusion is confirmed by Kant's *Prolegomena to Any Future Metaphysics*, written in 1783 (i.e., in between the first and the second editions of the *Critique of Pure Reason*). An anonymous, strikingly dense review of the first edition in the *Göttinger gelehrten Anzeigen* of 19 January 1782, "which, therefore, I would have left entirely unnoticed if it had not given me an opportunity for some elucidations which in some cases may save the reader of these *Prolegomena* from a mistaken interpretation [of transcendental idealism, PL],"[*173] had indeed lumped together Kant's *transcendental* or *critical* idealism with Berkeley's *subjective* or *psychological* idealism; and in a supplement to the *Prolegomena*, Kant set out to correct this misunderstanding. Since, however, at other places Kant continued to describe his own position as 'transcendental *idealism*',[†174] it seems at least exaggerated to speak of a fundamental shift in Kant's position, as Schopenhauer does.

It is true that the emphasis had shifted somewhat. For as a matter of fact the representations of our outer sense are now even being accorded some kind of *priority* as compared with those of our inner sense. The latter are of an exclusively *temporal* nature, which implies that they are in constant flux. But change has meaning only when it can be contrasted with something that does not change. In order to have self-awareness something unchanged (*beharrlich*) is therefore required; since I cannot find this in myself it must be found outside me, i.e., in

[*] "... daher ich diese Beurtheilung gänzlich übergangen sein würde, wenn sie mir nicht zu einigen Erläuterungen Anlaß gäbe, die den Leser dieser Prolegomenen in einigen Fällen vor Mißdeutung bewahren können." The review had been written by a certain Chr. Garve, the published text being a shortened version of his original text rewritten for the purpose by the editor, J.G.H. Feder.

[†] Especially so in the sixth section of the second chapter of the second book of 'Transcendental Dialectic', where transcendental idealism is presented as the key to the solution of the cosmological dialectic, i.e. of the antinomies of Reason.

those representations that are perceived in *space*.[175] The crux, however, is that in *both* editions Kant firmly holds that external objects (as phenomena, mark you, not as things-in-themselves!) can be directly perceived and need not be inferred as causes of our perception. But, as we have seen in the preceding chapter, this is precisely what, according to Schopenhauer, they *do* need. The fundamental difference is therefore not so much between Kant A and Kant B, but rather between Kant *tout court* and Schopenhauer. Indeed, the latter point was to be tacitly admitted by Schopenhauer himself in the second edition of the *Fourfold Root*, where he explicitly criticizes from this point of view Kant's *original* exposition of the fourth paralogism.[176] One might well wonder, therefore, what exactly made him insist so strongly upon calling Kant's revised exposition mutilated, if a ready answer did not present itself in Schopenhauer's evident desire to play down the fundamental difference between Kantian and Berkeleyan 'idealism'. This answer may seem odd, as Schopenhauer never failed to insist that he, too, was an *empirical realist* who never thought of denying the reality of the empirical world.[*177] But the question is: Was this claim of his really justified?

b. Schopenhauer's physiological epistemology: analysis

For an answer to that question we must consider once more the theory of perception first presented by Schopenhauer in *On Vision and Colours* and later, more extensively so, in the second edition of the *Fourfold Root*. Its basic point was, as we recall, the distinction between 'sensation' (*Empfindung*) and 'intuition' (*Anschauung*). The former is nothing but an event in a part of our organism which remains, so to speak, 'underneath the skin' and which, as such, contains nothing that lies outside us. Intuition, in contrast, while taking its raw material from sensation, makes something quite different out of this by transforming it into the representation of an outer world as distinguished from the 'I'. It is capable of doing so because our Understanding (*Verstand*), which is located in another part of our organism, the brain, unconsciously applies to our sensations the law of causality (its one and only category) which demands a *cause* for every single sensation — a cause which, according to the law of space (one of our two forms of intuition) is interpreted as an 'object' *outside* of our organism. "When Understanding ... begins its activity and proceeds to apply its one and only form, the law of causality, then a powerful alteration takes place, since subjective sensation is now transformed into objective intuition."[†178] This object, however, is not the 'thing-in-itself', but only a 'phenomenon', since it exists only for the subject; this

[*] In Section 5 of the first book of the first volume of *The World as Will and Representation* Schopenhauer speaks in this context of the "foolish dispute on the reality of the outer world" ("der törichte Streit über die Realität der Außenwelt").

[†] "Wenn der Verstand ... in Thätigkeit gerät und seine einzige und alleinige Form, das Gesetz der Kausalität, in Anwendung bringt, geht eine mächtige Verwandlung vor, indem aus der subjektiven Empfindung die objektive Anschauung wird."

is why we said that the difference between Kant and Schopenhauer is that, to the latter, external objects must be inferred already as *phenomena* in order to be causes of our perception. This leads to the strange paradox that the very thing which Schopenhauer calls the 'objective view of the intellect' (namely, its physiological, and to that extent 'materialist', interpretation as a mere function of the *brain*) leads here to a radically subjectivist, and to that extent 'idealist', epistemology! For remove the tiny piece of our organism called 'the brain', and the objective world passes into nothingness.

In our view, Schopenhauer is the victim of a delusion here. To the 'objective view of the intellect', the acts of thinking and of perceiving are nothing but a physiological function of the brain, Schopenhauer says, "so that one is justified in asserting that the whole objective world ... is really only a certain movement or affection of the pulpy brain-mass inside the skull;"*[179] and somewhat further on: "the nervous system ... is bounded by the skin; however, once advanced in the brain to the status of intellect, it crosses this boundary by means of its cognitive form of causality, and thus perception arises, as a consciousness of *other* things, as a picture of beings in space and time that change themselves by virtue of causality."†[180]

Now as a physiological fact it is of course true that the whole perceptual process takes place inside my skull and therefore underneath my skin; up to this point the world can indeed be said to be 'in my head'. But this would justify Schopenhauer's theory only if it followed from this *physiological* fact that there is also a *psychological* priority of the 'inside'; i.e., if in the beginning I were *conscious* of the fact of a process taking place underneath my skin, the initiator of which process would then subsequently, by an operation of my Understanding, be projected outside myself. But is this really how it goes? Take Schopenhauer's illustration of the young child we quoted at length on p. 67. The idea that its perception of the outer world is produced in its brain from sensations of its retina is something it learns only much later, in the biology lesson at school; the awakening of its consciousness of the outer world, which Schopenhauer so fascinatingly describes, is therefore likely to have its origin, not in some intellectual leap (however unconscious) from 'inside' to 'outside', but rather in frustrated vital impulses which teach the child to distinguish between its own body (the movements of which it can influence, and which is therefore seen now as 'belonging to itself' and, to that extent, as 'inside') and the objects of its desire (which do not automatically respond to its impulse, and are therefore seen now as 'outside'). It is only through such a confrontation with outer reality that the child

* " ... ja, daß man berechtigt ist, zu behaupten, die ganze objektive Welt ... sei eigentlich nur eine gewisse Bewegung oder Affektion der Breimasse im Hirnschädel."

† "Das Nervensystem ... ist durch seine Haut begränzt; jedoch im Gehirn bis zum Intellekt gesteigert, überschreitet es diese Gränze, mittelst seiner Erkenntnißform der Kausalität, und so entsteht ihm die Anschauung, als ein Bewußtseyn *anderer* Dinge, als ein Bild von Wesen in Zeit und Raum, die sich verändern gemäß der Kausalität."

becomes conscious, not only of this outer reality, but also *pari passu* of its own inner reality, thus gradually learning to distinguish between 'I' and 'world', whereas in the beginning — during the period when, in Schopenhauer's words, "it gazed stupidly into the world" — it lived in some kind of undifferentiated whole. In other words, not only from an *abstract, logical* point of view (this, we recall, Schopenhauer always acknowledged), but also (this being what Schopenhauer failed to see) from a *concrete, psychological* point of view there is a strict *correlativity* of subject and object.* This, then, does away with the third argument we made Schopenhauer bring up against Kant, namely, that the correlativity of knowing subject and object is true *in abstracto*, but that in concrete reality there is a priority of the subject *because* there is a priority of *consciousness*. For, although it may be admitted that the world is 'in consciousness', it is not 'in my head', since my head, too, is 'in consciousness', 'consciousness' being nothing but being aware of my head (or rather — and this is important, as we shall see — of my whole body) and the world as distinct, yet inseparably linked, entities.

The most striking point in Schopenhauer's theory of perception, as we have sketched it in the above, is of course the twofold new dimension he gives to the category of causality (the only remaining category!). First, the very fact that it plays a central role in a theory of *perception* implies that to Schopenhauer causality is not so much a category of *experience* in the Kantian sense (which, as we know, is for all practical purposes synonymous with Newtonian science) but rather, so to speak, a category of *intuition*, actually introduced by him as a third constitutive principle besides the pure forms of space and time, the *union* of which he elsewhere states it to be.†[181] Second, causality is to Schopenhauer more than a way of establishing law-dominated connections between objects of experience (or

* "Correlation", not "identity"! Cf. Chapter II, p. 44.

† "Now the law of causality derives its importance and necessity only from this, that the essence of change does not consist in mere change of the situations as such, but rather in the succession of *one* situation to *another at the same place*, and in the presence of this situation *here* and the other *there* at *one* and the same time: it is only this reciprocal limitation of space and time by each other which endows a rule, in accordance with which the change must take place, with importance and at the same time with necessity ... The change, i.e., the succession that takes place in agreement with the law of causality, therefore, always has to do with a specific part of space and a specific part of time *together* and in union. Consequently, causality unites space with time." ("Nun aber erhält das Gesetz der Kausalität seine Bedeutung und Nothwendigkeit allein dadurch, daß das Wesen der Veränderung nicht im bloßen Wechsel der Zustände an sich, sondern vielmehr darin besteht, daß an *demselben Ort* im Raum jetzt *ein* Zustand ist und darauf ein *anderer*, und zu *einer* und der selben bestimmten Zeit *hier* dieser Zustand und *dort* jener: nur diese gegenseitige Beschränkung der Zeit und des Raums durch einander giebt einer Regel, nach der die Veränderung vorgehn muß, Bedeutung und zugleich Nothwendigkeit ... Die Veränderung, d.h. der nach dem Kausalgesetz eintretende Wechsel, betrifft also jedesmal einen bestimmten Theil des Raumes und einen bestimmten Theil der Zeit *zugleich* und im Verein. Demzufolge vereinigt die Kausalitat den Raum mit der Zeit.").

rather between the *changes* these undergo). Of course it is that; but it is also — and primarily so — a way of *constituting* these objects in the first place. Or, in other words, the principle of causality is not only at work as an ordering and arranging agency *in the interior* of the phenomenal world; nay, it even *creates* this world (namely, by postulating external objects as causes of our sensations)! But this leads to insuperable difficulties. For, in spite of Schopenhauer's own repeated warnings that the law of causality (and the principle of sufficient reason in general), being the *a priori* of all objects, is valid between *objects* only and not between *subject and object*, his theory of perception implies all the same that the knowing subject actually creates (and thus 'causes') the object — an object which is at the same time said to have caused *previously* the sensation in the sense-organ![182] And these inferred objects are *not* things-in-themselves, but mere phenomena, since our brain, no less than our sense-organs, is an operator 'underneath the skin', which can therefore provide subjective knowledge only.

It follows from this that, for all Schopenhauer's recurrent protestations that he is an 'empirical realist' who has never thought of denying the reality of the external world, this external world is in fact reduced by him to the status of a *phantasm* unfolded before our eyes by our Understanding. Indeed, the very expression was used by Schopenhauer himself in a passage in his main work;[183] and elsewhere he spoke of the "dream-like nature of the whole world" and quoted with approval the title of one of Calderon de la Barca's plays: *La vida es sueño* ("*Life is a dream*").[184]

c. Schopenhauer's physiological epistemology: assessment

And yet, if we wish to do justice to Schopenhauer we ought to place his view of the nature of causality in its historical context, which is provided in particular by the Hume–Kant controversy; for only this will enable us to appreciate the indisputable merits of Schopenhauer's epistemological innovation.

"To refute him [Hume] has been, ever since he wrote, a favourite pastime among metaphysicians," writes Bertrand Russell.[185] But, although he finds none of their refutations convincing, even Russell feels compelled to admit that "where psychology is concerned, Hume allows himself to believe in causation in a sense, which, in general, he condemns."[186] Indeed, Hume says in effect that the often observed succession of A and B *causes* the impression of A to *cause* the idea of B: two unwarranted usages of the word in one sentence — according to Hume's own thesis. Hume's insistence on the nongivenness of causation among percepts was surely irrefutable from an empiricist point of view; but that is precisely why we have noted in our first chapter (p. 19) that traditional empiricism bore in itself the seeds of its own decay. For this meant that, even before Kant, an *intellectual* component (in the guise of the law of association) had tacitly to be admitted in addition to the raw material of sense-data. Kant, however, insisted that association is possible only by 'affinity' (*Affinität*) which solely can decide which perceptions are associable and which are not; and this *empirical* affinity in its turn is possible only as a consequence of *transcendental* affinity, i.e., the principle of the unity of apperception in relation to the cognitions that can be said to be 'mine'.*[187] This insight lies at the root of Kant's proof of the law of causality — a proof we have already met in our second chapter (p. 50) together with Schopenhauer's objections against it.[188] The heart of the matter is of course Kant's awareness (besides the circularity that arises from a strictly empiricist point of view) that even an admitted *psychological necessity* of the law of *association* is not enough to guarantee the *validity* of the law of *causality*. This is why he introduced the synthesizing activity of Understanding.

But there was still another possible objection to Hume, which was *not* met by Kant's transcendental argument. Recall Schopenhauer's objection that empirical phenomena may very well succeed one another in an objective manner without the first thereby becoming cause and the latter effect. The classic example is the regular succession of day and night, which nobody in his right senses had ever taken for a succession of cause and effect. Russell, wishing to get rid of what he considered the superstition of the law of causality, yet prepared to speak of a causal sequence in those cases in which a uniform sequence has been observed so often as to provide inductive probability for it to appear likewise in future, suggested henceforth calling *every* case of a sufficiently frequent sequence "causal", and thus no longer refusing to say that night is the cause of day.[189] But unfortunately this does not alter the fact that in those dark, pre-Russellian ages people *did* refuse it (or rather, did not hit upon the idea), so that the notion of causality, regardless of whether justified or not, cannot have its *origin* in regular succession. This is why Schopenhauer came up with his alternative demonstration of the *a priori* validity of the law of causality (a position to which he stuck in no lesser measure than Kant), the essence of his demonstration residing, as we have seen, in the conception that causality is necessary not only for the *regularity* of *some* (not of all) phenomena, but for the *constitution* of these phenomena in the first place. "It is only by the category of causality that we recognize objects as *real*, i.e., as *acting* upon us," says Schopenhauer with an etymological emphasis which alas gets lost in translation.†[190] But, apart from those problems arising from this which we have raised in the preceding section, this also meant of course that the function of the *a priori* became a different one from what it had been for Kant. Instead of, as its primary job, making *science* possible,‡ it was now invoked to

---

* "Also stehen alle Erscheinungen in einer durchgängigen Verknüpfung nach notwendigen Gesetzen und mithin in einer *transzendentalen* Affinität, woraus die *empirische* die bloße Folge ist ... Die erste liegt in dem Grundsatz von der Einheit der Apperzeption in Ansehung aller Erkenntnisse die mir angehören sollen."

† "Durch die Kategorie der Kausalität allererst erkennen wir die Objekte als *wirklich*, d.i. auf uns *wirkend* an." It is this etymological connection with the verb *wirken* (i.e., 'to act') which made Schopenhauer prefer German *Wirklichkeit* to the 'reality' and other variations of Latin *realitas* in the Latin-derived languages. Cf. also Chapter VIII, p. 212.

‡ The question (frequently answered in the negative nowadays under the influence of, above all, quantum-mechanics in its Copenhagen interpretation) of whether causality is indeed a

make *perception* (or, in Schopenhauer's term, *Anschauung* (*intuition*)) possible. Kant, had he lived to know it, would therefore in all likelihood have been as little pleased with Schopenhauer's mode of saving the *a priori* nature of causality as Goethe had been with Schopenhauer's mode of defending his theory of colours! Yet it is precisely here that Schopenhauer's originality and (in spite of all shortcomings so far acknowledged) his merits as an epistemologist reside. In order for us to gain a clearer understanding of this we must cast a more thorough eye than we have yet done upon Schopenhauer's 'Critique of Kantian Philosophy'.

After exuberant praise for Kant's 'Transcendental Aesthetic', the theses of which he regards as 'irrevocable truths', Schopenhauer nevertheless goes on to remark that, although nothing can be taken away from this doctrine, something could still be added to it. We read: "After the elaborate treatment of the general *forms* of all intuition in the 'Transcendental Aesthetic' one would expect to get some information about its *contents* — about the way *empirical* intuition enters our consciousness, that is, how our knowledge of this whole world, which is so real and important to us, arises in us. But on this topic Kant's entire doctrine really contains nothing but an expression he repeats as often as it is devoid of content: 'The empirical element of intuition is *given* to us'."*[191] Now Kant might of course have replied here, once again, that this reproach is beside the point since his *transcendental* investigation is indeed about forms only, not about contents. But let us listen first to the continuation of Schopenhauer's argument. Its core is that, by leaping immediately from the treatment of the pure forms of intuition in the 'Transcendental Aesthetic' to the treatment of *concepts* in the 'Transcendental Logic', Kant misses a link between *sensation* and *thought*. For at the beginning of the 'Transcendental Logic' Kant says that "our knowledge has two sources, namely, receptivity for impressions and spontaneity in forming concepts: the former being the faculty of receiving representations, the latter that of recognizing an object by these representations. By virtue of the former an *object* is given to us, by virtue of the latter it is thought." Whereupon Schopenhauer retorts: "This is wrong: for in that case the *impression*, for which alone we have mere receptivity,

---

necessary constitutive principle of *science* would require a chapter of its own. However, not only is it beyond our competence (which goes, too — it may be said in passing — for the parallel question of whether the mathematics of Riemann and Lobachevsky have done away with Kant's doctrine of space and time), but also the question is not of prime importance in the present context; in the following, we focus on Schopenhauer's theory that causality is a necessary constitutive principle of *intuition* (*Anschauung*).

* "Nach der in der transzendentalen Aesthetik gegebenen, ausführlichen Erörterung der allgemeinen *Formen* aller Anschauung muß man erwarten, doch einige Aufklärung zu erhalten über den *Inhalt* derselben, über die Art wie die *empirische* Anschauung in unser Bewußtseyn kommt, wie die Erkenntniß dieser ganzen, für uns so realen und so wichtigen Welt in uns entsteht. Allein, darüber enthält die ganze Lehre Kant's eigentlich nichts weiter, als die oft wiederholte, nichtssagende Ausdruck: 'Das Empirische der Anschauung ist uns *gegeben*.'"

and which comes from outside and is, strictly speaking, the only '*given*', would already be a representation, nay, already an *object*. But it is nothing more than a mere *sensation* in the sense-organ ...";*[192] and then follows a summary of the theory of perception with which we are already familiar.

Now the objections to this theory that we have raised in the above do not preclude most people's inclination nowadays (our own included) to recognize that Schopenhauer was basically right in insisting that not only *thinking* but also *perceiving* is an *active, synthesizing* process; or, in Kantian terms, that already at the perceptual level there is not only 'receptivity' but also 'spontaneity'.†[193] In fact, contrary to Schopenhauer's suggestion, Kant himself already recognized as much, by introducing the faculty of 'imagination' (*Einbildungskraft*) as a necessary link between 'sensuality' (*Sinnlichkeit*) and 'understanding' (*Verstand*).‡[194] The

* "'Unsere Erkenntniß', sagt er [Kant], 'hat zwei Quellen, nämlich Receptivität der Eindrücke und Spontaneität der Begriffe: die erste ist die Fähigkeit Vorstellungen zu empfangen, die zweite die, einen Gegenstand durch diese Vorstellungen zu erkennen: durch das erste wird uns ein *Gegenstand* gegeben, durch die zweite wird er gedacht'. Das ist falsch: denn danach wäre der *Eindruck*, für den allein wir bloße Receptivität haben, der also von außen kommt und allein eigentlich '*gegeben*' ist, schon eine *Vorstellung*, ja sogar schon ein *Gegenstand*. Er ist aber nichts weiter, als eine bloße *Empfindung* im Sinnesorgan ..."

† One may think, for example, of Popper's statement that — assuming a sufficiently wide sense of the term 'theory' — "all knowledge is theory-impregnated, including our observations"; or of Rudolf Arnheim's by now almost classical book *Visual Thinking*.

‡ "Daß die Einbildungskraft ein notwendiges Ingredienz der Wahrnehmung selbst sei, daran hat noch wohl kein Psychologe gedacht" ("that imagination is a necessary ingredient of perception itself has not, in all likelihood, occurred to any psychologist yet"). It is at first sight surprising that in this case Schopenhauer did not mention the 'mutilations' of the second edition (where — at least in the chapter 'On the deduction of the pure concepts of Understanding' ('Von der Deduktion der reinen Verstandesbegriffe') — the treatment of the faculty of imagination is much more cursory and where in fact it appears as a function of Understanding rather than as an autonomous faculty). It is well known that precisely this 'mutilation' was later to induce Heidegger to suggest that Kant 'recoiled' from the ultimate consequences of his own doctrine of the finitude of human knowledge. Heidegger's view, in its turn, provoked a hardly less well-known reaction from Cassirer. According to him, and rightly so in our opinion, it was rather Kant's wish, in view especially of Garve's review (cited by us in a footnote on p. 90), to draw a clear line between psychological idealism and his own transcendental idealism that was responsible for the alterations in the second version of the *Critique of Pure Reason*. Moreover, Cassirer rightly points to the fact that the core of Kant's doctrine of imagination, to wit the chapter 'On the schematism of the pure concepts of Understanding' ('Von dem Schematismus der reinen Verstandesbegriffe'), has been preserved unchanged in the second edition; Heidegger may therefore be said to have as much exaggerated the 'mutilations' of this edition compared with the first one as Schopenhauer did! Returning now to Schopenhauer, the reason for his not mentioning the faculty of imagination is, apart from the reasons given above, probably that Kant's introduction of this faculty does indeed not affect Schopenhauer's criticism of Kant's failure to perceive the constructive nature of perception *in general*, that is to say, at

fundamental difference between Kant and Schopenhauer is that, according to the latter, it is precisely the faculty of 'understanding' which makes this synthetic activity at the pre-conceptual level possible; a faculty, moreover, which we share with the animals. All *conceptual* knowledge (which alone is specifically human) belongs to the domain of 'reason', whereas to Kant the difference between Understanding and Reason was a difference *within* the domain of *conceptual* knowledge (a difference corresponding to the two books of which the 'Transcendental Logic' is composed — the 'Transcendental Analytic', which is about constitutive concepts, and the 'Transcendental Dialectic', which deals with regulative Ideas). By contrast, from Schopenhauer's point of view — he does not phrase it that way, but this is what it comes down to — Understanding ought to have been treated already in the 'Transcendental Aesthetic' (which would thus have expanded greatly) and Reason in the 'Transcendental Analytic' (which would have shrunk considerably in view of the drastic reduction of the categorial apparatus to the one remaining category of causality). The 'Transcendental Dialectic', furthermore, ought to have been dropped in its entirety,[*195] since the faculty that corresponds to it and which Kant calls Reason, in Schopenhauer's opinion does not exist at all.

Leaving aside, for now, the latter problem (we return to it in Chapter VI), there is certainly something to be said for Schopenhauer's attributing *Understanding* (and not only 'instinct') to animals as well; animals, after all, also live in a *meaningful* world! This of course serves to make the chasm between man and beast somewhat more relative, in a manner which reflects the essence of the biological revolution of the 19th century. Given Schopenhauer's background, it is not surprising to find him in this intellectual company. We recall from Chapter I, p. 39 that as a student at the university of Göttingen he had already become acquainted with works by contemporary physiologists like Cabanis (*Rapports du Physique et du Moral de l'Homme*) and Flourens, who discovered the functions of the great (*cerebrum*) and the little brain (*cerebellum*). Later Bichat (*Recherches Physiologiques sur la Vie et la Mort*) and others like Magendie, Charles Bell and Marshall Hall followed suit. This is what Wilhelm Gwinner had in mind when (cf.

---

the animal level as well. For Kant calls perception (of which imagination is a "necessary ingredient") "intuition of which one is conscious", thus confining it to the *human* level. That Schopenhauer, although rightly attributing intelligence to animals, in his turn goes too far by speaking of 'representations' at the animal level will concern us in the next section of the present chapter; in that context we shall have to make additional remarks on 'imagination', too.

* This is not literally true insofar as Schopenhauer did recognize as a merit of the 'Transcendental Dialectic' that it destroyed, for good, scholastic philosophy by eliminating theism from philosophy; this, however, he found more of "temporal and local" than of "general, everlasting, and purely philosophical" interest. The essential point is that Kant's 'Ideas of Reason' were not even accepted by Schopenhauer as regulative principles; to him Reason (*Vernunft*) was nothing but the human faculty to bring order to a variety of representations (*Vorstellungen*) by subsuming them under one general concept (*Begriff*).

Introduction, p. 1) he spoke of the "physical realism" that had contributed to Schopenhauer's philosophical development in an essential manner, and which had been amalgamated in his system with the "ethical realism" of Kant and the "aesthetic realism" of Goethe into a wholly new approach to the hoary problems of philosophy.

Indeed, Schopenhauer always scorned those philosophers who spoke of the human intellect as if it were a "winged angel's head without a body" ("geflügelter Engelskopf ohne Leib");[196] and he may therefore be said to have been the first important philosopher to philosophize "with the body for his guide" ("am Leitfaden des Leibes"), as his disciple Nietzsche was to call it later. He definitely did away in his philosophy with the old, cherished opposition between 'body' and 'soul'; both man and beast were always considered by him as a psychosomatic unity. He used to express this by saying that what is seen from within as an act of Will appears on the outside as an act of the body. It has been noticed that this starting point led him to interesting anticipations of modern biological insights into the adjustment of the organic specialization of each animal species (including man) to the demands of its instinct and to its conditions of life (especially so in his book *On Will in Nature*);[*197] and there are indeed significant similarities here, as we shall see later on. For all his originality Schopenhauer remained nonetheless conventional insofar as, like his sensualist and empiricist predecessors, he continued to treat the phenomenon of perception as isolated from the whole area of bodily-volitional functions. Certainly he was original here, too, in that the advantage he took of his knowledge of the results of contemporary physiology enabled him to recognize the crucial role of the *brain* in the perceptual process; a process regarded up until then by most philosophers as merely a matter of the sense-organs. But in Schopenhauer's account it looks as if the faculty of Understanding must, by means of its one and only category of causality, i.e. *by a purely intellectual leap* ('intellectual' not in the sense of 'Reason', but of 'Understanding', to be sure) *break through the wall of the body*, so to speak, in order to reach the outside world — a world which, consequently, is nothing but 'my representation' since space, time and causality are nothing but 'functions of the brain' and therefore 'subjective'.[†] Was the 'wall of the body' (to use our image) already somewhat astonishing in view of the fact that, although brain and nervous system certainly are located 'underneath the skin', the sense-organs are after all open gateways through which outer information enters the body — it is still much

---

[*] For example, by Arnold Gehlen, and also by Hübscher, who praises Schopenhauer for the same reason, with special reference to von Üxküll, whose influence on such representatives of the German school of 'philosophical anthropology' as, notably, Gehlen (just mentioned) and Plessner (soon to be mentioned as well) has been considerable, as we shall see.

[†] We may refer here once more to Schopenhauer's statement, quoted in the above on p. 92, about the nervous system which is bounded by the skin, but which, once advanced in the brain to the status of intellect, crosses this boundary by means of its cognitive form of causality.

more astonishing to watch Schopenhauer neglecting the fact that not only the *receptive* kind of feelings that are called sense-impressions, but at least as much those *propelling* feelings that may be called *vital urges* (pain, anguish, hunger, sex, and the like) provide the raw material out of which the perception of an outer world is construed! For are 'vital urges' not the core of what Schopenhauer calls "Will" — the key word of his entire philosophy?* How then could the metaphysician of "Will" as an *epistemologist* overlook this whole complex and present our becoming conscious of an 'outer world' as a purely intellectual leap? The more so since he always stressed that 'Intellect' is (at least originally) nothing but the servant of 'Will'. Is it not obvious, then, that the origin of our notion of causality should have something to do with the 'volitional' and not only with the 'intellectual' sphere?

The explanation of this amazing neglect is evidently to be found in Schopenhauer's wish to retain (in however un-Kantian a way) Kant's thesis of the *a priori* nature of the law of causality. This may be illustrated using his comment on Maine de Biran, who saw in "the willed effort" ("l'effort voulu") the origin of our notion of causality. Schopenhauer took him to task (while generously conceding that he came from a country where, just as in England, Kant's philosophy had not yet penetrated) for still not believing in the *a priori* nature of the law of causality,† but rather seeking it in the experience that, solely on the force of will as a cause, the movement of the body follows as its effect. But, Schopenhauer goes on, that is not so. For we do not experience the force of will proper as something *distinct* from our bodily action, such that the two have to be connected by causality. In fact, both are one and the same; between them there is no succession, since that which presents itself as an act of will to inner perception presents itself as an act of the body to outer perception.[198] And indeed, most people will concede (while adding that there is always a time lag between the two, however tiny) that in direct experience, though not perhaps on reflection, we are concerned, in this case, with identity rather than with causal connection, thus ruling out the hypothesis that here is where the origin of our belief in causality is to be sought.

But, our reader might ask, if we thus join Schopenhauer in judging Maine de Biran's search for the origin of the notion of causality in *inner* experience to be as mistaken as Hume's search for it in *outer* experience, what else is left to us but to recognize, with Kant and Schopenhauer, the *a priori* nature of causality? And indeed, we are inclined in a sense to agree with Schopenhauer insofar as he maintains that it is the notion of causality which makes *conscious perception* (or, as he would say, *intuition*) possible in the first place. But when we recall once again the passage on the awakening child (p. 66), the suggestion imposes itself that in the beginning there is a dim, animal feeling of frustrated vital impulses and of

---

* On this, highly controversial, Schopenhauerian terminology we have more to say in Chapters VII and VIII.

† Cf. on this Chapter III, p. 68.

forces working upon us, i.e., an *unconscious experience* of a *causal interaction* between ourselves and our surroundings.* It seems, therefore, that the notion of causality is originally imposed upon us by this primordial experience (and is to that extent *a posteriori*); yet we may concede to Schopenhauer that from then on it serves as a key-guide in the process of conscious perception of the world (and is to that extent *a priori*). To this extent, Schopenhauer was quite right with his thesis that "all intuition is an intellectual one"† — a thesis he illustrated by means of an array of acute observations on the operations which, along the road of perception guided by Understanding, lead to the construction of the image of an ordered external world.‡[199] It may be questionable in detail; yet, when we recall from the end of the preceding chapter Helmholtz' remark on "the wholly unclear and unjustified notion" of "Schopenhauer and his successors" and go on to consider the indisputable parallels between a notion so contemptuously dismissed and Helmholtz' own findings, we (in spite of our own critical remarks, and wholly passing by the question of possible plagiarism) cannot but think the remark

* In the animist world-view where man's inner experience of himself as 'willing' subject (the identity of action of the body and action of will, stressed by Schopenhauer against Maine de Biran) is projected, as it were, on nature around him, we may see this primary experience still at work ... but also on a higher level in Schopenhauer's own metaphysics where the world-riddle is explained by a bold analogy with the human body which, seen from the inside, shows itself as "Will"! And in a sense also still in the concept of 'force' in the physics of Schopenhauer's day, a circumstance which gives some credit to his claim that the concept of 'force' ought to be subsumed under the concept of 'will' rather than the other way round (more on this in Chapter VIII, p. 213).

† Indeed, the thesis, in one or another guise, of the 'intellectual' nature of 'intuition' has become rather a commonplace nowadays, as already remarked before (cf. a footnote on p. 97).

‡ The operations described already by Schopenhauer in *On Vision and Colours* (cf. Chapter III, p. 67) are presented in a much more elaborate way in the second version of the *Fourfold Root*, especially so in that here they are supplemented with digressions on perceiving three dimensions, on recognizing the correct distance of objects, and on seeing in perspective. Other important new additions are his arguments that (1) it is only the visual and the tactile sense (of which the former, while less perfect, has the advantage of much wider scope) that provide data for Understanding, since only these two make possible constructions in space, i.e., the constitution of an objective outer world (a blind man would *by himself* never be able to infer, from the sound of music, the presence of a musician or of instruments); (2) the intellectual nature of intuition is demonstrated by the fact that people born blind are often able nonetheless to attain a complete knowledge of spatial proportions, owing to their capacity to infer, e.g., from the resistance felt in fingering a cube, the presence of a solid body, and to construct the latter's form in *a priori* known *space* from the movements of their arms (this he illustrated by the example of a certain Saunderson, who, though born blind, studied mathematics, optics, and astronomy in Cambridge with success; also, conversely, by the fact that a child born without arms and legs (a Livonian girl called Eva Lauk is mentioned by Schopenhauer as an example) may perform the same feat.

ungenerous to say the least. Helmholtz' attitude probably had something to do, not only with the customary *pereant qui ante nos nostra dixerunt* ('let those perish who before us have said what we say', to use a phrase for which Schopenhauer had rather a predilection) but also with the 'no nonsense' intellectual climate of the time, which made scientists inclined to distrust ideas that came from such dinosaurs as philosophers with a metaphysical system (even if the system is as down-to-earth as Schopenhauer's).

Moreover, Schopenhauer's argument had the merit of demonstrating, in passing, that it is the *unexpected*, rather than — as Hume, and in a sense Kant, too, seemed to think — the *regular sequence*, that requires causal explanation. One example of this is his pointing to the reaction of his "very bright" ("sehr kluger") poodle to new bedroom curtains of the type that open in the middle when a rope is pulled: Atma at once began to search around for the other change that she *a priori* knew by necessity must have preceded this surprising change in her accustomed surroundings.[200] But one problem remains. We recall from the first section that the difference between Kant and Schopenhauer consisted in the former's holding that external objects (as phenomena, not as things-in-themselves, to be sure) can be directly perceived and need not be inferred as causes of our perception, which is precisely what they *do* need according to Schopenhauer. Now that we have come to reject this view of Schopenhauer's (at least in the guise in which he presented it), while acknowledging, in a sense, the role of causal inference in recognizing objects, have we not, then, condemned ourselves to the alternative that we infer to objects as *things-in-themselves* as causes of our perception? This is what Kant seemed to think; for in his treatment of the fourth paralogism he distinguished (but it must be repeated that he did so already in the *first* version so abundantly praised by Schopenhauer at the expense of the second!) between two meanings of 'external', namely, in the 'strict' (i.e., spatial and temporal) sense which he calls 'empirically external', and in another, 'transcendental', sense. Kant acknowledges that this is somewhat ambiguous in that the expression 'external' at one moment denotes something which, as a Thing-in-itself, exists independently of us, and at the other moment denotes something which belongs to the realm of (external) *phenomena*.[*201] But (even leaving aside such confusing terminology) when Kant thinks that we infer causally not to the latter (as Schopenhauer thinks) but to the former, is he not then committing indeed the "incredible inconsistency" rebuked by Schopenhauer in his 'Critique of Kantian philosophy',[†202] and noted before by

* "Weil indessen der Ausdruck: *außer uns*, eine nicht zu verneinende Zweideutigkeit bei sich führt, indem er bald etwas bedeutet, was als Ding an sich selbst von uns unterschieden existiert, bald was bloß zur äußeren *Erscheinung* gehört, so wollen wir *empirisch äußerliche* Gegenstände daher unterscheiden von denen die so in transzendentalem Sinne heißen."

† The "incredible inconsistency" ("unglaubliche Inkonsequenz") rests, of course, in applying the category of causality (the scope of which, according to Kant's own doctrine — precisely because of its being a 'category' — is restricted to the phenomenal world) to a supposed relationship between the phenomenal and the noumenal world.

Schopenhauer's teacher Schulze-Aenesidemus? That question brings us to the final problem to be treated in the present chapter.

d. The big question: is the world my representation?

We may sum up the findings of our preceding section in this formula: Kant and Schopenhauer were agreed that, for the recognition of objects, causal inference is needed one way or the other; the question that kept them divided was "what kind of objects?" To complicate the matter a little further, in his 'Refutation of Idealism' (to be found, as we saw, only in the second version of the *Critique of Pure Reason*) Kant distinguished between a *thing* outside me and the mere *representation* of a thing outside me.[203] Schopenhauer therefore accused Kant of placing an impossible "bastard thing" (*Zwitter*), namely, an 'object in itself' or an 'object of my representation', between Representation and Thing-in-itself: a 'bastard thing', since we are concerned here, in Schopenhauer's wording, with "an object which is not in need of a subject, a single thing and yet not in space and time — as it is not *anschaulich* —, an object of thought and yet not an abstract concept".[*204] According to Schopenhauer, nothing exists between 'representation' on the one hand, and the 'thing-in-itself' on the other. The reason is that it is impossible to differentiate between the representation and the object of the representation, since every object is *eo ipso* in need of a subject; hence, it can never be anything else but a representation.[205] Now this sounds plausible, but still ...

Let us look once more at Schopenhauer's polemic against Kant's assertion that by means of sense-impressions an object (*Gegenstand*) is *given* to us, whereas by means of concepts it is *thought*. In the previous section we have dealt with his objections against the first part of this statement; we now examine his objections against the second part. Schopenhauer's argument is here that thought, in being given a particular thing instead of a general concept as its object, is robbed thereby of its essential character of generality and abstraction.[206] But, one may ask, is it not the case that only the *concept* creates a real '*thing*', i.e., something *believed to exist independently* of the question of whether I perceive it or not? For it might be argued that animals, although certainly living in a structured and meaningful environment, lack the real notion of a 'thing' or a 'substance' in the sense of an identical, permanent substrate in which the objects of the different senses are unified, and of which the contents of our various sensations (sensations that are held to be *caused* by it) are then *interpreted* as so many 'qualities'. For this particular notion one needs "the faculty of representing an object that is not actually present", i.e., the faculty which Kant called 'imagination', and which presupposes the focus of the transcendental subject, which unifies the data of the

---

[*] "... ein Objekt, das keines Subjekts bedarf, [ist] ein einzelnes Ding, und doch nicht in Zeit und Raum, weil nicht aunschaulich, [ist] Gegenstand des Denkens, und doch nicht abstrakter Begriff."

different senses into one image (the 'synthetic unity of apperception').* And there we are back at Schopenhauer's theory of causality. In the preceding section we met Schopenhauer's poodle and we were indeed inclined to attribute to her, as her master did, a certain notion of causality. But we would suggest that the notion of causality appears in two distinct shapes which are expressed by two basic questions, namely: 'By what?' (German: '*wodurch*?') and 'Why?' (German: '*warum*?'). The first question was one that preoccupied Atma, too, as we recall (although, of course, in an unarticulated way); but the second question would not have bothered her at all, since it would not occur to her to cudgel her dog-brain about such questions as 'how far is our sensual experience reliable, i.e., how far does its information give us access to objective truth about the world?' For posing *such* questions not only 'Understanding' (in the Schopenhauerian sense of the word) but also 'Reason' is presupposed, since they betray the desire for an intelligibly coherent world; now the efforts to satisfy this eternal human longing for *explanation* have gone and still go through many different stages: 'ontogenetically' (in the child's mental development)† as well as 'phylogenetically' (in the succession of the various ontologies or world-pictures which the history of ideas presents to us, such as the Greeks overcoming the mythological world-picture, or the Scientific Revolution in its turn overcoming the world-picture of common sense which had found its classic representative in Aristotle). The question 'why?', in other words, always has *ontological* implications (which is why positivism — wrongly so, in our opinion — thinks it obsolete); well then, the 'things' or 'substances' of the world-picture of common sense and naive realism are precisely those 'objects in itself' or 'objects of my representation', so disdainfully rejected by Schopenhauer as 'bastard things' from 'representation' and 'thing-in-itself' (see above)! It seems, therefore, that these bastard things have a legitimate function after all in a theory that aims to give an accurate description of the ways of the human mind, even if only to demonstrate them in hindsight as fictitious. But that raises the question: what, then, ought we to understand by 'Representation' and by the 'Thing-in-itself'? As far as the latter and as far as Kant is concerned (it is otherwise with Schopenhauer, as we shall see) the answer, as far as the domain of *theoretical* reason is concerned, undoubtedly must (or at least ought to) be: a limiting concept (*Grenzbegriff*) which is symbolic for the never-ending quest of Understanding and Reason for an ever more comprehensive unity (of 'concepts' and 'ideas', respectively), which implies that it is constantly on the move like the ever receding horizon. Regarded this way it is 'imperative'

---

* On the faculty of imagination cf. footnote on p. 97. Adopting Schopenhauer's terminology, one may say that this faculty is not indeed a necessary link between 'intuition' and 'understanding' (as it was for Kant), but that it *is* a necessary link between 'understanding' and 'reason'.

† In this context the name of the French–Swiss pioneer in developmental psychology, Jean Piaget, should of course be mentioned before any other. Among his many works we point here especially to a book he published in 1937, *La construction du réel chez l'enfant*.

rather than 'indicative' in nature; it is therefore not by chance that, as Schopenhauer notices himself, it is only in his treatment of the third antinomy that Kant is forced to dwell at length on the Thing-in-itself.[207] The real bone of contention between Kant and Schopenhauer is indeed whether there exists such a thing as 'causality through freedom' as distinct from 'causality according to laws of nature'. This problem, with which we enter the *ethical* realm (the domain of 'practical reason') will concern us in our sixth and seventh chapters; to confine ourselves, for the moment, to the domain of 'theoretical reason' it must be said that in the case of the Thing-in-itself as in many more cases Kant's terminology is clumsy in the extreme. For, in view of what has been said in the above, the 'Thing-in-itself', whatever else it may be, is certainly not a 'thing'; yet, as we saw, in his treatment of the fourth paralogism Kant *did* speak about 'transcendentally external *objects*' as 'things-in-themselves', thus inevitably exposing himself to the objection made by Aenesidemus and Schopenhauer (see the end of Section c.).* It is indeed true, as Schopenhauer remarks, that Kant has nowhere made the Thing-in-itself into the subject matter of a special exposition or clear deduction and thus has given rise to much confusion.[208] We must therefore grant Schopenhauer that at times Kant was in danger of blurring his own distinction between transcendental *idealism* and transcendental *realism*. But it must be added at once that Schopenhauer himself was at times in danger of blurring the distinction between *transcendental* idealism and *psychological* idealism; indeed we hope to have made it plausible by now *that Schopenhauer's 'Kantianism' was rather a sophisticated Berkeleyanism*.†[209]

---

* They might perhaps be called *provisional* things-in-themselves (i.e., stages on the road to the 'real' Thing-in-itself, which however, will always remain elusive like the ever-receding horizon), if such a paradox is allowed! For the world-picture of classical or early modern science (for the terminology cf. Chapter I, footnote on p. 10) may be said to have been a 'thing-in-itself' compared with that of ancient science and of common sense (with which we associated in the above what Schopenhauer chose to call Kant's 'bastard things'); but the world-picture of modern science is in its turn a 'thing-in-itself' compared with that of classical or early modern science, as the world-picture of future science will be a 'thing-in-itself' compared with that of contemporary science ... and so on! With all this, however, we remain in the *phenomenal* sphere, i.e., in Kantian terms, in the realm of *necessity*; it is only as *moral* beings that we have access to the *noumenal* sphere, i.e., the realm of freedom. The 'real' thing-in-itself is therefore for Kant not the mysterious inner core of nature (as it was for Schopenhauer, as we shall see), but something which *transcends* nature: it does not belong to the domain of 'theoretical reason' but to that of 'practical reason'.

† 'Sophisticated' indeed, for of course the simple associationism of classic British empirical philosophy finds itself replaced in Schopenhauer's epistemology with Kant's synthetic vision (and in this regard Schopenhauer may be said to have remained true to *transcendental* idealism). That the formulation 'sophisticated Berkeleyanism' does not do any injustice to Schopenhauer appears from the fact that Schopenhauer himself always presented Kantianism (i.e., the unspoiled Kantianism of the first edition of the *Critique of Pure Reason*, to be sure!) as an improved Berkeleyanism. He marked his pertinent position in the clearest manner in the first chapter 'On the fundamental view of idealism' ('Zur

The latter blurring found its classic expression in the celebrated opening sentence of Schopenhauer's *magnum opus*: "The world is my representation". Let us look first at "representation" and then at "my". As to the former, we said in the above that Schopenhauer's refusal to distinguish between 'representation' and 'object of representation' sounded plausible, and we will go even further now and acknowledge that it is justified ... as long as it concerns indeed the 'object of *representation*'! For, although Schopenhauer was right in saying that "every object is *eo ipso* in need of a subject" (see above), it does not follow from this that an 'object' can never be anything else but an 'object of representation'. And as a matter of fact Schopenhauer knew this, witness the fact that at least one object, namely the object of the 'fourth class of the objects for the subject',* i.e., the object of inner experience, was for him explicitly *not* a representation;†[210] but we saw that Schopenhauer's wish to retain the *a priori* nature of causality made him exclude the vital impulses from his epistemology, so that (since they are not representations) they had to be upgraded towards the metaphysical realm.

The problems raised by his metaphysics ('world as will') will concern us in Chapter VIII; for now, we restrict ourselves to our assessment that, although there may be nothing *between* 'representation' and 'Thing-in-itself' as Schopenhauer taught, this does not yet mean that there is nothing *prior* to the 'world as representation'. Actually there is something prior to it, which we might call, with

---

idealistischen Grundansicht') of the second volume of his main work, where we read: "This [i.e., transcendental idealism, P.L.] leaves the *empirical* reality of the world unaffected, but insists that every *object*, i.e., the empirically real in general, is conditioned by the *subject* in a twofold way: first, *materially*, or as *object* in general, since an objective existence can be thought of only as facing a subject and as its representation; second, *formally*, because the *mode* of existence as an object, i.e., of being represented (space, time, causality), proceeds from the subject, is predisposed in the subject. Thus simple or *Berkeleyan* idealism, which concerns the *object in general*, is immediately joined by the *Kantian* one, which concerns the specifically given *mode* of being an object." ("Dieser läßt die *empirische* Realität der Welt unangetastet, hält aber fest, daß alles *Objekt*, also das empirisch Reale überhaupt, durch das *Subjekt* zwiefach bedingt ist: erstens *materiell*, oder als *Objekt* überhaupt, weil ein objektives Daseyn nur einem Subjekt gegenuber und als dessen Vorstellung denkbar ist; zweitens *formell*, indem die *Art und Weise* der Existenz des Objekts, d.h. des Vorgestelltwerdens (Raum, Zeit, Kausalität), vom Subjekt ausgeht, im Subjekt prädisponirt ist. Also an den einfachen oder *Berkeleyschen* Idealismus, welcher *das Objekt* überhaupt betrifft, schließt sich unmittelbar der *Kantische*, welche die speziell gegebene *Art und Weise* des Objektseyns betrifft."

* Cf. Chapter II. We return to this topic in the first section of Chapter VIII.

† "One is wholly wrong, however, if one calls pain and lust representations: they are not in the least that, but direct affections of the Will in its appearance, the body: a forced momentary willing or not-willing of the impression undergone by the latter." ("Man hat aber gänzlich Unrecht, wenn man Schmerz und Wollust Vorstellungen nennt: das sind sie keineswegs, sondern unmittelbare Affektionen des Willens, in seiner Erscheinung, dem Leibe: ein erzwungenes augenblickliches Wollen ode Nichtwollen des Eindrucks, den dieser erleidet.").

Üxküll (more on whom in the next section) the "world as environmental world (*Umwelt*)". For, although animals are able to act spontaneously and purposefully (guided not only — as Schopenhauer rightly stressed — by instinct but also by what in his wake we may call 'Understanding', considering this as the faculty to interpret sensations as messages from the outer world that are relevant to the animal's vital impulses), it seems highly doubtful whether one can appropriately speak of '*representation*' at the animal level;* and for exactly the same reason (namely, lack of 'imagination') which casts doubt in the above as to whether the real notion of 'thing' could be ascribed to animals. Certainly the animal's world is 'its' world (though in a species-determined rather than an individual sense), since it is a drastically reduced world, reduced not only by the limitations of its sense-apparatus but also by its vital impulses, for which only that which is of immediate interest can really be said to 'exist'. We might therefore say that the human subject has objects in a twofold way — in the sense of 'what does this mean to me at this moment?' (in this sense animals have objects, too) but also in the sense of 'what is it, judged by its own merits?'†[211] But this is possible only because the human subject is also a *subject* in a twofold way; or, more precisely, a *knowing* subject in a twofold way. For Schopenhauer, too, acknowledged the human subject to be a subject in a twofold way; but he distinguished only between a 'willing' and a 'knowing' subject, and to him this distinction also applied to the animal subject. Enlarging somewhat further on this theme, we might say that in a sense the plant is already a subject, since it has an urge to grow and procreate and since it communicates with its surroundings; using conventional terms, it already has a 'soul', i.e., an *inner* side, just as the human (and the animal) body has. The plant, in Schopenhauer's terminology, therefore, already may be said to be a 'willing' subject; but a *willing* subject only, whereas the animal is also a *knowing* subject, because it acts spontaneously, guided by "*motivation*, i.e., causality that passes through cognition" as Schopenhauer used to say.‡[212] And man? Man has not only 'Understanding', but also 'Reason', Schopenhauer acknowledged; but without fully recognizing that the essence of the latter resides in man's having not only *consciousness* (the animal has that, too, as Schopenhauer rightly stressed), but also that *consciousness of consciousness* which Kant expressed (in Section 16 'On the original synthetic unity of apperception' of the second chapter of the first book of the 'Transcendental Analytic') in the formula: "The '*I think*' must be *able* to accompany all my representations;"§[213] i.e., in having *self-consciousness*. It may be useful to recall here the two common-sense arguments mentioned by Schopenhauer in both versions of the *Fourfold Root* that go against his correct,

---

* To this we return in a footnote on p. 174.

† One might speak here, with Husserl, of the 'intentional' vs. the 'transcendent' object, except that the expression 'transcendent' seems ill-chosen here — something transcendent cannot be an object.

‡ "Motivation, d.h. die durch das Erkennen hindurchgehende Kausalität ...".

§ "Das '*Ich denke*' muß alle meine Vorstellungen begleiten *können*."

and also impeccably Kantian, thesis that the subject of knowledge is itself unknowable, because it is the condition of all representations (being their necessary correlate), so that it can never itself become representation or object. In our chapter on the *Fourfold Root* (p. 44) we have already dealt with one of these arguments, together with its equally correct, and impeccably Kantian, refutation by Schopenhauer; let us now look briefly at Schopenhauer's refutation of the other. We read: "To the objection: 'I do not only know, but I know also that I know', I would reply: Your knowing of your knowing is distinguished from your knowing only in wording. 'I know that I know' does not say more than 'I know', and this, without further qualification, does not say more than 'I'."[*][214] Therefore 'I know', according to Schopenhauer, is an analytical judgement, whereas 'I will' is a synthetic judgement; synthetic *a posteriori*, to be sure, in that it is derived from (inner) experience.[215] "*Known* is the subject only as a *willing* one, ... not as a *knowing* one," he says.[†][216] Here one might object, first, that in this case "Will" cannot be the 'Thing-in-itself' that Schopenhauer claimed it to be (the willing subject thus being an *object* for the knowing subject);[‡] and, second, that the knowledge of this willing subject is possible only thanks to a certain *distance* from it ... the same distance which lies at the root of the seemingly banal observation 'I know that I know' — an observation which implies in any case a certain distance from itself being taken by the 'knowing subject', which thus becomes what we called 'knowing subject in a twofold way' (better perhaps: 'knowing subject raised to the second power'), having not only consciousness, but also self-consciousness.

Now this distance is specifically *human*, even though it seems that in rudimentary form it can already be found in some other primates.[§] The observation: 'I know also that I know', is therefore crucially important after all, although it remains of course true that the knowing subject cannot *directly* become representation or object.[**] The animal is entirely wrapped up in the *hic et nunc*, the

---

[*] "Auf den Einwand: 'Ich erkenne nicht nur, sondern weiß doch auch, daß ich erkenne', würde ich antworten: Dein Wissen von deinem Erkennen ist von deinem Erkennen nur im Ausdruck unterschieden. 'Ich weiß daß ich erkenne', sagt nicht mehr als 'Ich erkenne', und dieses, so ohne weitere Bestimmung, sagt nicht mehr, als 'Ich'."

[†] "*Erkannt* wird das Subjekt nur als ein *Wollendes*, eine Spontaneität, nicht aber als ein *Erkennendes*"; also "Demnach erkennt das Subjekt sich nur als ein *Wollendes*, nicht aber als ein *Erkennendes*."

[‡] This problem is to concern us in Chapter VIII, which is about 'the world as will'; for the moment we confine ourselves to 'the world as representation'.

[§] We think here especially of the capacity for simulation that has been observed in apes.

[**] We recall here that Schopenhauer did recognize the possibility of *indirect* knowledge of the knowing subject insofar as its various faculties (intuition, understanding, reason) are known to us by way of the different classes of representations they stand for (cf. Chapter II, p. 43); indeed, without the capacity for critical reflection upon these, no epistemology would be possible at all. But this is precisely the deeper meaning of the seemingly banal observation: 'I know also that I know'!

here and now, as Schopenhauer himself observed; but it is precisely this lack of reflection, i.e., the absence of a certain *distance* from the facts of life, which causes it not to be able, in Wittgenstein's phrase, 'to make for himself images of the facts',*[217] that is, 'representations' in the strict sense of the word. For these (if referring to facts from the past) *can be reproduced more or less at will by memory*, or (if they refer to future possibilities) *they can be produced in advance*, and therefore presuppose a being *not* entirely wrapped up in the *hic et nunc*. But Schopenhauer's failure, described by us in the preceding section, to see that external 'reality' is immediately experienced by its resistance to our vital impulses and thus not in need of being intellectually 'inferred', made it necessary for him to introduce 'representation' even at the animal level and thus to assign to *genuine* (i.e., *human*) representations the demeaning title 'representations of representations'. And here lies the root of his virtual relapse (though on a 'higher level') from Kantian, transcendental or critical idealism back into neo-Berkeleyan, subjective or psychological idealism. For this meant that the problem of the *truth* of our judgements about the empirical world, i.e. the *real* problem of transcendental idealism which arises only at the genuinely representational level (representations giving — guided, as we may grant Schopenhauer, by the 'principle of sufficient reason' — a 'picture' or an 'image', i.e. always some kind of *explanation*, however primitive, of the world, which may be more or less 'true' or 'false'), came to be confused again (in spite of Schopenhauer's repeated assurances that he was an 'empirical realist') with the *pseudo*-problem of the *reality* of the empirical world.†

After all this we may be brief about "*my*" representation. The faculty of having representations produces the diverse, changing world-pictures we spoke of, and it will be noticed that their succession shows an increasing degree of abstraction. For we discern as the most striking characteristic of the scientific world-picture precisely its claim to eliminate all subjective, personal elements. Thus the modern scientific world-picture, although certainly a 'representation' and not a 'copy' or a 'mirror' of the world, can hardly be said to be 'my' representation.‡[218] Yet in

---

* "Wir machen uns Bilder der Tatsachen" ("We make for ourselves images of the facts").

† Recall here our imaginary conversation between Kant and Schopenhauer at the beginning of this chapter (p. 85)!

‡ Against this it might be argued that Schopenhauer talks here of the 'knowing subject', i.e., of Kant's transcendental subject, which is not the same as the individual knower. However, we fully agree with Janaway that Schopenhauer's favoured expression "The world is *my* representation" "is at best a highly misleading way of putting his point if — as we should — we take seriously Schopenhauer's remarks on the need to distinguish between 'my person' and 'I', i.e. about the difference between *individual* and [knowing] *subject*." In our opinion it is Schopenhauer's insufficient appreciation of the fact that the 'knowing' subject appears in man as 'knowing subject raised to the second power' (i.e., that only man is a subject in the Kantian, transcendental sense) which is responsible for the confusion.

another sense it is quite true that "the world is my representation" * ... but 'that is (at least for the moment) another story'.

e. Conclusions. Schopenhauer and Goethe once again

Returning now from our rather lengthy epistemological round-trip to our starting point, to wit, the antinomy as formulated by Schopenhauer, our conclusion must be that it is Schopenhauer's virtual abandonment of *transcendental* idealism in favour of *psychological* idealism which, in being combined at the same time with a rather drastic materialism ("a thinking creature without a brain is like a digesting creature without a stomach,"†[219] Schopenhauer said) is responsible for the way in which the antinomy could arise in the first place. For it must be repeated that, in spite of Schopenhauer's own frequent warnings against attributing any causal relationship to knowing subject and object, his theory of perception implies that the knowing subject actually creates (and thus 'causes') the object — an object which is at the same time said to have caused *previously* the sensation in the sense-organ! This is an antinomy, indeed, but one that, in our opinion, can be avoided in that it is based on a delusion.‡[220]

This does not mean, to be sure, that there is no problem here at all. For it is indeed as true that Kant's 'synthetic unity of apperception' (i.e., the faculty of saying 'I') is a necessary condition for having 'representations' in general and thus is a necessary condition for the constitution of 'science' in particular, as it is true that this very science demonstrates this 'synthetic unity of apperception' to be an organic function of the zoological species called 'man'. Therefore Schopenhauer

---

* To wit, in the sense of 'my person' (see the preceding footnote), since my personal view of the world is always coloured by subjective moods and preferences. For (as compared to the animal) man is not only 'subject raised to the second power', but also 'individual raised to the second power'; and thus at the same time more objective *and* more subjective than the latter ('subjective' this time not in the sense of 'knowing subject', but in the sense of 'willing subject'). This paradox will concern us later on in the final section of Chapter VIII, and in section b. of our final chapter.

† "Denn ein denkendes Wesen ohne Gehirn ist wie ein verdauendes Wesen ohne Magen." The aphorism calls to mind the assertion of Schopenhauer's favourite Cabanis that, just as the stomach is a machine for digesting food, so the brain is a machine for digesting impressions by the secretion of thoughts.

‡ We do not therefore agree with Volker Spierling, who holds that the inconsistencies in Schopenhauer's system are due to the incompatibility of transcendental idealism with the 'materialist' aspect of Schopenhauer's philosophy. These can in our opinion coexist peacefully since they are two different, yet equally legitimate, approaches to the problem of knowledge (although there will always remain a problem, as we shall see further down). The *unnecessary* difficulties into which Schopenhauer's epistemology ran are due rather to his virtual lapsing from *transcendental* into *psychological* idealism; and in this sense we do agree with Spierling that Schopenhauer, in considering himself a transcendental idealist, misunderstood himself.

was right in principle (albeit in a confused way) when he claimed that Kant's approach was one-sided and ought to be complemented. Only, it is not a 'subjective' approach (Kant's approach, as we saw, was not 'subjective' in the customary, psychological sense) that is to be complemented by an 'objective' one (for, paradoxically, Schopenhauer's very approach, as we saw likewise, is radically subjectivist in its consequences), but a 'transcendental' approach that is to be complemented by a 'genetic' one. 'Complemented', not 'substituted', for we are concerned here with two equally legitimate queries. The first is 'transcendental': 'What are the *logical* conditions that make our scientific understanding of the world possible?'. The second is 'genetic': 'What *biological* conditions have made possible a form of life that strives after such understanding in the first place?' The second query ought more strictly to be called 'phylogenetic', as a third, 'ontogenetic' query presents itself in addition: 'What *psychological* conditions make possible the rise of those logical conditions in the child's awakening mind?' Speaking even more strictly, the phylogenetic query ought to be differentiated in its turn, giving rise to the *biological* query just formulated, and a *historical* query which might run thus: 'What *historical* conditions have made possible the shift from one interpretation of the world to the other?'

Now the whole point here is that the transcendental and the biologically–genetic approaches in particular, if not carefully distinguished from one another, lead either to a *petitio principii* or to a *regressus ad infinitum.*

The *petitio principii* in question has been demonstrated clearly by Plessner in connection with Herbert Spencer's gentleman's solution to the problem of the *a priori.* Spencer had asserted that the Kantian forms of intuition and categories are *a priori* for the individual (being innate), but *a posteriori* for the species (being the results of a difficult process of adaptation). Here Plessner rightly observed that such a pragmatic view, which seeks the criteria for truth or justice in utility or life-furtherance, overlooks the fact that, if the conditions of our conception of the world have indeed come about by adjustment and selection, the world to which this adjustment has taken place must have possessed all these very features already by itself, since otherwise such a process could not possibly have started in the first place, let alone have come to a successful end. To explain categories as acquired adaptational achievements, thus turning them into innate forms of experience (i.e., comparing them with fins, feathers, claws or teeth) is tantamount to presupposing nature to be the very thing the existence of which is made possible by those selfsame categories to begin with (just as fins presuppose water and feathers presuppose air).*[221]

---

* As a more recent example of the 'Spencerian' attempt criticized by Plessner, the work of Konrad Lorenz may be mentioned here. He proposed a comparable gentleman's agreement, when he wrote that the 'spectacles' of our modes of thought and perception, such as causality, substance, quality, time, and place, are functions of a neurosensory organisation that has evolved in the service of survival. Speaking about Kant, who held that the forms of intuition and the categories are *a priori* (rightly so, says Lorenz, since they are not products of *individual* experience as Hume believed) and *for that reason* cannot have evolved, he

As regards the *regressus ad infinitum*, we would ourselves suggest the following comparison: if one compares the transcendental subject to a painter (to wit, the painter of a scientific world-picture), then the problem arises that a painting, if meant to give a really *complete* picture of the depicted situation, ought to include the artist at work himself. In the case under consideration this would mean that the transcendental subject should now appear in the guise of Man as a biological creature, as a part of the concrete, visible world; a part, however, that (by virtue of its being at the same time the transcendental subject) is necessarily engaged in making a picture of this world on the basis of certain *a priori* principles — a picture on which is to be seen again the painter himself ... and so on![*]

Plessner, therefore, in his desire to provide the young discipline of philosophical anthropology with a biological foundation, wisely confined himself to a phenomenological description of the various organizational levels of life, leaving aside the problem of how these appeared on the world stage. In this he was indebted to Jakob von Üxküll, who held that each animal (including man) lives in an "environmental world" (*Umwelt*) determined by its specific type of organization (its receptors and effectors, especially), in that this organization determines what is capable of becoming a stimulus or sign for the animal and what modes of response it is capable of carrying out. The sum of possible stimuli for each animal is called by Üxküll its "sign-world" (*Merkwelt*) and the sum of possible responses its "action-world" (*Wirkwelt*). These jointly constitute its specific *Umwelt*, to which a specific inner world (*Innenwelt*) corresponds.[†222] Now this Üxküll is the selfsame biologist of whom Schopenhauer (as we have observed in a footnote on p. 99) has

---

argues that in Kant's own time this was an unanswerable assertion, but that there is a response now. "The simple answer is that the system of sense organs and nerves that enables living things to survive and orientate themselves in the outer world has evolved phylogenetically through confrontation with an adaptation to that form of reality which we experience as phenomenal space. This system thus exists *a priori* to the extent that it is present before the individual experiences anything, and must be present if experience is to be possible. But its function is also historically evolved and in this respect not *a priori*".

[*] Lovers of art may be reminded here of Velasquez' famous picture 'Las meniñas'. On a more trivial level, elderly Dutch readers might think of a once well-known nurse on the tin of a certain brand of cocoa ...

[†] Plessner writes of Üxküll that he "is not interested in the genealogic origin of the building-scheme, but in the way it functions ... Organic building-schemes are functional styles with equal rights, not susceptible to being inserted into any hierarchical order ... Any organism has those complications that it merits and that are becoming to it ... The sign-world and the action-world of man is in quite the same way a monad around a living centre with its specific forms of intuition, categories, and further apriority." ("Üxküll interessiert nicht die stammesgeschichtliche Herkunft des Bauplans, sondern die Art und Weise wie es funktioniert ... Organische Baupläne sind gleichberechtigte Funktionsstile, in keiner Stufenordnung unterzubringen ... Jeder Organismus hat diejenigen Komplikationen die es verdient und die ihm zukommen ... Die Merk- und Wirkwelt des Menschen ist genauso eine Monade um ein lebendiges Zentrum, mit ihm spezifischen Anschauungsformen, Kategorien und sonstiger Apriorität.").

by some been held to be a forerunner. Certainly one should always be a little cautious in tracing such lineages; after all, there is a striking difference between Schopenhauer's metaphysical interpretation of the various stages of complexity in nature as Platonic Ideas which, in their turn, are stages of objectivation of the World-Will, and the much more down-to-earth interpretation given by Üxküll (and, for that matter, by Plessner). Yet they all share what may be called a 'morphological' approach, which (in spite of the evolutionist ring of expressions like "*stages* of objectivation") made both Schopenhauer and Üxküll reject Darwinism.*[223] The same approach, which in a sense constitutes a bridge between the transcendental and the genetic method by placing transcendental philosophy in the framework of the *Urphänomen* 'Life', may in its turn be traced back to

---

* Cf. footnote on p. 84. In *On Will in Nature* Schopenhauer makes the following comments on Lamarck: He saw rightly that will has determined the organization but, due to French (or, for that matter, English) metaphysical backwardness (to wit, ignorance of the Kantian doctrine of the ideal nature of space and time; cf. the remark on Maine de Biran we cited in Section c.), he did not see that this will lies beyond time and in this sense is more original than the animal itself. Thus he started from an animal without distinct organs or strivings, provided with perception only, through which it learns in the course of time under what circumstances it has to live; thence originate its will and its organs. But if these peculiar organs had developed over time, the species must have perished in the meantime! Schopenhauer then continues: "In truth, any organ should be regarded as the expression of a universal manifestation of will (universal, in that it has been made once and for ever), of a fixed longing; in short, of a volitional act, though not one of the individual but of the species. Any animal figuration is a longing of the will-to-life invoked by certain circumstances: e.g., this will-to-life was captured by a longing to live on trees, to hang from their branches, to eat from their leaves, without struggling with other animals or ever setting foot on the soil; this longing over time without end presents itself in the shape (Platonic Idea) of the sloth. It can barely walk since it is wholly oriented towards climbing: it is as adroit on the trees as it is helpless on the soil, and it looks itself like a moss-covered branch so that no persecutor may notice it." ("In Wahrheit ist jedes Organ anzusehen als den Ausdruck einer universalen, d.h. ein für alle Mal gemachten Willensäußerung, einer fixierten Sehnsucht, eines Willensaktes, nicht des Individuums, sondern der Species. Jede Thiergestalt ist eine von den Umständen hervorgerufene Sehnsucht des Willens zum Leben: z.B. ihn ergriff die Sehnsucht, auf Bäumen zu leben, an ihren Zweigen zu hängen, von ihren Blättern zu zehren, ohne Kampf mit anderen Thieren und ohne je den Boden zu betreten; dieses Sehnen stellt sich, endlose Zeit hindurch, dar in der Gestalt (Platonische Idee) des Faulthiers. Gehn kann es fast gar nich, weil es nur auf Klettern berechnet ist: hülflos auf dem Boden, ist es behend auf den Bäumen, und sieht selbst aus wie ein bemooster Ast, damit kein Verfolger seiner gewahr werde."). We now understand why, to him, Darwinism was only a variation of Lamarckism, in spite of the fact that Darwin's theory of natural selection differed from the latter in that it was less finalist (just as the wave-theory was to him only a variation of Newtonianism in spite of a fundamental difference which was, however, from his — and Goethe's — point of view irrelevant!). It is fair to grant, though, that the problem in the evolutionary explanation of the development of organs here signalized was (and indeed still is) a real one (as was the problem of giving a satisfactory explanation of pigments and their mixing before Helmholtz!).

Schopenhauer's other spiritual hero besides Kant, namely, the very man who coined the expression 'morphology' — Goethe.

After all, it was Goethe who wrote the *Metamorphosis of Plants* (his most important scientific work besides his *Theory of Colours*; cf. Chapter I, p. 23) and who for that reason was a partisan of Geoffroy Saint-Hilaire in the latter's well-known dispute with Georges Cuvier in the year 1830 (a dispute, incidentally, which he found much more interesting than the political revolution taking place in the same country in the same year). Moreover, it was Goethe who, in 1784, independently from Vicq d'Azyr (to whom priority belongs) had discovered the intermaxillary bone in humans, inspired by his persistent belief in unity in nature. From this stance, he made Faust call animals "my brethren",[*224] with which Schopenhauer would have agreed whole-heartedly. Therefore the emancipation of pre-human knowledge and also of human knowledge at the level of intuition (*Anschauung*), both of which are implicit in Schopenhauer's epistemology, must in principle have been congenial to Goethe. We may refer here to an interesting statement once made by Goethe. "In German philosophy," he remarked to Eckermann (to whom else?), "two major things were still to be done. Kant has written the *Critique of Pure Reason,* and with that infinitely much has happened, but the circle has not yet been closed. A competent man of importance should now write the critique of the *senses* and of human understanding; and, if this were done in an equally excellent manner, little would be left to desire in German philosophy."[†225] Might Goethe perhaps have thought here of his one-time disciple?

With some regret for an opportunity lost, one is tempted indeed to consider that Schopenhauer's scientific background qualified him more than anybody else for such a task. But in fact, as we have seen, his effective shift from transcendental to psychological idealism in Berkeley's vein not only made him misunderstand the quintessence of Kantianism, but also and at the same time created an unbridgeable gulf between Goethe and himself. For, while Schopenhauer may have been right in recognizing in Goethe's 'objectivist' mode of thought a kind of naive realism, Goethe was no less right in instinctively recognizing in Schopenhauer's subjectivist 'Kantianism', not only the heretical implications for the theory of colours treated at length in the previous chapter, but more generally the spleen and the morbid preoccupation with one's own self (known only too well to Goethe from his own *Werther* period, which he had managed to overcome only with

[*] "Du führst die Reihe der Lebendigen / Vor mir vorbei und lehrst mich meine Brüder / Im stillen Busch, in Luft und Wasser kennen ..." ("Thou makest the whole sequence of living beings / pass before my eyes and makest me know my brethren / in the silent wood, in the air and in the water ...").

[†] "In der deutschen Philosophie waren noch zwei große Dinge zu thun. Kant hat die 'Critik der reinen Vernunft' geschrieben, womit unendlich viel geschehen, aber der Kreis nicht abgeschlossen ist. Jetzt müßte ein Fähiger, ein Bedeutender die Kritik der *Sinne* und des Menschenverstandes schreiben, und wir würden, wenn dies gleich vortrefflich geschehen, in der deutschen Philosophie nicht viel mehr zu wünschen übrig haben."

difficulty in his 'classical' period, and which he disliked now in the Romantic movement). Schopenhauer may have felt this when, much later, he recalled the following episode: "'What?', [Goethe] once said to me, gazing at me with his Jupiter-eyes, 'Light would be there only insofar as you see it? No, *you* won't be there, if Light would not see you!'"*[226]

And so we are back again at the Goethe–Newton controversy. It is quite understandable that Goethe, who once said "The senses do not cheat, judgement cheats" ("Die Sinne trügen nicht, das Urteil trügt"),[227] was at first inclined to see a potential disciple in the young philosopher who — remarkably enough in his time — consistently defended the primacy of concrete *Anschauung* against abstract *Begriffe*. But we remember that, shortly after his encounter with Goethe, the young philosopher had also made the acquaintance — once again, in that famous maternal tea-circle — of the Orientalist Friedrich Majer, who had drawn his attention to the *Upanishads*. Or rather, his attention was drawn to the mutilated version of that work which had been published in 1801 by the Frenchman Anquetil-Duperron under the name of *Oupnekhat* (mutilated twice over, in fact, since this was a Latin translation of a Persian translation made from the original Sanskrit). Schopenhauer studied the book for the first time in the summer of 1814, and it was always to remain his bedside reading. The encounter with this Indian source of wisdom no doubt served to strengthen his already existing 'Kantian' conviction that, in the last resort, the "senses cheat", too. For, according to Indian thought the whole world of *Anschauung* is nothing but 'Veil of Maya' — freely translated: 'nothing but my representation'.† Schopenhauer would surely add at once that this is true *a fortiori* for the world constituted by our *Begriffe* (concepts), since these, in their turn, are 'representations of representations' and therefore *twice* deceitful! But this does not alter the fundamental fact of the *denigration* of empirical reality (*La Vida es Sueño!*); a denigration which, although philosophically legitimated by Schopenhauer's idealist epistemology, had its deepest roots in a profoundly pessimistic temperament to which his position in philosophy was wonderfully adapted.‡[228] It was precisely this temperament that was

---

* "'Was', sagte er mir einst, mit seinen Jupitersaugen mich anblickend, 'Das Licht sollte nur da seyn, insofern Sie es sehen? Nein, *Sie* wären nicht da, wenn das Licht *Sie* nicht sähe.'"

† Originally 'Maya' was the female aspect of divine creative power, turning later into the bewitching power which causes the semblance of the material world.

‡ It seems to us that Volker Spierling takes Schopenhauer a little too much at face-value in thinking that it is the other way round and that Schopenhauer's pessimism results with logical stringency from his analysis (meant to be transcendentally idealist) of the stuff of experience, since the *epistemological* foundation of Schopenhauer's pessimism cannot, in view of the necessity and universal validity claimed for it, owe its existence to inductive generalization of personally experienced or observed suffering (experience providing at most a *probable* validity). Perhaps it *ought* not, but *can* it not? In our opinion it is rather the opposite which seems hardly feasible. Or does Spierling *really* hold it feasible for a

responsible for an attitude towards life about which Goethe once warned the young philosopher in the following distich: “Willst Du Dich Deines Wertes freuen / So mußt der Welt Du Wert verleihen” (“If you wish to enjoy your own value / then you must bestow value upon the world”).[229]

In short, Schopenhauer’s pessimistic world-view could not fail to arouse the disapproval of Goethe, who never lost his belief in the ultimate harmony of Nature; and it is this difference in fundamental attitude towards Nature which lies at the root of Schopenhauer’s heretical interpretation of Goethe’s *Urphänomen* in the theory of colours which formed the subject of our Chapter III. It was therefore inevitable that Goethe was eventually drawn to the conclusion that “Dr Schopenhauer came to my side as a benevolent friend. We discussed many things in a spirit of agreement; but in the end a certain parting of the ways could not be avoided, as when two friends who so far have gone together, shake hands but, one of them wishing to go north and the other south, very quickly lose sight of each other.”[*230]

However, in the theory of colours still another man was to side with Goethe as a benevolent friend: Hegel! More than that, in his case no parting of the ways ever took place. Could it be that, after all, Hegel was more akin to Goethe than Schopenhauer had proven to be, in spite of the fact that, other than Schopenhauer and Goethe himself, Hegel was the very embodiment of abstract, non-intuitive philosophy? However this may be, an amazing coincidence such as this is interesting enough to deserve somewhat closer inspection, which is why we devote a special chapter to it.

‘cheerful and sanguine temperament’ (to use David Hume’s self-characterization) ever to ‘become’ a pessimist out of epistemological considerations?

* “Dr. *Schopenhauer* trat als wohlwollender Freund an meine Seite. Wir verhandelten manches übereinstimmend mit einander, doch ließ sich zuletzt eine gewisse Scheidung nicht vermeiden, wie wenn zwei Freunde, die bisher mit einander gegangen, sich die Hand geben, der eine jedoch nach Norden, der andere nach Süden will, da sie denn sehr schnell einander aus dem Gesicht kommen.”

V

BIEN ETONNÉS: HEGEL AND SCHOPENHAUER AS PARTISANS OF GOETHE'S THEORY OF COLOURS

a. Goethe and Hegel

Bien étonnés: quite astounded to find themselves back in each other's company, indeed! The French expression applies, not only, as is obvious, to the pair Hegel/Schopenhauer but, in a sense, to the pair Hegel/Goethe as well. We need only cite Mephistopheles' witty warning to Faust:[231]

Ich sag es dir: ein Kerl, der spekuliert,	"I tell you, a fellow who speculates
Ist wie ein Tier, auf dürrer Heide	is like a beast, over barren heath
Von einem bösen Geist im Kreis herum geführt,	carried around by an evil spirit
Und rings umher liegt schöne grüne Weide.	amidst fine green meadowland."

And, indeed, though the two men were rather frequently in touch during the period when Hegel was a *Privatdozent* at Jena (1802–1806) — a position obtained in part through Goethe as an intermediary — Goethe did not respond at all when Hegel sent him a copy of his *Phenomenology of Spirit*. An irate remark has even been preserved about a passage on botany in the Preface of this work (wrongly attributed by Goethe to the Preface of the *Science of Logic*). Hegel had written: "The bud disappears when the blossom breaks through, and we might say that the former is refuted by the latter; in the same way when the fruit comes the blossom may be explained to be a false form of the plant's existence, for the fruit appears as its true nature in place of the blossom ...".[*232] Goethe's reaction was: "It is scarcely possible to say anything more monstrous. It seems to me quite unworthy of a rational man to wish to annihilate the eternal reality of nature by means of a miserable piece of jocular sophistry." To do so might be forgiven a worldly minded empiricist, "but when a distinguished thinker, who penetrates an Idea and knows very well its intrinsic worth, and its still higher worth when it articulates a tremendous process of Nature — when he amuses himself with a sophist's caricature of that process, and with denying and annihilating it by means of

[*] "Die Knospe verschwindet in dem Hervortreten der Blüthe, und man könnte sagen, daß jene von dieser widerlegt wird; ebenso wird durch die Frucht die Blüthe für ein falsches Daseyn der Pflanze erklärt, und als ihre Wahrheit tritt jene an die Stelle von dieser."

ingenious yet mutually self-cancelling words and phrases, one knows not what to say."*[233]

Indeed, we need not wonder that this typical sample of 'dialectics', which seeks to 'refute' one real situation by another, was repugnant to a man like Goethe who, in his *Metamorphosis of Plants*, saw the plant as a living force that reveals itself with equal truthfulness at each stage of its growth. This negative attitude towards Hegel's philosophical pet child never left Goethe, as is apparent from faithful old Eckermann's record of a visit paid by Hegel to Weimar in the autumn of 1827. He informs us that Goethe had a high esteem for Hegel personally but did not greatly appreciate some of the fruits of his philosophy, meaning by this his dialectics in particular. Goethe prophetically expressed the fear that these might be abused for the purpose of turning falsehood into truth and truth into falsehood. Hegel answered, in a reply meant to be reassuring, that this sometimes happens, but only with spiritually sick people.[234] While Goethe's reaction is not recorded, he is unlikely to have stood convinced. Still, the mere fact of his kindly receiving Hegel in his house at the Frauenplan (which visit was the culmination of a renewal of their broken contact that dates back to 1817, leading to a friendly correspondence over the years 1817–29),[235] seems to indicate that Goethe's initial anger had meanwhile subsided. This may have had something to do with the fact that the same year (1817) had seen the publication of the first version of Hegel's *Encyclopaedia*, in which book Hegel publicly championed Goethe's theory of colours (as he was to do in a much more elaborate way in the second and third versions of 1827 and 1830, respectively, and in the oral supplements delivered in his lectures).

A detailed description of how Hegel accomplished this feat in the framework of his own philosophy of nature would carry us much too far away from our central concerns, the more so as we have done so elsewhere.[236] Suffice it here to summarize the essential points.

Although Hegel whole-heartedly joined Goethe in upholding, against Newton, the view that Light is homogeneous and that colour arises from an interaction of Light and Darkness, he differed from Goethe in thinking that the nature of Light can only be understood properly in connection with Gravity. Gravity should be distinguished from mere attraction; in fact (as Matter is heavy as such) Gravity is the essence of Matter since it contains both of its aspects, namely, attraction and repulsion (which, in their turn, represent the *internal* relationship between material objects, whereas space and time constitute the framework of their *external*

---

* "Es ist wohl nicht möglich, etwas Monstroseres zu sagen. Die ewige Realität der Natur durch einen schlechten sophistischen Spaß vernichten zu wollen, scheint mir eines vernünftigen Mannes ganz unwürdig ... Wenn aber ein vorzüglicher Denker, der eine Idee penetrirt und recht wohl weiß, was sie an und für sich werth ist, und welchen höheren Werth sie erhält, wenn sie ein ungeheures Naturverfahren ausspricht, wenn der sich einen Spaß daraus macht, sie sophistisch zu verfratzen und sie durch künstlich sich einander selbst aufhebende Worte und Wendungen zu verneinen und zu vernichten, so weiß man nicht, was man sagen soll."

relationship).[237] Now Gravity can be said to represent the eternal longing of Matter for the unity it never attains (for if it did, it would contract into one point).[238] Whereas Gravity thus reveals the unity of Nature in a vague, obscure form, Light does so in an ideal form, because it does not *seek* the unity but *manifests* it through its universal, uncorporeal nature. However, although *in principle* Gravity is overcome by Light, *in reality* the lower, mechanical level of nature continues to exist. Thus, there is an antithesis between Light and dark, heavy Matter, and this hinders Light's expansion and propagation. White is the corporeal fixation of Light, black that of Darkness; colour occurs between these two extremes.

Hence, we have here what Goethe had taught — the combination of Light and Darkness, and more particularly the specification of this combination, which first gives rise to colour.[239] Hegel seeks, however, to give a more elaborate explanation of this interaction, and of the role of the 'turbid medium' in particular. Of central importance here is the phenomenon of transparency. There are several transparent media of Light, differing from each other by their greater or lesser density (which has to do with their specific gravity). But there is also a difference between the transparent *elements* (i.e. air, water, and fire, as distinguished from opaque earth) and the equally transparent *crystals*, in that in the latter case the *externally* conditioned phenomenon of *refraction* gives way to the *internally* conditioned phenomenon of *dual refraction* (due to the *inner individualization*, i.e., the nuclear shape of crystals).[*240] Chemical colours are explained by *internal* darkening, due to their underlying *metallicism*; this is so because the process of inner individualization just mentioned (which in brittle glass has progressed from mere interior shape to actual interior splintering and therefore to a beginning of darkening) in the compactness of metallic being has progressed further to actual darkness.[241] Physical colours, in contrast, are explained by *external* darkening, i.e., by the dimming effect of a translucent yet turbid medium such as glass. The prism is the best means for producing such a dimming effect, since its specific shape (i.e., the inequality of the diameters of its profile) causes its dimming agency not to function uniformly but rather, in accordance with the different lengths of its diameter between the various parts through which lights passes;[242] therefore, it gives rise to a variety of colours.

We must refrain here from a critical assessment of this singular theory; the only fact that is of importance to our subject matter is the obvious one that, in contrast to Schopenhauer, Hegel treated the phenomenon of colour not on a (subjective) physiological basis, but rather on an (objective) physical (or what he believed to be physical) basis. Thus he must have impressed Goethe as a much less self-opinionated disciple than, a few years earlier, Schopenhauer had proved to be. However, there is no indication that this pleasant surprise ever roused Goethe to

---

* Hegel, during the period he lived in Nuremberg (1808–16), had made the acquaintance of Thomas Seebeck, who was also Goethe's chief adviser in optical matters; also, like Goethe, he had taken an active part in Seebeck's experiments on the phenomenon of 'entoptic colours'.

undertake a systematic study of Hegel's philosophy; in the last resort, the mode of thought of a man who saw Nature as nothing but an 'externalization' of the Absolute Idea and who therefore (in an essentially Gnostic way) treated Creation as some kind of dialectically necessary Fall (not the fall of man Genesis speaks of, but a Fall, so to speak, of God Himself) must have remained alien to him. And yet, the case of Hegel shows that, in principle, it was possible to arrive at a defence of Goethe's theory of colours from a philosophical position which, in its highly abstract panlogism and its consequent indifference to Nature's beauty,*[243] seems at first sight to be as un-Goethean as can be. What, then, did make it possible?

b. Hegel and Schopenhauer as positivist metaphysicians

Whatever sense of wonder may have arisen from watching Goethe in Hegel's company pales in comparison to the naming of Hegel in the same breath as Schopenhauer. The latter's life-long hatred of the "charlatan" and "nonsense-scribbler" (*Unsinnschmierer*; to select only some of the kindest epithets from Schopenhauer's inexhaustible chest of invectives) is too well known to require further demonstration. It is actually uncertain whether both thinkers ever acquainted themselves with each other's work on the theory of colours. The only thing we do know for sure is that Hegel was aware of Schopenhauer's contact with Goethe and of the existence of *On Vision and Colours*, as Schopenhauer had mentioned both in the *curriculum vitae* he presented in 1819 to the University of Berlin in order to obtain the *venia legendi*. (It is well-known that, as a young and wholly unknown *Privatdozent*, he had the conceit to set his lectures at the same hour as those of the elder professor — who around that very time made the breakthrough towards his great fame! — in the vain hope of luring the students away from the latter; this tragicomic episode was to remain his only attempt at an academic career, for, having fled from Berlin in 1830 because of the cholera epidemic that was indeed to fell his great rival, he settled in Frankfurt where he would pass the remaining thirty years of his life as a rentier). But while Hegel certainly read the *curriculum vitae*, he probably never read *On Vision and Colours*; nor is it likely that Schopenhauer ever made a serious study of his rival's contribution to this area of science (nor, for that matter, of his philosophical system in general). Yet the fact remains that, in the case of the Goethe–Newton controversy, at least once in their life-times they presented a common front. Is this mere coincidence, or does it point after all to some hidden, spiritual kinship?

Perhaps a clue is to be found in the Preface 'On scientific knowledge' ('Vom wissenschaftlichen Erkennen') to the *Phenomenology of Spirit*, where we read: "In

* Characteristic of the difference between Hegel and Schopenhauer in this respect is Schopenhauer's enthusiastic account of his climbing Mt. Pilatus in Switzerland on the 3rd of June 1804 as compared to notes in Hegel's diary, made during his trip to the Berner Oberland in 1796, which only bear witness to the profound boredom its majestic scenery aroused in him!

my view, which can justify itself only through the presentation of the system itself, all depends on grasping and expressing the True not only as *substance*, but also as *subject*."*[244] Freely translated: the Kantian revolution has rendered impossible a metaphysical system of the old type ('substance'); undaunted by that fact we proceed to construct a new type of metaphysics, taking as our starting-point Thinking itself, and thus turning its subject (Kant's 'transcendental I') from a merely functional entity into the inner core of the whole outer world. A glance in the engine-room of the cognitive process will thus at the same time provide us with the key to the riddle of the world!

Hegel would add that, in essentially the same enterprise, Fichte and Schelling had proceeded in too uncritical a way; the latter's 'philosophy of identity', in particular, is to be compared with a night in which all cows are black,[245] since it seeks to look immediately into the depths of the Absolute, thus producing nothing but artefacts that are neither fish nor fowl, neither poetry nor philosophy.[246] Philosophy should be scientific and articulate, and therefore it ought not to be satisfied with subjective intuition and vague feeling. Hegel therefore criticized Schelling's 'intellectual intuition', as Kant had already done by anticipation at the end of his 'Transcendental Aesthetic'. So, also, as we remember, had Schopenhauer already done in his dissertation. Later, too, the latter would repeatedly mock the "sixth sense of the bats" as something he did not claim to possess, though for him the reason resided rather in his wish to remain *anschaulich* in the workaday sense of the word, whereas Hegel insisted on a more stringent *conceptualization.* (Simplifying somewhat, it might be said that Schopenhauer objected more to the 'intellectual' and Hegel more to the 'intuition'.) To be sure, to Hegel as well as to Schelling, Being and Thought were basically identical. But Hegel rejected Schelling's more romantic holism, which seems to imply that he had a higher regard for empirical science than was true of romantic *Naturphilosophie.*

The inference is a correct one, as becomes clear from the — at first sight somewhat surprising — respect with which Bacon's work is treated in his lectures on the history of philosophy.[247] But, mark you, in the last resort empirical science does no more than transform unstable sense-appearances into a stable image, which, in Part 3 of the *Encyclopaedia*, Hegel calls "the steady picture of the inconstant phenomena" ("... dem beständigen Bilde der unsteten Erscheinungen") or (even more poetically) "a quiet Domain of Laws" ("ein ruhiges Reich der Gesetze").[248] It cannot therefore claim to give us more than an orderly *description* of the phenomena. Consequently, empirical science, with its coarse instrument of Understanding *(Verstand)*, should abstain from any claim to give us an *explanation* of the world; this greater task legitimately belongs only to speculative philosophy with its more refined instrument of Reason *(Vernunft).* This is why

* "Es kommt nach meiner Einsicht, welche sich nur durch die Darstellung des Systems selbst rechtfertigen muß, alles darauf an, das Wahre nicht als *Substanz*, sondern ebensosehr als *Subjekt* aufzufassen und auszudrücken."

Hegel praised Bacon while rejecting Newton — the empiricist tradition of 'Baconian' science at least has the merit of doing the preliminary work for speculative philosophy, whereas the mathematical tradition of 'classical' science tends to intrude into the domain of the latter, thus saddling physical science with a false metaphysics, as becomes most conspicuous with Newton.[*249]

Horribile dictu, something similar appears when we inspect Schopenhauer on this score. Certainly, to Schopenhauer Being and Thought were by no means identical, and consequently the faculty of *Vernunft* is far from occupying the same lofty status in his system that it is assigned in Hegel's; its domain is restricted precisely to what for Hegel (as it was for Kant) is the domain of *Verstand*. Moreover, although both rejected the Newton of the *Opticks*, the Newton of the *Principia* was also loathed by Hegel, but praised by Schopenhauer. According to the latter, Kepler had in his second and third laws only expressed the relationship actually obtaining between the radius vector of a planet to the sun and the velocity with which it traverses its orbit, whereas it was Newton who first demonstrated *that*, and *why*, things are arranged this way, "namely, because with *this* given distance to the central body the planet must have exactly *this* particular velocity in order not to fall into the sun or to run away from it."[†250] With Hegel it is otherwise. To him, all that is sound in Newton comes from Kepler, the difference between the two men being that "what Kepler expressed in a simple and sublime manner as *constituting* the *laws of celestial motion*, is changed by Newton into the *reflective form* of the *force of gravity*".[‡251] In neglecting the fundamental distinction between mathematical analysis and physical reality Newton saddled physical mechanics with a "*monstrous metaphysics* which, contrary to both experience and the Notion (*Begriff*), has its sole source in these mathematical demonstrations".[252] Here, in fairness to Hegel, it should be acknowledged that popular Newtonianism in his time tended indeed to attribute metaphysical status to the forces treated in the *Principia*. Still, Hegel could have known that Newton himself had been much more cautious in this respect. Though Newton certainly considered his 'forces' as really existing and not only as tools of measurement, he did not "frame hypotheses" about the ultimate reality behind them.[§253] To Schopenhauer this,

---

[*] Needless to say, the expression 'Baconian' vs. 'classical' science was not used by Hegel, but has been borrowed by us from Thomas Kuhn.

[†] "...weil nämlich, bei *solcher* Entfernung vom Centralkörper, der Planet gerade *solche* Geschwindigkeit haben müsse, um nicht entweder hineinzufallen, oder davonzufliegen."

[‡] We have seen already in the above that Hegel objected to this Newtonian concept because, according to him, gravity should be distinguished from mere attraction, since, being the essence of matter, it contains both attraction and repulsion.

[§] In the "General Scholium" which closes Book III of the *Principia* we read: "Hitherto we have explain'd the phaenomena of the heavens and of our sea, by the power of Gravity, but have not yet assign'd the cause of this power ... But hitherto I have not been able to discover the cause of these properties of gravity from phaenomena, and I frame no hypotheses. For whatever is not deduc'd from the phaenomena, is to be called an hypothesis; and hypotheses, whether metaphysical or physical, whether of occult qualities apparently, was modest enough. In the lectures which he had intended to hold at the Berlin university, but of which only a few were actually delivered because of lack of interest on the part of the students (who, to his great dismay, in overwhelming majority preferred to attend Hegel's lectures), he said that "the right path", shown by Bacon to lie between the "two errors of Scholasticism and Cartesianism", culminated in Newton — though not the Newton of the *Opticks*![254] Not so with Hegel. One wonders whether the latter would have preferred a return to the Ptolemaean situation of an almost complete separation between mathematical astronomy and physical astronomy?

For all this, there remains the shared praise of Bacon as against the alleged pretensions of a metaphysically conceived mechanicism (even though, to Schopenhauer, Descartes rather than Newton was the culprit here). It may be said therefore that, just like Hegel, Schopenhauer had an ambivalent attitude towards science in that he, too, distrusted its rationalism (which has not only methodological but also ontological pretensions and, hence, provides nothing but a poor, oversimplified world-view in which everything finds itself reduced to mechanics), while respecting its empiricism. The latter aspect of science, however, should be supplemented by a metaphysics of Nature better and richer than that offered as such by mechanicist philosophy. Now this, as we saw, is not basically different from Hegel's claim, although it was more modest insofar as Schopenhauer certainly did not believe in the possibility of 'absolute knowledge' as Hegel did. But no matter: Schopenhauer, too, was convinced that 'metaphysics of nature' grants us a deeper insight than mere 'aetiology of nature', as he calls it. The latter gives us no real *explanation* of the world (recall Hegel's identical remark!), for at the end of its voyage it always encounters a *qualitas occulta*, which it calls a 'force of nature', and of which it can only demonstrate the invariable constancy of its appearance once certain conditions are fulfilled; this constancy it calls a 'law of nature'. Therefore, even when aetiology has achieved all it can achieve, a philosophically minded scientist still feels as would somebody "who, he knows not how, has turned up in a company completely unknown to him, whose members, each in his turn, present the next one as his friend or his cousin, as if in so doing he has sufficiently introduced the other; he himself, all the while reiterating again and again his pleasure at meeting the introduced person, would always have the unexpressed question on his lips: 'but how, the deuce, have I arrived in this company in the first place?'"*[255]

---

or mechanical, have no place in experimental philosophy ... And to us it is enough, that gravity does really exist, and act according to the laws which we have explained, and abundantly serves to account for all the motions of the celestical bodies, and of our sea."

* "Bei der vollendeten Aetiologie der ganzen Natur müßte dem philosophischen Forscher doch immer so zu Muthe seyn, wie Jemandem der, er wüßte gar nicht wie, in eine ihm gänzlich unbekannte Gesellschaft gerathen wäre, von derer Mitglieder, der Reihe nach, ihm immer eines das andere als seinen Freund und Vetter präsentirte und so hinlänglich bekannt machte: er selbst aber hätte unterdessen, indem er jedesmal sich über den Präsentirten zu

In short: Hegel and Schopenhauer were agreed that with the formulation of *laws* of nature the task of science is completed, and that it is up to the philosopher to look further, like Faust, so "that I may detect the inmost force / which binds the world, and guides its course" ("daß ich erkenne was die Welt / im Innersten zusammenhält").* Science must do the preliminary work for philosophy, and in this resides both its justification and its limitation.

Surprisingly, this view of science as an enterprise essentially confined to the description of functional relationships between natural phenomena brings our two metaphysicians into close proximity to 19th-century positivism. The difference is, of course, that positivism makes a virtue out of necessity by proudly confining itself to the 'how' while dismissing any search for the 'what' as being senseless and irrelevant. Schopenhauer and Hegel, by contrast, present us with a kind of two-layer model in which science understood positivistically is enriched with a metaphysical superstructure. No doubt Schopenhauer's superstructure is quite different from Hegel's, yet it is the "two-layer model" which, on the one hand, distinguishes both men from earlier romantic *Naturphilosophie*, while, on the other hand, leading them to distrusting what they considered to be the trespass of science beyond its legitimate domain; to both men, Newton's *Opticks* was the classic exemplar of this fatal tendency.

Here one may ask why, for Schopenhauer, it was otherwise with Newton's *Principia*. After all, Hegel was right insofar as Newton had given a theory whereas Kepler had only given laws (even though the concept of gravitation was not ontologized by Newton himself), so that in this respect his rejection of the *Principia* was more consistent than Schopenhauer's praise for it. But 'forces' acting at a distance might be held to announce a kind of science more 'dynamic' (a term in high favour with romantic *Naturphilosophie*) than merely 'mechanical' science. Alfred Schmidt has rightly pointed out that Schopenhauer linked the empirical and the metaphysical aspects of his conception of nature (i.e., nature as it appears to discursive understanding and nature as Thing-in-itself) by presupposing a "dynamic" interpretation already at the level of phenomenal nature, just as Kant had done in his '*Metaphysical foundations of natural science*'.[256] And indeed, more thoroughgoing mechanicists initially had some difficulties with the Newtonian concept of gravitation, which may be precisely why Schopenhauer felt at ease with it for as long as it was presented by scientists with the restraint with which Newton himself had presented it, thus leaving to the metaphysician the task of revealing its true nature as far as this is possible for the human intellect. The *Opticks*, in contrast, dealing, as the book does, with a subject matter of much more direct appeal to the senses seems, in the way the topic is dealt with, to present a

---

freuen versicherte, stets die Frage auf den Lippen: 'Aber wie Teufel komme ich denn zu der ganzen Gesellschaft?'"

* cf. Chapter I, footnote on p. 25.

typical example of the violation of 'natural nature' in favour of 'artificial' nature,* and therefore to be the prototype of 'mechanical' science.

The latter feature would indeed account for Hegel's and Schopenhauer's shared rejection of the *Opticks*; but still not for their different attitudes towards the *Principia*. After all, the more 'dynamic' interpretation of nature Schmidt hinted at must have been grist to Hegel's mill, too. Perhaps part of the answer is revealed by Hegel's remark that "the Germans have often looked on impassively while the English have assumed authority in this way".[257] Might German nationalism (of which Schopenhauer was conspicuously free) have played tricks on Hegel here? Maybe so; but Hegel's distinction between 'gravity' and 'attraction' already showed that he also had more basic reasons for quarreling with the Newton of the *Principia* no less than with the Newton of the *Opticks*.† And we must repeat that, in so doing, Hegel showed himself to be more consistent, in a sense, than Schopenhauer as far as their shared 'two-layer model' is concerned. This is so because not only the Newton of the *Opticks*, but also the Newton of the *Principia* fails quite to conform to the strict division of tasks implied in that 'two-layer model' ... from which fact, however, we may infer the inherent difficulty of the model and the quarrels over competence it is bound to provoke! For science, notwithstanding positivism, seeks to explain, not just to describe; it pretends to provide a picture of reality more adequate than that of common sense, even if (enlightened by Kant's critical philosophy) it concedes that the 'Thing-in-itself' is beyond its reach.‡

That Hegel showed himself, in a sense, more consistent than Schopenhauer as far as their shared 'two-layer model' is concerned, probably has to do with Schopenhauer's having, in practice, a greater affinity with empirical science than Hegel,§[258] which is also one of the reasons for his unexpected popularity in the

---

* cf. Chapter I, p. 14.

† In the preface to the first edition of *The Two Basic Problems of Ethics* (a work that shall keep us occupied in the next chapter) Hegel's considerations on the phenomenon of gravitation and his critique of Newton's concept of general attraction are mentioned by Schopenhauer as illustrations of the absurdities with which the works of the *summus philosophus* abound. The disrespectful remarks in 'On the foundation of morals' — the second of the two essays collected in the volume — on *summi philosophi* like Hegel had been one of the arguments given by the Royal Danish Society for not awarding its prize to Schopenhauer's essay, although he was the only entrant; hence the qualification *summus philosophus*, which, of course, was adopted by Schopenhauer in an ironic sense.

‡ cf. Chapter I, p. 20, and IV, p. 104.

§ We think here especially of his life-long interest in physiology; it was precisely this expertise which enabled him, to his great joy, to ward off a critical remark by Hegel on his use of the term 'motive' during the dispute at the occasion of Schopenhauer's habilitation on 23 March 1820. He did so by signalizing in his turn Hegel's wrong usage of the term 'animal functions': "Here Mr Nitwit quite revealed himself" ("Hier zeigte sich ganz Monsieur Nichtwisser") he still reveled long after the event. Whether this was really such a terrible blunder on Hegel's part as Schopenhauer suggested we do not venture to decide.

second half of the 19th century (which saw a corresponding decline of Hegel's influence). Let us meanwhile dwell a little longer on the paradoxical closeness to positivism in our two thinkers — a closeness which, at least in Hegel's case, was noticed long ago by Meyerson and somewhat more recently by Hayek.[259] According to the founder of positivism, Auguste Comte, astronomy could become the first fully fledged science because its problems are the most general and simple ones. Each subsequent science (physics, chemistry, biology) depends, it is true, on its predecessor; but, since the phenomena concerned become steadily more complex, the same must hold for the methods by which these phenomena are treated. Although in principle all phenomena might be subject to mathematical treatment (Comte himself was originally a mathematician), in fact those phenomena that are further up in the scale of complexity are not, in his view, amenable to such an approach. We observe something quite comparable in our two German philosophers.* In discussing mathematics, Schopenhauer writes: "Applied mathematics, i.e. mechanics, hydraulics etc., is concerned with the lowest levels of the objectivation of Will, where most still belongs to the realm of mere Representation, yet it already has an empirical element which obscures the complete intelligibility and transparency [of pure mathematics] and with which enters the inexplicable. For the same reason, only a few parts of physics and chemistry still lend themselves to mathematical treatment; but this ceases higher up in the chain of Being, precisely because the content of the phenomenon here outweighs the form."†[260] What is suggested here is the conception of a *hierarchy of the sciences* based upon the richness of their subject matter and corresponding to a similar hierarchy in nature in accordance with the different levels of 'objectivation of Will', that is to say, with the different levels of complexity of natural phenomena.‡[261] In the second volume of *The World as Will and Representation*

---

* The difference is of course that, to the 'positivist' Comte, the hierarchy of the sciences cannot culminate in metaphysics. Yet his 'sociology' (the one and final science still to be developed) has all the characteristics of a meta-science — one, moreover, that should pave the way for nothing less than a novel, state-sponsored, atheist religion: *la religion de l'humanité*.

† "Die angewandte Mathematik, also Mechanik, Hydraulik u.s.w., betrachtet die niedrigsten Stufen der Objektivation des Willens, wo noch das Meiste auf dem Gebiete der bloßen Vorstellung liegt, hat aber doch schon ein empirisches Element [NB a reader surprised by this juxtaposition of *empirisch* and *Vorstellung* should keep in mind that Schopenhauer, in speaking of *Vorstellung*, has in mind here the formal, *a priori* component of 'the world as representation', PL], an welchem die ganze Faßlichkeit, Durchsichtigkeit, sich trübt und mit welchem das Unerklärliche eintritt. Nur einige Theile der Physik und Chemie vertragen, aus dem selben Grunde, noch die mathematische Behandlung: höher hinauf in der Wesenleiter fällt sie ganz weg; gerade weil der Gehalt der Erscheinung die Form überwiegt."

‡ The contemporary reader, to his surprise, is reminded here of Joseph Needham's conception of "integrative levels". For this he believed to have detected a parallel in dialectical materialism, invoking a certain Zavadovsky who wrote: "The true task of scientific research is not the violent identification of the biological and the physical, but the

this suggestion is elaborated into a schema that calls to mind the architecture of Hegel's *Encyclopaedia*. According to this schema, science proceeds from mathematics and logic *via* the sciences of dead nature (dominated by the notion of *cause* in the strict sense of the word) to the sciences of living nature (dominated by the notion of *stimulus*) and finally to the sciences of the human mind and of culture (dominated by the notion of *motive*). The parallel with Hegel's *Encyclopaedia*, which starts with the *Science of Logic*, proceeding from there to the *Philosophy of Nature* (which starts with 'mechanics' and ends with 'organic physics'), and finally culminating in the *Philosophy of Spirit*, is striking. Both start from the most abstract form and look next to the progressively richer and more complex forms in which their metaphysical principle ("Will" and "Absolute Idea", respectively) objectivates itself in the concrete world. Still, these very expressions "Will" and "Absolute Idea" should serve to warn us that by now we have carried our parallel as far as it goes and that we should refrain from pushing it beyond that point.

c. Goethe, Kant, Schopenhauer, Hegel, and history

For of course the difference between "Will" and "Absolute Idea", far from being merely a difference of words, rather points to a fundamental difference in the nature of Hegel's and Schopenhauer's metaphysics. Of the former it might be said that it hypostasizes the transcendental subject, thus *ontologizing Logic* and presenting a metaphysics of *Spirit*, in which the difference between phenomenon and Thing-in-itself has been abolished (*aufgehoben*). Schopenhauer, in contrast, firmly sticks to the latter difference, but, since the Thing-in-itself is not to him (as it was to Kant) a boundary concept (*Grenzbegriff*), but is rather ontologized by him into "Will", he presents instead a metaphysics of *Life* (biology being for Schopenhauer as important as history for Hegel).

Though from a strictly Kantian point of view both performances are illicit, we admit with some reluctance that, at first sight, Hegel's heresy has one strong point as compared to Schopenhauer's. This is the former's awareness of what in Chapter IV (p. 86) we have called in Popperian terms the 'World 3' nature of the subject matter of theoretical reason. As 'objective idealism' it is essentially a philosophy of *culture*, and as such — because of the *historical* character of culture — it is a philosophy of history. For such a thing there was no place in Schopenhauer's

---

discovery of the qualitatively specific controlling principles which characterise the main features of every phenomenon studied", and also "Biological phenomena, historically connected with physical phenomena in inorganic nature, are none the less not only not reducible to physico-chemical or mechanical laws, but within their own limits as biological processes display different and qualitatively distinct laws. But biological laws do not in the least lose thereby their material quality and cognisability, requiring only in each case methods of research appropriate to the phenomenon studied." Needham also claimed to have found an even more significant parallel in the 'organic' — as distinct from Western 'mechanical' — science which, according to him, had been practiced by Chinese neo-Confucian thinkers.

philosophy. According to Schopenhauer, any 'historicist' philosophy amounts to a return to times prior to Kantian critical philosophy, since it misunderstands *time* as constitutive for the Thing-in-itself rather than for the phenomenal world only. Strictly speaking, history cannot even be regarded as a true science in Schopenhauer's view, since it deals with the unique and the individual rather than the recurrent and the general (which categories enable empirical natural science at least to establish *laws*, which is impossible in history). The only *philosophically* relevant meaning the study of history can have is to find out that which remains identical under all change, i.e., the basic qualities of the human heart and brain (mostly bad in either case). "The motto of history in general should be: *Eadem, sed aliter* [i.e., the same, yet otherwise]. Who has read Herodotus has already studied enough history, at least from a philosophical point of view. For there one finds everything that constitutes the subsequent history of the world: humanity's strivings, doings, sufferings, and fate, such as they emerge from those human qualities and from the physical conditions of life on earth".[*262]

In this respect Schopenhauer, much more than Hegel, appears as a spiritual kinsman of Goethe [†263] who, in a well-known conversation with the historian Luden, quoted his own Faust's statement (adressed by Faust to his pedantic *famulus* Wagner): "What you call the spirit of the times is in fact nothing but the gentlemen's own spirit, in which the times reflect themselves."[‡264] Yet, for all his denigration of history as a whole (a "mixture of error and violence") and his scepticism towards the possibility of acquiring secure historical knowledge ("past times are to us a book with seven seals"),[§] Goethe nevertheless, in the notes and essays added as supplements to his *West-östlicher Divan*, believed he had recognized one recurrent, dominating theme in history, namely, "the conflict of

---

[*] "Die Devise der Geschichte überhaupt müßte lauten: *Eadem, sed aliter*. Hat Einer den Herodot gelesen, so hat er, in philosophischer Absicht, schon genug Geschichte studirt. Denn da steht Alles, was die folgende Weltgeschichte ausmacht: das Treiben, Thun, Leiden und Schicksal des Menschengeschlechts, wie es aus den besagten Eigenschaften und dem physischen Erdenloose hervorgeht."

[†] Hübscher rightly says that Schopenhauer shares with Goethe not only a distrust of mathematics, which (in Schopenhauer's wording) "teaches only the 'how much' and 'how big', but never the 'what' of phenomena" ("die nur das Wieviel und Wiegroß, aber nie das Was der Erscheinungen lehrt") — this particular distrust is shared by Hegel as well — but also a distrust of history and its inconclusiveness, which never comes to an end and never arrives at a definitive end (whereas mathematics, according to Goethe, can claim at least the merit that it is sensible enough to occupy itself only with things about which one can gain certainty ("die sich klüglich nur mit Dingen abgibt, über die man gewiß werden kann")). Hence, too, their shared rejection of all constructions of a belief in progress.

[‡] "Was ihr den Geist der Zeiten heißt, das ist im Grund der Herren eigner Geist, in dem die Zeiten sich bespiegeln."

[§] "... Mischung von Irrtum und Gewalt"; "... vergangene Zeiten sind für uns ein Buch mit sieben Siegeln".

unbelief with belief'.*[265] But this is quite unlike history considered — as Hegel did — as a planned development towards a future end. It is more like Schopenhauer's *eadem sed aliter*. Schopenhauer himself attributed at least sufficient value to history to regard it as "the rational self-awareness of mankind" ("das vernünftige Selbstbewußtsein des menschlichen Geschlechts"): just as individual man is, by force of his reason, not tied to the *hic et nunc* but lives at the same time in the future and the past, so a nation becomes conscious of itself through history.[266]

In practice, this means that both Goethe and Schopenhauer allow only for empirical (or 'evenemental', in Braudel's latter-day terminology) history, not for any philosophy of history, let alone for any 'historicism' as Popper was later to call it.†[267] Characteristically, both stress the importance of the *individual*. As Goethe remarked in the same conversation: "Mankind? That is an abstraction" ("Die Menschheit? Das ist ein Abstractum").[268] And Schopenhauer, for all his denigration of the *principium individuationis* as 'Veil of Maya', nevertheless finds in the individual's maxim of action the core of morality. This links him not only to Kant but also to Kierkegaard's 'ethical subjectivity', as Alfred Schmidt has rightly remarked.[269] After all, 'Will', as opposed to 'Intellect', and in spite of its Schopenhauerian apotheosis as metaphysical ground of the world, is always intrinsically personal ('existential', to use the fashionable term of a few decades ago). This very point was noted already by the well-known Jewish thinker Rosenzweig, who favourably opposed Schopenhauer's concern with the irreparable anguish and sufferings of the concrete individual to Hegel's fascination with the all-reconciling, impersonal Whole of Absolute Spirit, and who therefore wrote about the former: "Schopenhauer was the first among the great thinkers to ask, not for the essence, but for the value of the world ... Here stood a man who no longer philosophized in the context of the history of philosophy, so to speak as its representative or as an heir to the actual state of its problems, but as one who had set himself the task 'to think about life', because life is 'a miserable affair'."‡[270] This interest in the individual living subject has something to do, in its turn, with Schopenhauer's emphasis upon concrete *Anschauung,* which he shared with Goethe and which distinguished both men from Hegel; and, although the fact that Hegel, too, defended Goethe's theory of colours shows that no necessary condition

---

* "Das eigentliche, einzige und tiefe Thema der Welt- und Menschengeschichte, dem alle übrigen untergeordnet sind, bleibt der Conflict des Unglaubens und Glaubens."

† 'Historicism' in this Popperian sense is not to be confused with 'historism' (more on which below).

‡ "Schopenhauer fragte als erster unter den großen Denkern nicht nach dem Wesen, sondern nach dem Wert der Welt ... Hier stand ein Mensch, der nicht mehr im Zusammenhang der Philosophiegeschichte und gewissermaßen als ihr Beauftragter, als Erbe des jeweiligen Standes ihrer Probleme philosophierte, sondern der sich vorgesetzt hatte, 'über das Leben nachzudenken', weil es — das Leben — 'eine mißliche Sache ist'." (Incidentally, Schopenhauer made this statement in a conversation, not with Goethe (as Rosenzweig wrongly thought) but with the poet Wieland during a visit paid to the latter by young Schopenhauer in April 1811).

(nor, of course, a sufficient reason) for such partisanship rested here, it certainly provided an additional motive for Schopenhauer's siding with Goethe in the matter (the prime motive, shared with Hegel, being the wish to show positive science its subordinate — albeit legitimate — place as compared to metaphysics).

These are intricate matters and, in whatever manner one wishes to draw the various alignments and crossroads, we may with justice regard it as a regrettable limitation of Schopenhauer's philosophy that it is very eloquent about *nature* while nowhere offering even an outline of a philosophy of *culture*. Up to this point in the argument one might grant Hegel an advantage over Schopenhauer, yet two things should not be forgotten here.

First, we must raise the issue of *progress* in history. Hegel's insight in the importance of history was not so original as is sometimes suggested — one need only recall Vico and Herder to acknowledge that the allegedly 'unhistorical' 18th century had already paved the way here. According to Vico, we can know only what we produce ourselves, such as mathematics; however, the latter yields only formal knowledge, as Nature has been made, not by man, but by God. History, on the contrary, is a work of man; therefore, the most secure knowledge to be gained is of history. A wide gulf yawns, of course, between Vico's philosophy, in which the human spirit knows the historical reality it has itself produced *sub specie divinae veritatis*, and Kant's allegedly unhistorical, critical philosophy; and indeed Kant's apparent conception of Euclidian geometry and Aristotelian logic as definitive, suggest a rather unhistorical view of human reason. Yet the fact remains that Hegel did not even acknowledge Newton's significance as a *stage* in the development of science (even less so than did 'unhistorical' Schopenhauer as we have seen); moreover, in a footnote on p. 20 we already gave as our opinion that there is a significant parallel after all between Kant's 'Copernican revolution' and Vico's *verum factum*. For, in order to trace the principles that govern human understanding, Kant proceeded in an indirect way by turning his attention to its actual achievements, i.e., to the fact of existing science, i.e., *to a branch of culture*. Hence, there is a tacit historical dimension to Kant's mode of thought.[*271] Moreover, in some of his lesser works Kant explicitly granted an important role to history. Not only did he introduce, some decades prior to Laplace, the notion of development on the cosmic level in his '*General Natural History and Theory of*

[*] A similar observation was made long ago by Max Scheler. He says of Kant that, at least in the domain of epistemology, a present system of historical intellectual actions becomes the criterion of reason, thus opening for the first time the possibility of gauging reason against reality instead of reality against dogmatic reason; this, Scheler thought, betrays a much more profoundly fundamental relationship to the historical mode of thought than Kant is customarily given credit for ("Wenigstens auf dem Gebiete der theoretischen Erkenntnistheorie wird ein vorliegendes System geschichtlicher Denkhandlungen zum Maßstab der Vernunft; und damit erscheint zum erstenmal die Möglichkeit, die Vernunft an einer *Realität* und nicht umgekehrt alle Realität an einer dogmatischen Vernunft zu messen. Das verrät, trotz Kant's vielgeschmähtem unhistorischem Sinn, eine tiefliegende und *prinzipielle sachliche* Beziehung zu einer geschichtlichen Denkart.").

the Heavens' (*Allgemeine Naturgeschichte und Theorie des Himmels*; 1755), but in the sphere of history proper there is also his '*Ideas for a General History in a Cosmopolitan Sense*' (*Ideen zu einer allgemeinen Geschichte in weltbürgerlicher Absicht*) — an essay outlining an overall view of the history of mankind with the typical Enlightenment concept of progress as its clue. Since the establishment of an international legal system is proclaimed to be the final aim of human history, it might even seem at first sight that Kant was no less of an 'historicist' than Hegel; and when Kant speaks of human progress as a "purpose of nature" (*Naturzweck*), since all natural propensities of any creature are destined to arrive in the end at full development, then one is reminded of Herder's *Ideen zu einer Philosophie der Geschichte der Menschheit*, the unfavourable reception of which by Kant we mentioned in our first chapter (p. 30). And indeed the establishment of a righteous civil constitution (with an international legal system as its necessary capstone) is proclaimed here by Kant to be the "biggest problem forced by nature upon mankind" ("das größte Problem für die Menschengattung zu dessen Auflösung die Natur ihn zwingt"). But not by chance does Kant call this problem at the same time the most difficult of all for mankind to solve. For, although it is true that nature endowed man to this purpose with reason and with freedom of will, it is no less true that man is inclined to misuse this freedom; and, although as a reasonable being he desires a law which puts limits on the freedom of *all* people with a view to general welfare, he will always, because of his selfish animal inclinations, be inclined to make an exception for himself. Therefore he needs a master who can compel him to break his own will and obey a generally valid will; but, alas, this master will be a man, too! Therefore the complete realization of this idea is impossible ("out of such crooked timber as from which man has been made nothing wholly straight can be carved," he said): only its *approximation* has been imposed on man by nature.[*272] Thus for Kant one could speak of *possibilities* and of a *task* (*Aufgabe*) rather than of relentless progress bound to take place in any case. Hence his rejection of Herder's naturalist determinism, especially when this is presented as leading to an imagined ideal state in which 'happiness' is realized; for under the aspect of happiness not even a comparison between different human societies or between generations in favour of one or the other is possible since everybody creates for himself his own image of happiness.[273]

This, then, brings us to our second point, which is that any historicism is really (to phrase the matter a little paradoxically) bound to be of a rather unhistorical nature. The Dutch essayist Jacques de Kadt once wittily remarked that, in Hegel's system, unfortunately Spirit is not put into history but rather History into Spirit.[274] And indeed, Hegel's interpretation of history as Absolute Idea arriving at an awareness of itself as Absolute Spirit by the detour of its self-externalization in Nature came down to constructing, out of freshly awakened historical consciousness, not so much a science, but rather a mythology. Moreover, it is one

* "... aus so krummem Holze, als woraus der Mensch gemacht ist, kann nichts ganz Gerades gezimmert werden."

thing to recognize the importance of the irrational in history (as every historian is bound to do), but quite another to join Hegel in incorporating — by means of a brand-new 'dialectical' logic — the irrational into the concept of Reason itself.*[275] Finally, Hegel, in stressing the reason-imbued nature of history while at the same time altering reason beyond recognition by incorporating the irrational into it (a *tour de force* which was in its turn a logical consequence of his attempt to reconcile his belief in Reason as the ultimate reality with the evident presence of the irrational in nature and history), managed simultaneously to fall into two opposite traps that the historian must at all costs seek to avoid. To explain what we mean by this, we must point to still another parallel between Comtean positivism and *Hegelian* metaphysics (yet not, for a change, between Comtean positivism and *Schopenhauerian* metaphysics!). What parallel is this?

It consists of the fact that both Hegel and Comte regarded History teleologically, as leading to a predetermined end. *Pace* Francis Fukuyama in his sensational book *The End of History and the Last Man*,†[276] this led to two fatal consequences, namely, the denigration of the concrete human individual in favour of abstract collectives,‡ and the obliteration of the difference between 'fact' and 'value' (or between 'is' and 'ought'), since everything, however morally objectionable, can be seen as a necessary stage in the development of Spirit towards its ultimate self-realization. Indeed, if we consider as the 'Scylla' of the historian a '*historism*' that carries the historian's duty to enter with an open mind into the ideas and circumstances of other ages to the extreme of a complete

---

* The actual sterility of this, at first sight brilliant, brain-wave has been clearly demonstrated in Meyerson's critique of Hegel.

† Fukuyama defends Hegel against the charge that he had been a reactionary apologist of the Prussian monarchy, or even a precursor of 20th-century totalitarianism, by suggesting that Hegel really had the modern constitutional state (i.e., liberal democracy) in mind when he spoke of the gradual growth of mankind to full rationality. This may or may not be so, yet it does not alter the fact that, to Hegel, man as an ethical being finds his ultimate destination in the political sphere.

‡ This aspect of Hegelianism was, as we know, the main target of Kierkegaard in his *Concluding unscientific afterword*. For the appreciation of the philosopher's role in society this leads to the curious consequence of what we may call a 'presumptuous modesty': for on the one hand the individual thinker can never be anything more than a mouthpiece of the spirit of the age, but on the other hand a Hegelian philosopher can, precisely because of this, claim to speak 'on behalf of Time' (this being a book title chosen by a Dutch historian, Jan Romein, infected with a liberal dosage of Marxist Hegelianism). And because of that he may even, as reportedly (and *se non è vero è ben trovato*) Hegel did himself from his conviction that in his time at least in principle history had come to an end (politically speaking in the conception of the modern state, philosophically speaking in his own system in which the World Spirit has arrived at full self-consciousness!), venture to say: "Gentlemen, I dare to say: I do not only tell the truth, I *am* the truth" ("Meine Herren, ich darf wohl sagen: ich rede nicht nur die Wahrheit, ich bin die Wahrheit") — a pronouncement that one would be inclined to call blasphemous if it were not so utterly ridiculous.

relativism, and as his 'Charybdis' the inclination to judge (most often, to condemn) the past in the name of some present-day ideal, then Hegel's '*historicism*' (which claims to reconcile the idea of progress with historical consciousness by way of the 'dialectical' method), instead of cautiously manoeuvering between the two extremes, manages to combine the former's 'a-moralism' with the latter's subordination of the past to some preconceived destination. As one consequence among others, Hegel actually had no use for individual ethics ('*Moralität*' as he calls it disparagingly) but only for a social ethics ('*Sittlichkeit*' as he calls it), culminating *via* family and civil society in the *state*, where, therefore, all ethical aspirations of mankind crystallize.*

To Schopenhauer this doctrine constituted an "apotheosis of philistinism" ("Apotheose der Philisterei");[277] and this judgement certainly provides one reason among others for the lack of appeal exerted by Schopenhauer's philosophy before 1848 as well as for the temporarily fashionable status it began to enjoy after that fateful year. But one might well ask with Thomas Mann whether Schopenhauer's own position towards the political sphere could not be called an 'apotheosis of philistinism' as well, though in the opposite way?[278] "The worst government is that which one is forced to praise" ("Le pire gouvernement est celui dont on est forcé de dire du bien") he once remarked to a French visitor, Frédéric Morin;[279] and as a far-sighted hint to the bee-hive happiness of modern totalitarian systems it is to the point. Yet one might expect from the philosopher who always stressed that (cf. Chapter IV, p. 99) man is not a "winged angel's head without a body" that he would have been aware that 'justice' and 'humaneness' (the two cardinal virtues according to his own moral philosophy as we shall see in the next chapter) need a 'body' — which in this case means a secured piece of territory, however small, in order not to remain pure abstractions that commit one to nothing. Certainly, their concretization will always be very imperfect; and the state, even considered as a voluntary association of men, remains in itself just one aspect of human finitude. To see with Hegel in the notion of citizenship the apogee of human civilization is therefore, of course, an inadmissible exaggeration; on the other hand, from Schopenhauer, who always prided himself on his Hanseatic–republican backgrounds, one might have expected a somewhat greater appreciation of the fact that German *Bürger* not only means *bourgeois*, but also *citoyen* ...†

Hegel's overestimation of the legitimate scope and possibilities of the political sphere was to Schopenhauer for the rest just another symptom of what he used to call Hegel's "ruthless optimism" ("ruchloser Optimismus"). Yet here, too, we must

* It is of course true that, although Hegel takes the state for the culmination of the realm of 'Objective Spirit', this realm is transcended in its turn by the realm of 'Absolute Spirit' (culminating *via* art and religion in — Hegelian — philosophy); so that one cannot strictly speak of Hegel's 'worship' of the state. The problem, however, is that, to Hegel, *ethics* belongs only to the realm of 'Objective Spirit'.

† We shall briefly return to this topic in Chapter IX (p. 231) when discussing Schopenhauer's attitude in the year of revolution, 1848.

take care to differentiate, insofar as Hegel was not just the foul optimist depicted by Schopenhauer. After all there is room for the negative and for pain and death in his system, and he was of course quite aware that the real course of history presents us with anything but an idyll. However, in the last resort the Hegelian philosopher stands above all these things, since he is able to regard them as inevitable moments in the life of the Spirit, culminating eventually in that great identity in which all antagonisms are reconciled. This is why a thinker like Rosenzweig preferred Schopenhauer over Hegel, as we saw in the above. The fact that Schopenhauer's system, other than Hegel's, *did* have a use for individual ethics is what, more than anything else, lends a certain plausibility to his claim of being Kant's only worthy successor. For we must reiterate that ethics was Kant's ultimate concern, and that this accounts for the central place the problem of freedom occupies in his philosophy — a freedom which is for him the *proprium* of the noumenal realm, just as necessity is the *proprium* of the phenomenal realm. To Schopenhauer this was likewise a crucial issue,* and he was bold enough, in this respect as well, to proclaim himself not only Kant's legitimate heir, but also the man to improve upon him and to complete him. An even more consequential claim, this, than the similar claim we saw Schopenhauer raise in the domain of epistemology, and which we have assessed so far; an even more acutely needed, careful examination of that second claim will be our concern in the two chapters that follow.

* Recall that Schopenhauer considered the problem of moral freedom to be the second great topic (besides that of the relationship between the ideal and the real) which distinguished modern philosophy, culminating in Kant (cf. Chapter IV, footnote on p. 85).

VI

ETHICS

a. Introduction: are there two Kants?

Heinrich Heine, in a book still very much worth reading, entitled '*On the History of Religion and Philosophy in Germany*' (*Zur Geschichte der Religion und Philosophie in Deutschland*), imagines Kant's faithful old servant Lampe, umbrella under arm, watching sadly his master's merciless destruction of all the ancient certainties; and "then Immanuel Kant has mercy upon him and shows that he is not only a great philosopher but also a good man; he considers the matter and says to himself, half good-naturedly, half ironically: 'Old Lampe needs a God, otherwise the poor man cannot be happy; man, however, should be happy on earth, as is taught by Practical Reason; for all I care, well then, may Practical Reason guarantee the existence of God'. Consequently, Kant sets out to distinguish between Theoretical Reason and Practical Reason, and with the latter, as if with a magic wand, he called back to life again the corpse of Deism killed by Theoretical Reason."[*280]

All this is very witty; yet, although we can highly appreciate a light touch all too rarely found in books on philosophy, we cannot quite abstain from playing the schoolmaster here. For, apart from Heine's mistaken use of the term 'Deism' where 'Theism' is meant,[†281] and — more importantly — from the circumstance that considerations about "being happy on earth" were precisely *not* the core of

---

* "Da erbarmt sich Immanuel Kant und zeigt, daß er nicht bloß ein großer Philosoph, sondern auch ein guter Mensch ist, und er überlegt, und halb gutmütig und halb ironisch spricht er: 'Der alte Lampe muß einen Gott haben, sonst kann der arme Mensch nicht glücklich sein — der Mensch soll aber auf der Welt glücklich sein — das sagt die praktische Vernunft — meinetwegen — so mag auch die praktische Vernunft die Existenz Gottes verbürgen'. Infolge dieses Arguments unterscheidet Kant zwischen der theoretischen und der praktischen Vernunft, und mit dieser, wie mit einem Zauberstäbchen, belebte er wieder den Leichnam des Deismus, den die theoretische Vernunft getötet."

† The term 'Deism' is normally used only for the Enlightenment conception of God as the great 'watch-maker', i.e., as mere Creator, not as Providence. Kant in his *Prolegomena* carefully distinguishes between this 'purely transcendent' conception and the more full-blooded 'theism' of traditional Christian faith.

"practical reason" in Kant's view,* it should be remembered that already in the *Critique of Pure Reason* 'God' appears as one of the three regulative ideas of Reason ('Soul' and 'World' — viewed as a comprehensive whole — being the other two). True, He appears here in the guise of a theological 'ideal', to be translated with only a little licence as an 'as if'; but this virtually remains so in the *Critique of Practical Reason*, as we shall see; so in this sense no contradiction is involved. In the second part of the *Critique of Pure Reason*, entitled 'Methodology of Pure Reason' ('Methodenlehre der reinen Vernunft'), and more elaborately so in the *Prolegomena*, Kant made some highly instructive remarks on the subject. Though it is absurd, he says, to claim any knowledge of that which is not an object of possible experience, it would be still more absurd to regard our intuition of space and time as the only possible intuition and our discursive reason as the prototype of any possible reason; for in that case our principles, which confine the usage of reason to possible experience, would become transcendent themselves in proclaiming the limits of our reason to be the limits of the things-in-themselves. One should distinguish between "limits" (*Schranken*) and "boundaries" (*Grenzen*). In mathematics and natural science, human reason is subject to "limits", but not to "boundaries", i.e., it does recognize that there is something which is forever beyond its reach, but not that it will ever be completed in its inner progress. Metaphysics, on the other hand, displays its "boundaries" in the dialectical attempts of Pure Reason; it is precisely the function of the transcendental ideas (because we can neither abstain from them nor ever realize them) not only to *show* the boundaries of a pure usage of Reason but also to point out how to *determine* them. For in all boundaries there is also something positive (for example, a plane — while the boundary of corporeal space — is a space itself; a line is a space that is the boundary of the plane, and a point is the boundary of the line, yet still a place in space), whereas "limits" contain negations only.[282]

Moreover, neither of the two *Critiques* is about God in the first place. In the first *Critique* the confusion that comes from the illegitimate usage of the categories of Understanding is demonstrated in order to vindicate, so much the more strongly, its legitimate usage; it is therefore about the foundations of *science* in the first place. It is not against faith as such (quite the contrary: "so I had to get rid of knowledge in order to gain room for faith," says Kant in the Preface to the second edition of the *Critique of Pure Reason*!),†[283] but against faith presenting itself as science that Kant's critique is directed. In the second *Critique*, it is not the postulate of 'God', but the postulate of 'Freedom' which forms the core of the argument, even though it is true that 'God' and 'Immortality' are presented as the

* The question of whether considerations of this kind were not allowed by Kant to enter 'through the back door' after all will concern us later on; the *essence* of Kantian ethics, as Schopenhauer rightly — and with approval — remarked, lies in his banishing eudaemonism from its temple.

† "... ich mußte also das Wissen aufheben um zum Glauben Platz zu bekommen."

end points at which the entire enterprise culminates.* This is stated explicitly in the Preface to the *Critique of Practical Reason*, where we read: "Freedom, however, is also the only one among all Ideas of Speculative Reason, of which we *know* the possibility *a priori* (without, however, understanding it), because it is the condition of the moral law that we do know. But the ideas of *God* and of *Immortality* are not constitutive factors of the moral law, but only of the necessary object of a will determined by this law, i.e., of the merely practical usage of our Pure Reason; of these ideas we can therefore not even pretend that we can *know* and *recognize* their possibility, let alone their reality."†[284] This is not to be wondered at, since already in the *Critique of Pure Reason* there is a difference between, on the one hand, the 'paralogisms' of rational psychology and the pseudo-proofs of the existence of God in rational theology, and, on the other, the 'antinomies' of rational cosmology. In the first two cases, Kant argued, we deal with simple sophisms (i.e., the proofs of the existence of the soul and of God are presented by logically false arguments), whereas in the third case we are concerned with pairs of mutually conflicting arguments which are both irrefutable, that is to say, they are logically sound.[285] And since 'Freedom' enters the stage for the first time in the treatment of the third antinomy of rational cosmology, it might be maintained that, right from the start, it had better philosophical credentials than 'God' and 'Immortality'!

The point is, however, that its appearance on the stage already in the *Critique of Pure Reason* puts 'Freedom' in the position of providing the link between the two *Critiques*, thus ensuring the *continuity* of both, rather than introducing an *opposition* between the two. And this is indeed obvious once we consider that, in an important sense, Kant meant the *Critique of Pure Reason* only to do the preliminary work for the *Critique of Practical Reason*.‡ A necessary preliminary

---

* The 'immortality' of the *Critique of Practical Reason* was in fact a closer specification of the 'soul' of the *Critique of Pure Reason,* whereas the 'freedom' of the *Critique of Practical Reason* appears already in the treatment of the third antinomy in the *Critique of Pure Reason* (more on this below).

† "Freiheit ist aber auch die einzige unter allen Ideen der speculativen Vernunft, wovon wir die Möglichkeit a priori *wissen*, ohne sie doch einzusehen, weil sie die Bedingung des moralischen Gesetzes ist, welches wir wissen. Die Ideen von *Gott* und *Unsterblichkeit* sind aber nicht Bedingungen des moralischen Gesetzes, sondern nur Bedingungen des nothwendigen Objects eines durch dieses Gesetz bestimmten Willens, d.i. des bloß praktischen Gebrauchs unserer reinen Vernunft, also können wir von jenen Ideen auch, ich will nicht bloß sagen, nicht die Wirklichkeit, sondern auch nicht einmal die Möglichkeit zu *erkennen* und *einzusehen* behaupten."

‡ The juxtaposition of 'pure' and 'practical' reason is misleading insofar as both *Critiques* present us with a *transcendental* investigation, i.e., with the pure forms of Reason. More appropriate titles would therefore have been *Critique of Pure Theoretical Reason* and *Critique of Pure Practical Reason*, respectively. In the preface to the *Critique of Practical Reason* Kant anticipated this point by arguing that the investigation had only to *demonstrate* that there is indeed a "pure practical reason", whereas it had to *criticize* only the *practical* aspect of the latter (unlike the case of speculative reason, where the pretensions of the *pure* aspect had to be cut down to size). The "pure" aspect of pure

work, to be sure, since the question 'What can I know?' must be answered before one can make significant theories in any domain whatsoever. But this does not alter the fact that the problems of ethics ('What ought I to do?'), culminating in the idea of human freedom, always stood at the centre of his thought. And we should like to add, against any positivist distortion of Kantianism, that this primacy of 'Practical Reason' is quite justified, too: first because of Kant's constructivist view of science as an *activity*, and second because of the fact that science (while of course fitted out with its own internal logic) derives its ultimate justification from an ethical principle — the obligation to seek for objective Truth.* *Why* we should seek for this is a question to which science itself can provide no answer. Quite another question is, of course, whether Kant was sufficiently successful in his attempt to lay down the transcendental foundations of ethics. We shall have ample opportunity to deal with that question in the course of this (and the next) chapter, but let us first dwell a little longer on the continuity of Kant's thought by taking a somewhat closer look at the role played by the idea of Freedom in the *Critique of Pure Reason.*

As mentioned already, freedom comes in here in connection with Kant's treatment of the third antinomy, as follows: "Causality according to natural laws is not the only [form of causality] from which the phenomena of the world can in their entirety be deduced. To explain them requires, in addition, the assumption of causality by freedom" (*thesis*) vs. "There is no freedom, but everything in the world occurs solely according to natural laws" (*antithesis*).†[286] We shall abstain

---

practical reason, Kant held, need not be criticized, for if pure Reason can be really practical, then it proves its reality and the reality of its concepts sufficiently through its *actions*. Against this it might be argued that, even granted this, the objection remains that 'theoretical' (or 'speculative') and 'practical', not 'pure' and 'practical' are antithetical. Moreover, as we shall see, the problem is that not everyone, upon having read Kant's exposition, will be convinced that the pure concepts of 'practical reason' by themselves provide a sufficient motor driving moral action.

* From this point of view the truth-commandment may be regarded as the link between theoretical and practical reason.

† "Die Kausalität nach Gesetzen der Natur ist nicht die einzige, aus welcher Erscheinungen der Welt insgesamt abgeleitet werden können. Es ist noch eine Kausalität durch Freiheit zur Erklärung derselben anzunehmen nothwendig", vs. "Es ist keine Freiheit, sondern alles in der Welt geschieht lediglich nach Gesetzen der Natur." The other antinomies are of no specific concern to us; but, for completeness' sake, here they are. The first runs: "The world has a beginning in time, and is spatially closed into boundaries as well" ("Die Welt hat einen Anfang in der Zeit, und ist dem Raum nach auch in Grenzen eingeschlossen") vs. "The world has no beginning, and no boundary in space, but is infinite both in time and in space ("Die Welt hat keinen Anfang, und keine Grenzen im Raume, sondern ist, sowohl in Ansehung der Zeit, als des Raumes, unendlich"); the second: "Every compounded substance in the world consists of simple parts, and nothing exists but the simple, or that which is compounded out of the latter" ("Eine jede zusammengesetzte Substanz in der Welt besteht aus einfachen Teilen, und es existiert überall nichts als das Einfache, oder das, was aus diesem zusammengesetzt ist") vs. "No compounded thing in the world consists of

from presenting Kant's argument in detail; of interest to us here is what Schopenhauer had to say about it.

Let us recall, first and foremost, that (speaking in general) Schopenhauer wished to have nothing to do with the faculty of "Reason" (*Vernunft*) as presented by Kant in the 'Transcendental Dialectic', that is, as the "faculty of principles" (*Vermögen der Prinzipien*), to be distinguished from the faculty of "Understanding" (*Verstand*) which has only to do with "rules" (*Regeln*), to wit, the categories that make possible the unity of experience in mathematics and science. According to Kant, reason aims to find the "Unconditioned" (*das Unbedingte*) behind the conditioned pieces of knowledge provided by Understanding, thus completing the unity of the latter. This aim is called by Kant a "logical maxim", which may become a "principle of pure reason" on the following assumption only: "If the conditioned is given, let then also be given the whole sequence of hierarchically ordered conditions, which therefore is itself unconditioned."[*][287] Schopenhauer thought this formulation unclear and paraphrased it as follows: "If the conditioned is given, then the totality of its conditions, and therefore also the *Unconditioned*, through which alone this totality becomes complete, must be given as well."[†][288] But according to Schopenhauer this principle is not only objectively invalid but not even subjectively necessary, since reason as a faculty of knowledge has to do with objects only, which are necessarily subject to the principle of sufficient reason, *a parte ante* as well as *a parte post*, an "unconditioned" thus being an un-thing.[289] This is why, as noted before in Chapter IV, p. 83, he considers Kant's antinomies to be "sham fights". Only the antitheses are truly based on the forms of our faculty of knowledge, that is (speaking objectively) on the necessary, *a priori* certain, most general laws of nature, whereas the theses have only a subjective grounding in that they are based merely on the weakness of the rationalizing individual whose imaginative power becomes tired by an infinite regress.[290]

---

simple parts, and there exists everywhere nothing simple in it" ("Kein zusammengesetztes Ding in der Welt besteht aus einfachen Teilen, und es existiert überall nichts Einfaches in derselben"); the fourth: "To the world belongs something which, whether as a part of it or as its cause, is a simply *necessary* being" ("Zu der Welt gehört etwas, das, entweder als ihr Teil, oder ihre Ursache, ein schlechthin nothwendiges Wesen ist") vs. "Nowhere exists a simply necessary being, whether in the world or outside it, as its cause" ("Es existiert überall kein schlechthin nothwendiges Wesen, weder in der Welt, noch außer der Welt, als ihre Ursache"). According to Schopenhauer (to whom, as we noted in Chapter IV, all these antinomies were nothing but sham fights) the fourth is equivalent to the third, since with the "freedom" of the third nothing else can in the given context be meant but the idea of the cause of the world.

* "Wenn das Bedingte gegeben ist, so sei auch die ganze Reihe einander untergeordneter Bedingungen, die mithin selbst unbedingt ist, gegeben."

† "Wenn das Bedingte gegeben ist, so muß auch die Totalität seiner Bedingungen, mithin auch das Unbedingte, dadurch jene Totalität allein vollzählig wird, gegeben seyn."

Such is Schopenhauer's objection, and one may well ask whether he is not himself engaged here in a 'sham fight' against Kant. For Kant, in his solution of the antinomies (specifically, of nos. 3 and 4, which Kant called the 'dynamic' antinomies, as distinct from the 'mathematical' nos. 1 and 2) says precisely that, yes, *in the phenomenal world* only the antitheses are true, whereas the theses are true *in the noumenal world* only.* There is all the more reason to ask this because, as is well known, Schopenhauer wholeheartedly accepted the distinction between 'appearance' (*Erscheinung*) and 'thing-in-itself', telling us at the same time that "Kant's intended solution of the third purported antinomy gave him an opportunity to express in a very beautiful way the deepest thoughts of his entire philosophy, above all, his exposition of the distinction between the empirical and the intelligible character, which I count among the most outstanding things ever said by a human being."†[291] It is this distinction, indeed, which enabled Kant, as we shall see, to save the idea of freedom while accepting determinism in the empirical world, and thus (in his own opinion) to lay a secure foundation for ethics; Schopenhauer did (or believed he did) the same. What, then, *did* make him oppose Kant's antinomy? The reason is that, to him, the 'Thing-in-itself' is *not* of a 'noumenal' nature! Later on, in the next chapter, we shall dwell at length on this fundamental difference between Kant and Schopenhauer; but here already we wish to point out a highly characteristic remark made by Schopenhauer. In his view, Kant failed to see that the concept of freedom has its origin in man's immediate knowledge of his own will — a will which, as the world-creating Thing-in-itself, is indeed exempt from the principle of sufficient reason and therefore from all necessity, thus being wholly independent and free, nay, all-powerful. Kant's failure to see this is what caused him to seek the origin of the concept of freedom in a subtle speculation, making Reason, in its eternal striving after the Unconditioned, posit the transcendent Idea of Freedom, from which the practical concept of freedom is allegedly derived.[292] In other words, Schopenhauer, like Kant, cannot, for ethical reasons, renounce freedom, but unlike Kant he refuses to accept any

---

* The distinction between appearance and thing-in-itself serves Kant as a means of solving the antinomies in this sense that, regarded thus, in the third and the fourth (called by Kant 'dynamic') thesis and antithesis can peacefully coexist, whereas in the first and the second (called by Kant 'mathematical') thesis and antithesis both appear to be false. The distinction 'mathematical' vs. 'dynamic' is used also by Kant in his table of categories, where they designate the first two classes ('quantity' and 'quality') and the last two classes ('reality' and 'modality'), respectively. The difference is that in the first case it is about objects of our intuition as such and in the second case about the existence of those objects; in the last two classes the categories therefore appear as correlative pairs (e.g. 'cause–effect' or 'possibility–impossibility').

† "Übrigens ist es die beabsichtigte Auflösung des vorgeblichen dritten Widerstreits, welche Kanten Gelegenheit giebt, die tiefsten Gedanken seiner ganzen Philosophie sehr schön auszusprechen, vor allem aber die Auseinandersetzung des Gegensatzes zwischen empirischem und intelligibelen Charakter, welche ich dem Vortrefflichsten beizähle, das je von Menschen gesagt wurde."

connection between freedom and *reason*. This is so because, to him, the latter is only the awareness *a priori* (on the abstract, conceptual level) of the necessarily valid, general forms of all experience and, hence, is bound up *eo ipso* with a complete determinism. Indeed, as we have already seen in the preceding chapters, he calls 'Reason' what Kant calls 'Understanding', the latter word designating for Schopenhauer the faculty of pre-conceptual knowledge that we share with the animals.

Even if one accepts Schopenhauer's usage of the term 'Understanding', and, consequently, *up to a certain point*, of the term 'Reason', one need not in our opinion reject wholly Kant's usage of the latter term. Schopenhauer's usage stems from Reason being to him only that which Horkheimer called "instrumental reason";[*293] therefore it is, so to speak, 'all of a piece'. The possibility is alien to his mind that reason might in some sense be *divided against itself*, insofar as it is guided by two opposite maxims, one of which says that the conditioned should be reduced to the unconditioned and the other that each condition should be regarded as conditioned in its turn. Certainly one may say that it is precisely the task of the latter maxim to unmask the former as an illusion, but this does not detract either from the fact that *both* are tendencies of human reason, or from the fruitfulness of certain cognitive illusions as incentives for marching forward on the road leading to the greatest amount of possible human knowledge (the point being only that they ought to be abandoned as soon as they are perceived as such!). Schopenhauer seemed to forget that, in this context, Kant did not leave it at that passage we quoted, and on the Unconditioned that must be "given" (*gegeben*), and which Schopenhauer rejected. For in another passage Kant acknowledged that no maximal number in the sequence of conditions is *given* in the world of sense, but only *set as a task* (*aufgegeben*) in the *regression* of the latter. This is why Kant's principle of reason is valid, not as an axiom, but only as a problem for our understanding, which means that it is not of a *constitutive*, but of a *regulative* nature.[†294] And this is also why the theses of the 'dynamic' antinomies, which can

---

[*] According to Horkheimer "objective reason" (i.e., philosophical systems which believe that the fundamental structure of Being can be discovered and a conception of human destiny be derived from it) has in the modern age more and more been superseded by "instrumental reason" (i.e., a mathematically *formalized* reason, which, precisely because of its purely formal nature, is at the same time *subjective*, since its rationality has only to do with the question of which means fit best to certain ends and leaves aside all questions regarding the rationality of these ends themselves). Medieval Scholasticism was an example of 'objective reason'; and German Idealism may be said to have been a last desperate attempt to restore it in a new shape. It is interesting to note that such representatives of the neo-Marxist *Frankfurter Schule* as Horkheimer and Adorno held views in this respect that are very close to those of the 'reactionary' Heidegger (cf. Chapter I, p. 15).

[†] Quite another question is, of course, whether the three Ideas of Reason, given by Kant as specifications of the Unconditioned, can really be considered as regulatively necessary to human thinking. For example, Schopenhauer is certainly right when he says that neither Brahmanism and Buddhism nor the philosophers of Antiquity knew anything of God as a

peacefully coexist with their antitheses, are considered by Kant to be more reasonable in their claims than those of the 'mathematical' antinomies (in which thesis and antithesis appear both to be false), since in the former the unconditioned is placed beyond the range of homogeneous, sense-subjected conditions. In the dynamic antinomies, the unconditioned is taken for something different in kind from the conditioned, a force wholly beyond the senses that has nothing in common with the conditioned. Thus the demands of both reason and understanding are satisfied: since reason demands that there should be an unconditioned and since the understanding demands that nothing in the world of sense should be unconditioned, a dynamic idea, in asserting that the unconditioned belongs to a sphere with which understanding is not concerned, does not come into conflict with its own principles. One is sometimes almost inclined to say (preferably beyond the hearing of true Hegelians, who would all too readily take advantage of the suggestion) that Schopenhauer was a little *too* undialectical a thinker ...

However this may be, our little excursion into Schopenhauer's 'Critique of Kantian philosophy' has served not only as an illustration of the continuity between Kant's two *Critiques* (for it should be reiterated that the 'Transcendental Dialectic' from the *Critique of Pure Reason* was at issue here throughout), but also as a preliminary hint as to where the agreements as well as the differences between Kant's and Schopenhauer's respective moral philosophies are likely to be encountered. We shall now proceed to sketch an outline of the former, following, in the main, what we join Schopenhauer in considering to be its clearest and most elegant exposition, which is in Kant's '*Groundwork of the Metaphysic of Morals*' (*Grundlegung zur Metaphysik der Sitten*). This treatise was written in 1785, three years before the *Critique of Practical Reason* and twelve years before the *Metaphysic of Morals* (intended to sketch the positive applications of the preceding "critical" works); and it already contains most of the essentials of both.

b. Kant's moral philosophy

"Nowhere in the world, nor, for that matter, outside it is anything at all conceivable that may unconditionally be held to be good but only a *good will*" — with these robust words Kant opens his rightly famous book.[*295] Talents or gifts of fortune can become malign and miserable when the will, that is, the character that uses them, is not good. This sounds simple enough; yet, Kant says, there is something strange about it. Nature has endowed all creatures with those instruments that are most apt and adapted to one or the other purpose; now, if survival and well-being (in short, happiness) were the real purpose of a reasonable

first cause or an *ens realissimum*; in his invariably amiable manner he sums up these critical notes of his by remarking that "one should not identify Judaism and Reason" ("man soll nicht Judenthum und Vernunft identificiren").

[*] "Es ist überall nichts in der Welt, ja überhaupt auch außer derselben zu denken möglich, was ohne Einschränkung für gut könnte gehalten werden, als allein ein *guter Wille*."

and willing creature, reason would be a very inappropriate instrument; instinct would do much better. Yet reason is given to us as a practical faculty, i.e., as a faculty that is able to influence our *will*. Therefore it must be the true destiny of reason to create a will that is good in itself and not a will that serves as a *means* for something else.[296]

In order to develop this idea further Kant introduces the concept of *duty*. A man's action may of course go against his duty, but even if this is not the case the action may still arise from a selfish intention (e.g., a shopkeeper treating his clients honestly since otherwise they would go elsewhere). More than that, Kant argues, even the absence of any motive based on vanity or selfishness fails to guarantee the moral value of an action. For there are souls of so compassionate a nature that it gives them intense pleasure to bestow joy on others; yet such an action has no real moral value, for it arises, not from a sense of duty, but from *inclination* (*Neigung*). As opposed to this, a cold-hearted man, indifferent as such to other people's suffering, may do good towards them out of his sense of duty; this would therefore have real moral value.[297]

This particular assertion of Kant's has always sounded highly problematic, even to many convinced Kantians, as Schiller's well-known epigram testifies:[298]

Gerne dien ich den Freunden,	"I like to serve my friends,
Doch tue ich es leider mit Neigung, /	but unfortunately I do it with inclination,
Und so wurmt es mir oft,	and thus I often reproach myself
Daß ich nicht tugendhaft bin. /	for not being virtuous.
Da ist kein anderer Rat,	There is no other way,
Du mußt suchen sie zu verachten, /	you must seek to despise them
Und mit Abscheu alsdann tun,	and then do it in disgust,
Wie die Pflicht dir gebeut. /	as duty commands you."

This, too, is very witty, though in a somewhat more heavy-handed way. And it is also somewhat of a caricature, since Kant of course never condemned the inclination to serve one's friends as such; he says only that this is not enough to make an action morally valuable. And everybody will agree that it is indeed easier (and thus morally less meritorious) to be kind to people you like than to people you cannot stand. But such a truism would, no doubt, be despised by Kant. Instead, he clarifies his position by means of a closer analysis of the concept of duty.

Duty is defined by Kant as the "necessity of my actions out of sheer respect for the law".* For, if we abstract from all possible objects of the will, no other conditions for a dutiful action remain but, objectively, the *law*, and, subjectively, sheer *respect* for the latter.†[299] From what, then, does the concept of duty derive?

---

* "... die Notwendigkeit meiner Handlungen aus *reiner* Achtung fürs praktische Gesetz".

† Kant was aware that, in using the word 'respect', he introduced a sentiment, so that one might reproach him for introducing psychological instead of transcendental categories after all. He defended himself against this possible charge by pointing out that, though respect be a sentiment, it has not been *received* through influence but is *self-produced* by a concept of reason, which makes it fundamentally different from sentiments of the former kind.

Not from experience; for behind apparently dutiful actions an impartial observer of human conduct will always see the dear self inspiring the actions, rather than the commandment of duty, which would too often require self-negation. What remains is only the conviction that even if no actions from true virtue ever occurred they are nevertheless inexorably demanded by *Reason*.[300]

How does Reason proceed here? Kant points out that everything in nature works according to laws. But only a reasonable being is able to act according to the *idea* of law; this ability is called, by Kant, *Will*. Now if reason determined the will infallibly, the actions of such a being would be necessary not only objectively but also subjectively. But where this is not the case, so that the will is still subject to subjective motives not always in agreement with the objective laws of reason (as is the case with man), then objective law must *compel* the will, and the idea of law will therefore take the character of a commandment, which is to be expressed in the formula of an *imperative*.[301] Now imperatives are of two kinds only; they are either *hypothetical* or *categorical*. The former indicates the practical necessity of a possible action as a means of obtaining something other that that which one wants; the second demonstrates an action, apart from any relation to another purpose, as objectively necessary in itself.[302] The goodness of such an action consists of the disposition out of which it is performed, regardless of the result. In the first case we are concerned with what Kant (by analogy with the 'analytical judgements' of the *Critique of Pure Reason*) calls an "analytical–practical statement (*Satz*)"; for, from the pure notion of the willing of a certain purpose follows the necessity to act in such a way as to achieve that purpose. In the second case we are concerned with a "synthetic–practical statement *a priori*". It is "*a priori*" because no condition derived from some *inclination* is presupposed. It is "synthetic" because the willing of an action cannot be analytically deduced from another, already presupposed, willing (as we do not have such a perfect will), but is immediately linked to the concept of the will of a reasonable creature — a concept which does not already contain within itself the action concerned.[303] It is evident that the moral law is of the second type; hence, it is a Categorical Imperative. How can it be expressed?

This problem is easy to solve, since, besides the idea of law itself, a Categorical Imperative can only contain the necessity that the 'maxim' (i.e., the subjective principle of action) must be in accordance with the law. So there is but one Categorical Imperative, and it runs: "Act only according to that maxim which you can at the same time want to become a general law." Or, since the generality of that law refers to what is called 'nature' in the most general terms (i.e., the existence of things as far as they are determined by general laws): "act in such a way as if the maxim of your action would by your will become a general law of nature."[*304]

---

[*] "Handle nur nach denjenigen Maximen, durch die du zugleich wollen kannst, daß sie ein allgemeines Gesetz werden"; and "Handle so, als ob die Maxime deiner Handlung durch deinen Willen zum *allgemeinen Naturgesetze* werden sollte."

Kant has often been reproached for the purely formal nature of his categorical imperative; he would certainly have replied that here, too, we are dealing with a *transcendental* investigation, i.e., not with the question 'what is the material content of ethics?', but with the question 'what are the *a priori* principles that make ethics possible in the first place?' Moreover, Kant presents us with still another formulation, in which it is recognized that there is one *absolute* material value, i.e., a good that cannot become a means in the search for some other purpose. *Man* (or, more generally speaking, any reasonable being, as the moral law ought not to be dependent on the specific nature of *human* reason, but should be valid for *any* reasonable being) exists as a purpose in itself, and should always be regarded as such. Thus the categorical imperative would run: "act in such a way that you use mankind, in your own person as well as in the person of everybody else, never as a means only, but always at the same time as an end."[*305]

One further step may still be taken. From the fact that every reasonable being is an end in itself it follows that all reasonable beings must be systematically linked together through shared, objective laws; this means that the will of any reasonable being must, of necessity, be thought of as a *general, lawgiving* will. His will is not simply subjected to law; for such subjection is rooted in the fact that he himself must be seen as a lawgiver. Precisely *because* he may regard himself as *creator* of the law he must regard himself as *subject* to it. This is called by Kant the principle of the *autonomy* of will; in his own opinion it distinguishes the result of his inquiry from all previous accounts of ethics, because these were always founded on some interest outside pure Will and therefore always *heteronomous* in nature.[306]

Furthermore, the circumstance of every reasonable being finding himself subject to the *law* which says that each should treat himself and all others *never as a means only* but always *at the same time as an end in itself* yields a community of reasonable beings systematically held together by shared, objective laws. Since the laws are about the mutual relationships of these beings as ends and means, such a community may be called a Realm of Ends (*Reich der Zwecke*), the citizens of which are lawgivers as well as being subject to the law. Every reasonable creature belongs to it as a member, and God is its supreme head. However, this realm can be thought of only as an *ideal* of Reason.[†307]

This, then, is how Kant claims to have discovered the true foundation of ethics. But he adds at once that this result of his investigation can be wholly satisfactory only if we are able to prove that a good will, as defined in the foregoing, is indeed *possible*, i.e., that morality is not a delusion. So we have now arrived at the crucial question which, *mutatis mutandis*, is the same which was at the centre of the *Critique of Pure Reason*, namely, "how are synthetic judgements *a priori*

---

* "Handle so, daß du die Menschheit sowohl in deiner Person, als in der Person eines jeden andern jederzeit zugleich als Zweck, niemals bloß als Mittel brauchest."

† This is why we said in our introductory section that the 'as if' nature characteristic of the Deity of Theoretical Reason is characteristic of the Deity of Practical Reason as well.

possible?"* Translated into the language proper to the realm of Practical Reason, this becomes "how are synthetic-practical statements *a priori* possible?"

To answer the question, another Critique of Reason is required; in the present case, a Critique of Practical Reason. In the final section of the *Grundlegung* Kant outlines such a Critique, thus anticipating the larger work. Its conclusion (surprising at first sight) must be that, in this case, the answer is the exact opposite of its counterpart in the domain of theoretical reason. The categories of the latter can function only if used as conditions for possible *experience*; practical reason, instead, can only demonstrate its validity by *opposing* that which is given in experience (such as selfishness). Practical reason is concerned with Will, which is a kind of causality proper to living creatures as far as these are reasonable, and the capacity of this particular kind of causality to operate independently from *alien* causes that might determine it is called *freedom*. Similarly, *natural necessity* is the propensity of all creatures that lack reason to be induced to activity under the very influence of such alien causes. *Both* kinds of causality have to do with laws. For freedom is not lawless; the point is only that natural necessity implies *heteronomy*, whereas freedom of will implies *autonomy*, that is, the faculty of Will to be a law to itself.[308] The concept of a reasonable creature presupposes that of a free will, for it cannot act but under the idea of freedom, even if the latter were not demonstrable theoretically.

Now this seems to be circular, and Kant indeed admitted as much. "We conceive of ourselves as free in the order of acting causes, so as to be able to think of ourselves as subject to moral laws in the order of ends; and afterwards we think of ourselves as subject to these laws because we have attributed to ourselves freedom of will, for both freedom and will's own legislation represent autonomy, hence, they are synonymous."†[309] So we arrive here once again at the distinction made previously between the 'phenomenal' and the 'noumenal' worlds, which now appear as the world of conditioning causes and the world of ends, respectively. Man belongs to both worlds. If he belonged only to the latter, all his actions would

---

* In the present book this famous (and controversial) Kantian actor has not previously made his appearance. But, as the reader knows, this new class of judgements has been introduced by Kant as fundamental for the domains of pure mathematics and pure natural science (the latter, too, was believed by him to exist) because they combine the indubitable certainty of analytical judgements (which are always and by definition *a priori*, their certainty being based on the fact that the predicate is already included in the subject, i.e., on their being tautological — an expression not used by Kant) with the informational relevance of synthetical judgements (which, however, in their usual — and, as was held by all philosophers before Kant, only — form are *a posteriori*, since they are derived from experience and are thus always liable to 'falsification', to use Popper's term).

† "Wir nehmen uns in der Ordnung der wirkenden Ursachen als frei an, um uns in der Ordnung der Zwecke unter sittlichen Gesetzen zu denken, und wir denken uns nachher als diesen Gesetzen unterworfen, weil wir uns die Freiheit des Willens beigelegt haben; denn Freiheit und eigene Gesetzgebung des Willens sind beides Autonomie, mithin Wechselbegriffe."

automatically be in accordance with the principle of action of pure Will; if only to the former, all his actions would be determined by desires and inclinations, i.e., by heteronomy. The Categorical Imperative is the expression of man's being a citizen of both worlds; paradoxically, therefore, it is in the word 'ought' (*Sollen*) that human freedom reveals itself.

As noted already, in the *Critique of Practical Reason* the final section of the *Groundwork* is presented in a more elaborate way. Especially important here is the second section of the third chapter of the first book, entitled 'Critical illumination of the analytic of pure practical reason' ('Kritische Beleuchtung der Analytik der reinen praktischen Vernunft'). We read there that the concept of causality, considered as *natural necessity*, and in contrast with the same concept considered as *freedom*, has to do only with the existence of things as far as such existence can be *determined in time*; in other words, it has to do with appearances, not with things-in-themselves. Now it is evident that causality would be incompatible with, nay, would flatly contradict, any notion of freedom if (as is usually done) the determining factors for the existence of things in time are taken to be determining factors for the things in themselves. For causality implies that any event (hence, any action, too) which takes place in time takes place *necessarily*, since it is determined by that which took place previously. Now since the past is forever beyond my power, it follows that any action of mine is necessarily determined by factors beyond my power; in other words, I can never be free at the moment of my acting. And since the law of causality is inevitably valid for all things as far as their existence is determinable in time, it follows that freedom ought to be rejected as a meaningless and impossible concept if the existence of things as things-in-themselves were to be conceived in the same way. Thus, if one wishes to save freedom, there is no other way out than to attribute the existence of a thing as far as it is determinable in time — and, consequently, to attribute causality-considered-as-natural-necessity — *to the appearance only*, but freedom to the same entity as *thing-in-itself*.

This distinction is inevitable if one wants to maintain the two notions of causality and freedom, which would otherwise be mutually exclusive.[*310] It makes no difference whatsoever whether the determining factors of causality-as-natural-necessity rest inside or outside the human subject, nor whether (if they rest inside) they are of an instinct-like nature or rather result from rational considerations; for such considerations would owe their existence to *a preceding situation* in any case, and this in its turn to a situation preceding this situation, and so on. Causality,

---

* Kant's refusal to allow any difference between nature as appearing as an object of the inner sense only and nature as an object of the outer senses as well, beautifully illustrates the gap that separates his way of thinking from Schopenhauer's, to whom the former (because of its appearing in time only, rather than in both time and space) appeared as fundamentally different from the latter, so different in fact as actually to provide the hidden clue to the Thing-in-itself! More on this in our Chapter VIII, section a.

no matter whether it is of a psychological or of a mechanical nature, is about determining factors of the past, which therefore, at the moment when the human subject has to take action, are *beyond his power*. If one absolutely wants to, one may speak of psychological freedom in those cases in which we are concerned with a purely inner series of ideas of the soul, but that does not detract from their natural necessity. Natural necessity, however, is precisely that which leaves no room for *transcendental freedom*. For the latter must be conceived as independent of all that is empirical, hence, as independent of nature as such, no matter whether the latter appears as an object of the inner sense only (i.e., in time only) or of the outer senses as well (i.e., both in time and space). Without such a transcendental freedom no moral law nor any human responsibility on account of it would be possible.[311]

In order to resolve the apparent contradiction between natural mechanisms and freedom, in one and the same action, one must keep in mind that only the acting subject *as phenomenon* is subject to natural necessity, insofar as each of its actions is determined by factors that belong to the past and are therefore beyond its power. But the same subject *as noumenon* always considers itself as determinable only by the laws which it gives to itself through Reason, as is shown by the verdicts of that remarkable organ in us that we call 'conscience' (*Gewissen*). However much a man may try to depict an unlawful act he has committed as something towards which he has been driven by the stream of natural necessity, and thus to clear himself of any guilt, he will find that this plea held, however eloquently, by the lawyer inside him is unable to silence the plaintiff inside him, at least for as long as he feels that at the moment when he committed his injustice he was in full possession of his faculties and therefore of his freedom. Here, too, is to be found the root of the repentance of a deed committed long ago as soon as one thinks of it again; a painful sentiment, which is senseless insofar as it cannot undo what happened, but which is justified because Reason is indifferent to the flow of time when the moral law of our intelligible existence is at stake. Reason only asks whether what happened was a deed of mine, going on without fail morally to connect it with the very same sentiment, no matter whether the deed was committed in the present or long ago.[312] One may therefore concede that a man's future behaviour could be predicted with the same exactness as a solar or lunar eclipse if only we had an insight into even his most secret motives — and yet maintain that man is free. For if we were capable of intellectual perception (*intellektuelle Anschauung*) of the same human being (as a matter of fact we lack this property, for which rational understanding is given to us by way of compensation), we would perceive the whole chain of his behaviour as dependent on the spontaneity of the subject as thing-in-itself; and of such a determination no physical explanation could possibly be given. But as we lack this kind of perception,* it is the moral law which assures us of the difference between the

---

* Recall here Schopenhauer's comment on this alleged faculty (Chapter II, footnote on p. 43, and Chapter III, p. 65).

relationship of our empirical actions to our empirical subject on the one hand, and the relationship of this empirical subject to its intelligible substrate, on the other.[313]

It is this doctrine of the difference between the empirical and the intelligible subject that, in tandem with the 'Transcendental Aesthetics', was hailed by Schopenhauer as "the two great diamonds in the crown of Kant's fame that will never fade away".*[314] Such a eulogy, however, prevented him no more from criticizing severely Kant's moral philosophy than his eulogy of the 'Transcendental Aesthetics' with its doctrine of the ideal nature of space and time prevented him from objecting to many another aspect of Kant's epistemology, as we have seen in Chapter IV. We turn now to Schopenhauer's critique and to how it was connected to Schopenhauer's own moral philosophy.

c. Schopenhauer's moral philosophy

Although the core of Schopenhauer's objections against Kant's moral philosophy may of course be found already in his 'Critique of Kantian Philosophy', a much more thorough-going treatment is offered in the essay 'On the Foundation of Morals' ('Über die Grundlage der Moral'). Written in 1840 for a prize competition organized by the Royal Danish Society, Schopenhauer used the occasion to unfold his own ethical doctrine. This fascinating essay was published in Germany together with an other essay, called 'On the Freedom of Will' ('Über die Freiheit des Willens'), under the covering title '*The two fundamental problems of ethics*' (*Die beiden Grundprobleme der Ethik)*. This other essay, too, had been written for a prize competition, one set up by the Royal Norwegian Society of Sciences in 1839, and, as Schopenhauer proudly mentioned on the title-page, it brought him the prize. Not so 'On the Foundation of Morals', which was *not* awarded the prize, as Schopenhauer equally explicitly mentioned on the title-page (probably with the unspoken addition '*credite lectores*!').† Although chronologically, therefore, 'On the Freedom of Will' comes first, 'On the Foundation of Morals' is the treatise which contains Schopenhauer's systematic effort to come to grips with Kantian ethics; we shall begin, therefore, with a brief exposition of the arguments unfolded in this essay.

Schopenhauer recognizes as Kant's greatest merit his purging of all *eudaemonism* from ethics. Plato alone excepted, all previous thinkers, whether ancient or modern, had treated virtue only as a means to an end.[315] But there are also grave defects in Kant's ethics. The first defect is the *imperative* form given to it. Kant had said: "In a practical philosophy the purpose is not to demonstrate reasons for what happens, but laws for what should happen, even if it never happens". On the contrary, Schopenhauer retorted. The moral philosopher (and the philosopher in

---

* "Sie, nebst der transcendentalen Aesthetik, sind die zwei großen Diamanten in der Krone des Kantischen Ruhmes, der nie verhallen wird."

† For the background to all this cf. Chapter V, footnote on p. 125.

general) must content himself with the explanation and interpretation of what is given, i.e., of what really is or really happens, going on to seek to understand it.[316] Kant makes an unwarranted usage of the concept of 'law'. Originally this was confined to civil law (a human institution and, as such, arbitrary); only later was the concept metaphorically applied to nature. Insofar as man belongs to nature, there is indeed a law for human will, too, namely, the law of motivation. This can be demonstrated in reality, since it is nothing but one specific manifestation of the general law of causality. But the existence of moral laws may not *a priori* be taken for granted; in doing so, Kant commits a *petitio principii.*[317] Nay, more: by speaking of an *absolute* 'ought' (*Sollen*) and of an *unconditioned* duty, he saddles the reader with a *contradictio in adiecto*, for all *Sollen* is necessarily conditioned by punishment or reward, hence, it is inevitably *hypothetical*, never *categorical*. Later, in his *Critique of Practical Reason*, Kant had unvoluntarily acknowledged as much by postulating a condition after all, or even two conditions — a reward and a rewarder. True, the postulated reward — the immortality of the rewarded — is presented under the name of the 'Highest Good', which is the union of virtue and happiness. But this should not deter us from noticing that something else is actually going on here, in that eudaemonism is covertly being reintroduced.[318] Moreover, the archaic form 'Thou shalt' (*Du sollt*) used by Kant already betrays the origin of the imperative form. This is the Mosaic Decalogue, which deserves no place of its own in philosophical ethics.

Kant's fundamental error, according to Schopenhauer, is that, upon very meritoriously separating the *a priori* element in human knowledge from the *a posteriori* element, he was so impressed with his success that he introduced it in ethics, too, where it is out of place.[319] Consequently, the principle of morals must for Kant be a synthetic judgement *a priori* of a merely formal nature. Moreover, he thinks it must be valid, not for man only, but for 'all possible reasonable beings', which is like talking of 'heavy beings apart from bodies', as reason is known to us only as a human faculty. "One suspects that Kant perhaps thought here of the dear little angels," Schopenhauer mischievously suggests.[*320] The absurdity of applying the concept of knowledge *a priori*, which legitimately belongs to theoretical reason only, to what Kant calls 'practical reason' as well, is, to Schopenhauer, demonstrated already by the fact that in theoretical philosophy, according to Kant himself, apriority is confined to the *phenomenon*, which should logically, therefore, be the case as well in moral philosophy. But actually the moral thing, the Will, is considered by Kant (and, if taken by itself, quite rightly so!) as

* "Man kann sich des Verdachts nicht erwehren, daß Kant dabey ein wenig an die lieben Engelein gedacht." Cassirer, in his essay on Heidegger's interpretation of Kant, remarks at this point that Kant was of course not thinking here of the "dear little angels" and that it was only in order to avoid an anthropological interpretation that Kant spoke of "all possible reasonable beings"; for as a transcendental philosopher he was anxious neatly to separate the tasks of the disciplines of *ethics* and of *anthropology*. This is no doubt correct; but Schopenhauer has a point in so far as the picture that Kant gives of "reasonable beings" has indeed angelic rather than human traits (cf. our comment in Section b. of the next chapter).

intrinsically related to the Thing-in-itself — nay, as the very way in which this mysterious thing presents itself in us.[321] The consequences of this confusion become clear when Kant even goes so far as to say that the mere concept of duty is at the same time the reason for its fulfillment, since an action has moral value only if it is done purely out of duty without any inclination towards it (of course Schopenhauer uses the occasion to refer here to Schiller's epigram, without literally quoting it). To Schopenhauer the assertion that the value of a character begins only in the situation where somebody without sympathy of heart, cold and indifferent towards other people's sufferings and in fact not born to be a friend of mankind, nonetheless renders benefits to mankind, but only from acid duty, is revolting to any really moral sentiment; it is an "apotheosis of lovelessness" ("Apotheose der Lieblosigkeit"), and as such diametrically opposed to Christian ethics, which places love above everything and admits nothing without it.

Schopenhauer goes on to quote the *Critique of Practical Reason*: "The disposition with which man must follow moral law is such as to follow it out of *duty*, not out of *voluntary inclination*, not even out of some *uncommanded* striving, spontaneously arisen and gladly undertaken."[*322] And now his scorn hardly knows bounds: so, a *command* is required; what slave morals! Indeed, when the cold-hearted man called to the stage by Kant does good to other people, this can really be out of *fear* only (unless he has additional objectives), argues Schopenhauer.[323] But, he goes on, the point is that, since Kant despised all empirical motives of Will, indeed no other material for this law of his has been left to him but its own form. This form is 'law-likeness'; that is, its essence is that it be valid for all; consequently, the material content of the law contains nothing but this general validity itself. Hence Kant's formulation: "Act only according to the maxim, whereby you can wish at the same time that it should become a general law for all reasonable beings."

This, then, is the foundation of Kant's entire ethics. It should be objected against it that it cannot possibly happen to a man spontaneously to arrive at the idea of seeking a law to which his will must submit and ply itself. Such an event cannot begin to happen until another powerful moral incentive has given rise to it in the first place.[324] "For morality is concerned with man's *real* actions, not with aprioristic sand castles, for the outcomes of which nobody would care a bit amidst the seriousness and the constraints of life, and the effect of which upon the tempest of the passions would consequently be like that of a purgative syringe upon a conflagration," says Schopenhauer in his characteristically plastic way.[†325] Pure,

---

[*] "Die Gesinnung, die dem Menschen, das moralische Gesetz zu befolgen, obliegt, ist, es aus *Pflicht*, nicht aus *freiwilliger Zuneigung* und allenfalls *unbefohlener*, von selbst gern unternommener Bestrebung zu befolgen."

[†] "Denn die Moral hat es mit dem *wirklichen* Handeln des Menschen und nicht mit apriorischen Kartenhäusern zu thun, an dessen Ergebnisse sich im Ernst und Drange des Lebens kein Mensch kehren würde, deren Wirkung daher, dem Sturm der Leidenschaften gegenüber, so viel seyn würde, wie die einer Klystierspritze bei einer Feuersbrunst."

abstract concepts *a priori* can never motivate people ("Concerning any other reasonable beings I cannot join in the conversation", Schopenhauer ironically adds).*[326] Thus the second fault of the Kantian foundation of morality is its lack of real content. This is not to be wondered at, since Kant seeks the foundation of ethics in *reason*, whereas in fact a man may very well act in a well-considered, thoughtful and methodical way (i.e., be a reasonable man), yet follow the most self-centred, unjust or even ruthless maxims; reasonableness' and 'viciousness' are therefore quite compatible, at least as long as one sticks to a linguistic usage common to all nations.[327]

Having thus criticized the *foundations* of Kantian ethics, Schopenhauer now turns to its supreme *principle*. This, he says, requires another look at its formulation: "Act only according to the maxim whereby you can wish at the same time that it be valid as a general law for all reasonable beings." This implies tacitly that, because the law for my *actions* has been made general, it also becomes the law for my *passivity*: I cannot want injustice and lovelessness as such; yet if I could be confident that I would always be the active part, injustice and lovelessness might very well be the general maxim. Thus Kant's categorical imperative actually reveals itself to be a hypothetical imperative, as indeed any imperative, according to Schopenhauer, of necessity must be.[328] It tacitly presupposes the calculation that, if I act in accordance with it, it is likely to be eventually reciprocated. This might be of use as a foundation for the principle of political order, but not for the principle of morals, Schopenhauer thinks.†[329]

Now what about Kant's alternative formulation, which says that Man — and in general any reasonable being — exists *as an end in itself* and represents an *absolute value*? Both italicized expressions can be seen to contain a *contradictio in adiecto*. 'Being an end' means 'being willed', so that any end is an end only in relation to a will, of which it is the end, i.e., the direct motive. Furthermore, a 'value' is nothing but the appreciation of one thing *compared* with another, hence it is *eo ipso* relative. To speak of an 'end in itself' is as senseless therefore as it is to speak of a 'friend in itself' or an 'uncle in itself'; and the notion of man as an 'absolute value' leads only to high-sounding, yet empty phrases about 'human dignity'. Everything considered, Kant's second formulation is just a clumsy expression of the Golden Rule: *Quod tibi fieri non vis, alteri ne feceris*. Moreover, it must be objected that a criminal about to be executed is very well — and rightly so — used only as a means, namely, as a means of deterrence.[330] But the principal objection is that Kant's formulation insults authentic moral sentiment in that it lowers unreasonable beings, namely, animals, to the status of mere things, that is, of *means* that are not at the same time *ends*. True, in another book Kant rejects cruelty to animals; but only because acting thus violates the duty of man against *himself*.[331]

* "Von andern vernünftigen Wesen kann ich nicht mitreden."

† "Zur Begründung des Princips des Staatsvereins wäre das tauglich, aber nicht zu der des Moralprinzips."

Against all this Schopenhauer posits his own maxim: *Neminem laede, imo omnes, quantum potes, juva* ("do not harm anybody, but rather help everybody as much as you can").[332] Indeed, Schopenhauer concedes, he, too, seems thus to arrive at an imperative. But he warns that in ethics, as in all sciences, one should distinguish carefully between 'ο τι *(ho ti)* and διοτι *(dioti)*, i.e., between the *principle* and the *foundation* of ethics.[333] In Schopenhauer's view, Kant has made a mess of both. Schopenhauer's 'imperative' claims no more than to be the most concise expression of that conduct to which we attach moral value. But the really big problem is to find out what this prescription is based upon, that is, to reveal what gives it its strength. For from the other side egoism loudly screams *neminem juva, imo omnes, si forte conducit, laede* ("do not help anybody, but rather harm everybody if it happens to suit you"); to this, malice even adds the variation *imo omnes, quantum potes, laede* ("but rather harm everybody as much as you can"). Against *these* two maxims to posit a warrior just as strong or even superior — that is the problem of all ethics. *Heic Rhodus, heic salta*![334]

Well, then, any inquiry into the foundation of morals must take as its starting-point man's *real* conduct and the motives by which it is inspired. Now, in observing this conduct, we see that man, in the overwhelming majority of cases, acts out of egoism; yet sometimes (alas, all too rarely!) we meet people who act out of voluntary justice and unselfish humaneness. To such conduct we spontaneously attach moral value; and thus we have found out the true motives of all actions of genuine moral value. No action can take place without a sufficient motive; now what moves the will is only weal and woe in the widest sense. As long as the weal and woe of the actor himself are concerned there can be no question of moral value. An action can have moral value (or, for that matter, moral reprehensibility) only in relation to other people; it can then be an action committed out of justice or humaneness, or out of its opposite.[335]

The question therefore is: how can the weal or woe of other people motivate my own will? Obviously, because in a sense I identify myself with another person. This is what we are wont to call *compassion*, which therefore forms the true and only foundation of morals.[336] Generally speaking, there are only three basic motives for human actions: egoism, which aims at one's own weal; malice (*Bosheit*), which aims at another's woe; and compassion (*Mitleid*), which aims at another's weal.[337] The latter can take two forms, one negative, which is called 'justice' (*Gerechtigkeit*) and lies behind the *neminem laede*, and one positive, which is called "humaneness" (*Menschenliebe*) and lies behind the *omnes quantum potes juva*. The latter, however, should be applied to other living creatures as well![338]

In the end, compassion is rooted in metaphysics, that is, in a profound awareness of the ultimate lack of reality of the *principium individuationis*, which awareness in Indian wisdom goes by the expression *tat tvam asi* (i.e., 'thou art that' — 'that' meaning all kinds of being external to yourself).[339] In European thought, Kant's doctrine of the ideal nature of space and time has provided the theoretical framework for this insight. This is one of Kant's two greatest merits, as ought in justice to be stressed after so much criticism![340] His other principal merit comes to the fore when we ask ourselves 'Why do people really act in such very different ways?' Obviously, because they are different in character. And here Schopenhauer introduces Kant's distinction between the empirical and the intelligible character, which is where Schopenhauer held Kant's "greatest and most brilliant merit as regards ethics" resided.[341] At this point the problem of freedom comes in, and this, we recall, was made the principal subject of Schopenhauer's twin essay in ethics, 'On the Freedom of Will', which we begin to examine now.

According to Schopenhauer, freedom in the strict sense of the word is a *negative* concept, since we mean by it only the *absence* of inhibiting factors, which are themselves positive in that they appear as *forces*. Depending on the specific nature of the inhibiting force in question, freedom comes in three distinct varieties, as physical, intellectual, or moral freedom.[342] Passing by Schopenhauer's treatment of the first two, we concentrate on the variety that matters most to us here, namely, moral freedom. For here the concept of freedom is not related to the question of what we *can* do, but of what we *want* to do; and then the problem arises as to whether our will *itself* is free, that is to say, unrestrained by any necessity.[343] Such a free will would be a will not subject to the principle of sufficient reason; speaking concretely, this means that a man in possession of such a will would be able, under certain external circumstances, to commit either of two diametrically opposed actions.[344]

Here we must distinguish carefully between *wish* (*Wunsch*), i.e., the act of volition in a state of becoming, and *decision* (*Entschluß*), i.e., the act of volition in a state of readiness; whether the latter state has been reached, can only be demonstrated by the resulting *deed*. The philosophically uneducated believe that in a given case two opposed volitional acts are possible, thus conflating 'wishing something' and 'willing something'. One can indeed *wish* two opposite things at a time, but one can *will* only one of them; and which of the two this is becomes apparent only from the deed. Now the question is why, of two opposing wishes, the one grows into a volitional act and into a deed, and the other does not. Consciousness cannot by itself answer this question, since it does not know the result *a priori*, but only experiences it *a posteriori*.[345] Let us turn therefore, not indeed to our consciousness of ourselves, but to our consciousness of other things. This faculty certainly suffers from the drawback that the Will itself cannot be an object of its direct perception (this it is only for consciousness of ourselves, which, however, we have just found lacks the competence to answer our question); only willing *beings* can be that. But this liability is outweighed by the asset which resides in our having at our service a much more perfect organ — Understanding (*Verstand*).[346]

The most general and essential form of Understanding is the law of causality, since it is only by means of this that our perception of the outer world comes into being in the first place, which proves that it is *a priori* valid, being the necessary condition of all possibility of experience (cf. Chapter III, p. 66, and Chapter IV,

p. 94). Now the objects of experience can be classified into inorganic and organic ones, and the latter in their turn into plants and animals (man being the most perfect representative of the latter category). Consequently, the law of causality, too, has three corresponding modifications. These are: (1) *cause* in the narrowest sense of the word, by virtue of which all mechanical, physical, and chemical changes of the objects of experience take place; (2) *stimulus*, by virtue of which all changes of the organisms *as such* take place; and (3) *motivation*, that is, causality passing through the medium of *knowledge*. The latter appears on the stage at the very moment at which that more complicated being, the animal (which, owing to its relative complexity, has more differentiated needs), is no longer able to satisfy these needs by simply expecting the stimulus (as plants do) but must be able to choose, to take hold of, or even to go in search of the means to satisfy these. Therefore it needs an intellect (in an infinite range of degrees of perfection), which presents itself materially as the nervous system and the brain; mere receptivity for, and reaction to, stimuli is replaced by a receptivity for, and a response to, motives, i.e., by consciousness and spontaneous movement.[347] Human consciousness (called 'Reason') differs from animal consciousness in that it is capable not only of perceptual knowledge of the external world but also of deducing abstract, general concepts (*notiones universales*) from such knowledge; these are designated by means of words. Man therefore has a much wider horizon than the animal. He is able to call to mind his motives and the influence these exert on his will, and he can do so in arbitrary order, alternately, and repeatedly.

This, then, is *relative freedom*. Superficial people take this to be that freedom of will which, they say, evidently distinguishes man from animal. But the difference is only one of degree. Just as with perception, thought, too, becomes a *motive* as soon as it acts upon the will; but motives are causes and all causality implies necessity. Consequently, only a very superficial mind can take this relative, comparative freedom to be absolute, to be a *liberum arbitrium indifferentiae.* For the ability to deliberate, which is its fruit, actually provides us with nothing else but an often painful *conflict of motives*, the battlefield of which constitutes the whole human mind. "For it enables the motives repeatedly to test one another's strength using its will, thus putting it in the same situation as a body upon which different forces act in opposite directions, until in the end the decidedly strongest motive gets the upper hand over the others and determines the will — a result called decision, and appearing as the outcome of the struggle with complete *necessity*."[*348]

---

[*] "Er läßt nämlich die Motive wiederholt ihre Kraft gegen einander an seinem Willen versuchen, wodurch dieser in die selbe Lage geräth, in der ein Körper ist, auf welchen verschiedene Kräfte in entgegengesetzten Richtungen wirken, — bis zuletzt das entschieden stärkste Motiv die andern aus dem Felde schlägt und den Willen bestimmt; welcher Ausgang Entschluß heißt und als Resultat des Kampfes mit völliger *Nothwendigkeit* eintritt."

It is true, Schopenhauer admits, that on the scale of different forms of causality we perceive an increasing heterogeneity between cause and effect, such that the connection between them loses more and more of its originally immediate intelligibility. This heterogeneity reaches its peak in motivation; yet, just as with stimulus, the relationship remains a causal one and therefore it has lost nothing of its necessity.

We must remember, though, what a cause is in the first place. It is the preceding change which makes the succeeding one necessary. No cause in the world produces its effect out of nothing; it only produces, at a certain time, at a certain place, and in a certain being, a change which is always in conformity with its peculiar nature, and the potentiality for which must already be present, therefore, in this particular being. Hence, we see in physics and in chemistry that natural forces are presupposed everywhere, so that every explanation (meaning, to be sure, explanation from causality) consists of a reduction to such forces. Forces of nature are not subjects of explanation, but rather the principles of all explanation; nor are they subject to causality, as a force of nature is the very thing which furnishes any cause with the capacity to act. The explanations of celestial mechanics presuppose gravitation; the explanations of chemistry presuppose the secret forces that appear as affinities; all explanations of physiology presuppose the life-force, and so on. The same is true for the kind of causes we call motives; these, too, presuppose a natural force, the name of which is Will. This force, however, is known to us, not only from without (as is the case with the other natural forces), but also (by virtue of our consciousness of self) from the inside, i.e., directly. The very special, individually conditioned nature of the Will, by virtue of which the way it responds to the *same* motives is *different* for each individual human being, constitutes what is called one or another human being's *character* or, to be more exact, his *empirical character*, in that it is known, not *a priori*, but from experience.[349]

In animals the empirical character is different in each species, even though, as every owner of a cat (not Schopenhauer's example) will readily grant, the highest and most intelligent animals already show a remarkably individual character, however much dominated in the end always by species character. In man it is different for each individual. That is to say, human character is *individual*; moreover, it is *empirical*. This is why no one can know in advance how another person — or, for that matter, how he himself — will act in a given situation. Only after having stood the test one can be certain of the other as well as of oneself, but *then* one is indeed certain: one can always rely upon proven friends and well-tried servants.[350] Furthermore, empirical character is *constant*, that is, it remains the same throughout one's whole life. Finally, it is *innate*, for it is not a work of art or of circumstances and therefore subject to chance; rather, it is a work of Nature itself. It follows from all this that freedom of will, such that it were to allow the possibility of opposite acts for every man in every situation, is wholly inconceivable. This is so because in that case his character ought to be a *tabula rasa* (such as Locke held the intellect to be) without any innate propensity in this or any other direction, as such a propensity would disturb the perfect equilibrium presupposed in the *libero arbitrio indifferentiae*.[351]

Still, this is not Schopenhauer's last word on the matter. For in the final pages of the essay, when the sympathetic reader has long been wholly convinced of the hopelessness of freedom's cause, there appears, like the rabbit out of the magician's hat, Kant's distinction between the empirical and the intelligible character. Schopenhauer points here to "a fact of consciousness which I have left unremarked so far in order not to disturb the course of my investigation. This is the perfectly clear and secure sense of *responsibility* for what we do; of *accountability* for our actions, which rest on the unshakeable certainty that we are ourselves the doers of our deeds."*[352] Everybody feels that, however *subjectively* necessary his actions are, nevertheless, under the same circumstances (i.e., acting from the same conditioning motives) *objectively* a quite different or even opposite action would have been possible — had he only been an other man than he is! Since, however, he is not, and since he has the character he has, no other action was possible *for him*; the responsibility of which he is aware concerns therefore (everything considered) not the action, but his character — not in the ephemeral, offensive act, but in the permanent qualities of the offender, that is, in his character, rests his guilt, and, consequently, his responsibility. And since the latter is the only datum from which one may infer moral freedom, it follows that freedom must reside there, too. The problem is, however, that we have seen this character to be innate and immutable.

Here, then, Kant comes in once again, to wit, with his distinction between empirical and intelligible character, "which is among the finest and most profound thoughts ever produced by this great mind, nay, by humanity as such."†[353] It is from here only that we can begin to understand — insofar as this is at all possible for human capabilities — how the strict necessity of our actions may nonetheless go together with that particular freedom to which our sense of responsibility bears testimony. The relationship between the empirical character and the intelligible character, Schopenhauer goes on, rests in a straightforward manner upon the very foundation stone of Kant's entire philosophy, which is his distinction between phenomenon and Thing-in-itself. Just as, to him, the perfect *empirical reality* of the world of experience goes together with its *transcendental ideality* (cf. Chapter IV, p. 88), just so the strictly *empirical necessity* of acting goes together with its *transcendental freedom*. Man's empirical character, in that it is an object of experience, is bound to time, space, and causality, and is subject to their laws. But

---

* "Es giebt nämlich noch eine Thatsache des Bewußtseyns, von welcher ich bisher, um den Gang der Untersuchung nicht zu stören, gänzlich abgesehen habe. Diese ist das völlig deutliche und sichere Gefühl der *Verantwortlichkeit* für Das was wir thun, der *Zurechnungsfähigkeit* für unsere Handlungen, beruhend auf der unerschütterlichen Gewißheit, daß wit selbst *die Thäter unserer* Thaten sind."

† " ... welche zum Schönsten und Tiefgedachtesten gehört, was dieser große Geist, ja, was Menschen jemals hervorgebracht haben."

his intelligible character, that is, his Will as Thing-in-itself, is absolutely free, that is to say, it is independent of the law of causality, which is a mere form of phenomena.

So we must seek the work of our freedom, no longer in our separate actions, but in the whole being and essence (*existentia et essentia*) of man, which must be conceived as his free deed. Only, to the faculty of knowledge (bound as it is to time, space, and causality) this free deed necessarily presents itself as a plurality and variety of actions — actions which, precisely because of the original unity of what stands expressed in them, must all be of one and the same nature and therefore appear as strictly necessitated by those present motives which provoke them and condition them. Thus for the world of experience the principle of *operari sequitur esse* ('action follows being') is valid without exception. Hence, that *freedom* which is not to be found in the *Operari* cannot but reside in the *Esse*. The basic error committed by all previous thinkers on the subject resided in this: necessity was attributed to the *Esse* and freedom to the *Operari*, whereas it is really the other way round, in that *from what we do we recognize what we are.*[354]

d. Some incidental reflections on the present-day significance of Kant's and Schopenhauer's moral philosophies

Julien Benda in 1927 prefixed his well-known, polemical book *La trahison des clercs* ('*The treason of the intellectuals*') with a remark by the 19th-century French philosopher Renouvier to the effect that "the world suffers from a lack of faith in transcendent truth" ("Le monde souffre du manque de foi en une vérité transcendante").[355] The patient reader, who has followed our exposition of Kant's and of Schopenhauer's respective moral philosophies, will perhaps with some nostalgia observe likewise that, whatever the differences between the two thinkers, both men still firmly believed in universally valid, eternal moral principles that in one way or another can be known by man. To be sure, Kant's distinction between the domains of theoretical and practical reason, i.e., between man as knower of the phenomenal world and man as moral agent, might have conveyed the impression that we are concerned here with the famous distinction between 'fact' and 'value' so dear to neo-positivist thinkers of every kind. However, neo-positivism has been inspired here, not so much by Kant himself, but rather by the man who had 'awakened [him] from his dogmatic slumbers' and who had explicitly stated for the first time that it is a logical mistake to infer what *ought to be* from statements that tell us only what *is*.[356] To Hume this implied that reason, since it cannot distinguish between good and evil, "is and ought to be the slave of the passions".[357] And here, too, Hume's scepticism about reason served Kant as a challenge to be overcome without relapsing into dogmatism. "Surely, world, science and morals must depend upon some deeper and more *necessary* structure. Kant's portrayal of a radically divided self contrasts with Hume's tolerant compromise and his worldly calm. For Hume similar habits and movements of imagination set in order both the factual and the moral. For Kant the clarified justified account of scientific and factual knowledge has the intended effect of displaying as *totally different* the free activity of mind in moral choice and discernment. Kant's man as knower of the phenomenal world (exercising theoretical reason) is to be distinguished from his man as moral agent (exercising practical reason)," the late Iris Murdoch wrote in her study on *Metaphysics as a Guide to Morals* (which is quite as profound as it is brilliant, and in which not only Kant but — *mirabile dictu* — also Schopenhauer is fully given his due).[358]

According to Murdoch, this difference between Hume and Kant accounts for the fact that for Kant (as for Wittgenstein still) the distinction between 'facts' and 'values' was meant to *segregate* value so as to keep it pure and untainted, neither derived from nor mixed up with empirical facts. By contrast, with thinkers inspired rather by Hume the very same move resulted in a reduced or even perfunctory account of morality, leading in its turn (with the increasing prestige of science) to a marginalization of the 'ethical' ("big world of facts, little peripheral area of value"). It seems indeed hardly an exaggeration to say that, to representatives of the Vienna Circle (like Carnap) or of British analytical philosophy (like Ayer and Russell), the difference between factual world and morality turns into one between 'meaningful' and 'meaningless' propositions. Thus to them the latter category finds itself reduced to a personal affair (a matter of taste, so to speak) not susceptible to strict philosophical treatment.*

But one is inclined to say that the ethical consequences of this anti-metaphysical tradition (with its radical *separation* between the world of fact and the world of value) seem relatively harmless compared to those of the post-Kantian metaphysical tradition (with its complete *merger* of both worlds). 'Post-Kantian', certainly; but let it be noted that, in Schopenhauer's view, Kant willy-nilly paved the way for the deification of Reason by German Idealism, in that he proclaimed Reason — in the guise of 'practical' Reason — to be the source of all virtue while at the same time proclaiming Theoretical Reason to be a faculty directed towards the 'Unconditioned'. Schopenhauer expressed the point thus: "That is how nothing but Kant's mistakenly assuming ... a practical reason caused doctrines to arise out of stern and sober, critical philosophy which are quite heterogeneous to it ... in that any dreamer could now make his figments pass for the 'absolute' pronouncements and revelations of those doctrines. Here, then, resides the origin of that philosophical method which came up right after Kant and which seeks to mystify, to impress, to deceive, to throw dust in one's eyes, to puff up."†[359] And this much

---

* To avoid a possible misunderstanding: a man like Russell who, over his entire life-time, was passionately engaged in (among other things) pacifist movements (whether he showed here much political sense is another question), of course never denied the profound importance of moral problems. But to him, in the last resort, these had their origin in our emotional rather than in our intellectual propensities.

† "So hatte jene einzige von Kant gemachte falsche Annahme einer ... praktischen Vernunft zur Folge daß aus der strengen, nüchternen kritischen Philosophie die ihr heterogensten Lehren entsprangen ... für deren 'absolute' Aussprüche und Offenbarungen jetzt jeder Phantast seine Träumereien ausgeben konnte ... Hier also liegt der Ursprung jener

must be granted to Schopenhauer: without Kantianism, no Hegelianism, insofar as Kant's lack of clarity over the 'Thing-in-itself' * undoubtedly paved the way for the attempt undertaken by German Idealism to get rid of so seemingly awkward a concept by abolishing the difference of subject and object by means of an all-absorbing universal Self, i.e., by hypostasizing Kant's 'transcendental subject'.

That this is indeed what happened, has already been suggested here several times; but our very choice of the word 'hypostasizing' is quite deliberately meant to make clear, once again, the fundamental difference between Kant and his German Idealist successors. For the point is that, whereas Kant's transcendental subject was nothing but the functional correlate of expressed (either true or false) judgements about the world, and was also static in nature (the fixed 'categories'), it was ontologized by Hegel into a dynamic 'Absolute Idea' that 'externalizes' itself into the world so as, in the process of history, to arrive at a full awareness of itself as 'Absolute Spirit'. Thus the hoary, scholastic metaphysics of Being had no sooner disappeared from the stage than a new metaphysics arose, which might be called neo-Heraclitean insofar as it was a metaphysics of Becoming dialectically conceived (albeit in a very peculiar sense, in that Heraclitus' ever-changing stream has now been adorned with a final purpose!). The dualism of fact and value has now been happily overcome; but, to quote Murdoch again: "Hegel joins fact to value, but thereby ultimately kills value and loses the individual. Value is, in the totality, overcome."[360]

It is indeed true, as we have suggested in the preceding chapter, that Hegelianism leaves no room for individual responsibility, nor, therefore, for a true ethics. At its best it leaves room for a *Rechtsphilosophie* (philosophy of law), and at its worst (in the case of the Marxist interpretation of Hegelianism) for a 'class-determined' ethics, i.e., for full ethical relativism.

Marxism, which allegedly 'put Hegel on his feet' by defining consciousness (Hegel's ultimate reality, that is) as class-consciousness, seems by now, after the downfall of communism in Eastern Europe, to have had its innings; but (if we look at the other end of the metaphysical spectrum) what about structuralism and its 'poststructuralist' and 'postmodernist' successors? Ferdinand de Saussure's theory that a sign derives its identity, neither from the object represented nor from the consciousness of the subject, but rather from the internal relationships within the language system, so that language constitutes an autonomous reality, was certainly an interesting scientific hypothesis. It may or may not have been fruitful in its proper domain; whether it was, we must leave to the experts. Yet in any case this is a far cry from (as it tended to become later on) a metaphysical vision of language as a huge, transcendent structure in which both the individual user of language and extra-linguistic reality find themselves submerged.

---

unmittelbar nach Kants Lehre auftretenden philosophischen Methode, die im Mystifizieren, Imponieren, Täuschen, Sand in die Augen streuen und Windbeuteln besteht."

* Cf. Chapter IV, p. 105.

In one particular sense it might be said that structuralism presents us with another version of objective idealism — 'idealism' insofar as language is to be considered as an expression of the structure of the human *mind*; and 'objective', in that structuralism does not take the consciousness of the *individual subject* for its point of departure. It is certainly true that structuralism (unlike the finalist Hegelian brand of objective idealism, and its 'materialist' Marxist counterpart) appears to lack totalitarian implications. It is more 'tolerant' — *too* tolerant, really, to judge by its cultural–anthropological variety. Lévi-Strauss, in his (by itself) praiseworthy effort to avoid the kind of ethnocentrism which (whether openly or tacitly) assumes Western man to be superior to so-called primitive man, seems to fall into the opposite (yet really no less ethnocentric) extreme of depicting cultures (not languages, in this case) as transcendent structures leaving no room for the autonomous individual. His view is therefore determinist; moreover, in that all 'symbolic systems' are held to be equal, this, too, leads of necessity to moral relativism. "Alongside its Hegelian–Marxist interpretation in which [history] appears as relentlessly directed towards an end, there exists an interpretation according to which it would lead nowhere: all civilizations would be equal in value. Modern atheism does not rest in the denial of God, but in the absolute indifferentism of *Tristes Tropiques* [the title of one of Lévi-Strauss' most famous works, PL]", writes Lévinas with much justice.[*361]

No doubt is possible that structuralism, like German Idealism, might with some justice claim to stand in the tradition of Kant, this time precisely *because of* Kant's purely functional conception of his 'transcendental subject', the *a priori* forms of which he deduced from the conditions for the possibility of being an *object*. Kant's transcendental subject is nothing but the formal condition of my thoughts and of their concatenation, not a 'Self'. And indeed, historically speaking our conceptual dominion over the external world — due in particular to the mathematization of reality — entails the reduction, not only of the world, but also of the subject. Once the world has grown abstract through being reduced to its mathematical properties, the subject grows abstract, too; as Cathérine Chalier rightly remarks, it becomes pure representational activity.[362] Only, for Kant there remained something else beyond the transcendental subject (which is indeed no more than *actus purus*) and the empirical subject (which is indeed, as Hume demonstrated, contingent, i.e., a psychological event, a pure phenomenon), this 'something else' being the 'intelligible character', that is, man as a morally responsible subject![†363] And it is this very belief for which no room has been left in much contemporary thought.

---

* "A coté de son interprétation hégélienne et marxiste où elle apparait dirigée inéluctablement vers un but, il existe une interprétation d'après laquelle elle n'irait nulle part: toutes les civilisations se vaudraient. L'athéisme moderne, ce n'est pas la négation de Dieu. C'est l'indifférentisme absolu des Tristes Tropiques."

† That is also what Kant calls "person" or "personality" (*Persönlichkeit*). In the *Critique of Pure Reason* the term appears in the third paralogism where it is still being used as really synonymous with the transcendental subject (the thesis of the "personality of the soul" being a "purely identical thesis of self-consciousness", since it means nothing more than

Still under the influence of structuralism, but already on the way towards its even more radical, 'poststructuralist' and 'postmodern' successors, Foucault proclaims 'Man' to be a recent invention of Western civilization from the 16th century onward and, as such, to be bound to disappear again, "like a face drawn in the sand at the border of the sea" ("comme à la limite de la mer une visage de sable").[364] Here Kant's 'transcendental subject' has not so much been hypostasized (as in German Idealism), but rather dissolved, together with its objects, into an array of successive *epistèmè*'s, each of which is a product of history, language, and society, in short, of the prevailing *discours* of a given period. Foucault's concern here is with the history of the human sciences; but the historian of natural science may easily detect a parallel with Kuhn's concept of successive 'paradigms'. True, Kuhn and Foucault always remained different in that the former at least tried to explain how the switch from one paradigm to another comes about, and was therefore more of a historian than his French counterpart, who, it must be admitted, preferred for his own enterprise the term 'archeology'. Still, emphasizing discontinuity is what unites both men. So does an inherent relativism, in that both were agreed that one can no longer meaningfully speak of any 'progress' — not even in the domain of science where (just as with technology) the concept of progress (so questionable, even dangerous in any other domain, as we shall have an opportunity to underscore in our final chapter) would at least seem in the right place in view of its essentially cumulative nature denied, for that very reason, by both Foucault and Kuhn.

From there, only one more step is needed to arrive, on the one hand, at Feyerabend's 'anarchist' epistemology ("anything goes"), or, on the other, at Derrida's 'deconstructionist' literary criticism, in which the very possibility of

---

that, "during the whole time that I am conscious of myself, I am conscious of this time as belonging to the unity of my Self"), but in the *Groundwork* it is defined as "reasonable beings that are ends in themselves": "Those beings, whose existence does not depend upon our will but upon nature, have nevertheless, if they are not reasonable beings, only a relative value, to wit as a means, because of which they are called *things*, whereas reasonable beings are called *persons*, since their nature already defines them as ends in themselves, i.e., as something which may not be used as means only, thus prohibiting any arbitrariness, and as an object of respect." ("Die Wesen, deren Dasein zwar nicht auf unserm Willen, sondern der Natur beruht, haben dennoch, wenn sie vernunftlose Wesen sind, nur einen relativen Werth, als Mittel, und heißen daher *Sachen*, dagegen vernünftige Wesen *Personen* genannt werden, weil ihre Natur sie schon als Zwecke an sich selbst, d.i. als etwas, das nicht bloß als Mittel gebraucht werden darf, auszeichnet, mithin so fern alle Willkür einschränkt (und ein Gegenstand der Achtung ist).)" In the Introduction to the *Metaphysics of Morals* Kant differentiates between "moral" and "psychological" personality; the latter meaning the faculty of being aware of one's identity in the different situations of one's existence (and which is therefore, as in the third paralogism, virtually synonymous with the transcendental subject), the former meaning the freedom of a reasonable being under moral laws: "person" is defined, therefore, as "that subject whose actions are fit to be imputed to him" ("Person ist dasjenige Subjekt, dessen Handlungen einer *Zurechnung* fähig sind").

defining the correct meaning of a text finds itself denied. The latter feat was accomplished by Derrida turning De Saussure's thesis that the signifiers can only be identified by their mutual differences — i.e., by other, momentarily absent, signifiers — into the even more radical thesis that, in view of the finite number of *signifiants* in a language and the infinite number of *signifiés*, there can be no fixed relation whatsoever between the two. Just as Foucault made Man disappear, Derrida makes the author and his intentions disappear. The reader should no longer try to detect the proper meaning of a text, but leave free room for the play of association.

It might be objected here that, with Foucault and Derrida in particular, we have left the domain of *ethical* relativism. But that is not so; witness, for example, the case of Richard Rorty. Although stemming from a quite different philosophical tradition rooted in American pragmatism and analytical philosophy, and not primarily interested in the analysis of texts, Rorty nonetheless shares Derrida's conviction that we live in an essentially chaotic, 'contingent' world in which there can no longer be any question of transcendent 'truth' nor, consequently, room for any rationally founded ethics. It is true that Rorty seeks to avoid ethical nihilism by distinguishing between the 'private' and the 'public' sphere, and by insisting on the demand of *solidarity* in the latter, which culminates in his absolute rejection of *cruelty* to our fellow man.[*365] But what other justification for this demand and for this rejection is to be found in the premises of his own thought, than the purely pragmatic consideration that we seem too much attached to the former to be capable of abolishing it, and too much spontaneously repelled by the latter for us to be able to accept it ?

Nothing in the above is meant to suggest some simple-minded 'back to Kant', to be sure. When cultural relativists enjoin us not to forget the historical context of Kantian ethics they are of course quite right. For it is certainly true that Kant's work on the subject could arise only in Europe, and only in the 18th century. Quite obviously it was conditioned historically (speaking broadly) by a need felt after the 17th-century Wars of Religion to find some neutral, common ground which the universal human faculty of Reason — so successful recently in unveiling the laws of nature — seemed most fitted to provide. However, none of these truisms excludes in the least the possibility that, for all their historical situatedness, Kant's ethical teachings might have universal validity after all. Speaking not about Kant but about Plato (but the point is the same), Lévinas remarks rightly that the overcoming of 'Platonism' in modern thought (in the sense that we have learned to understand that the meanings of philosophical concepts are culturally and historically conditioned) could only have taken place thanks to the very generosity of Western thought — a generosity that has discovered *abstract* man and thus has

* It is true that in more recent publications Rorty has distanced himself from postmodern relativism in social theory and has pleaded for a new political engagement, but, as far as we can see, without abandoning his pragmatic starting-point.

proclaimed the absolute value of the person, including in this respectful attitude even the *cultures* in which these persons live and express themselves. "Platonism has been overcome thanks to the very means that have been provided by the universal way of thinking which goes back to Plato, i.e., by that much abused Western civilization that has managed to understand those particular cultures which have never understood anything of themselves."*[366]

A far more serious objection to a 'back to Kant' programme would be to consider the simple fact that so acute a thinker as Schopenhauer, while filled with all due admiration, felt entitled to put into question the very foundations of Kantian ethics. Is not the conclusion inevitable that, in the domain of 'practical reason', Kant had been much less successful than he had been in the domain of 'theoretical reason' in laying those solid foundations which (in the latter case) were meant to banish confusion from metaphysics for all time by raising it into a fully fledged science as Galileo and Newton had done for natural science?

To pose our question about the possible, universal validity of Kantian ethics thus makes it incumbent upon us to try, first, to lay bare the core of the difference between Kant's and Schopenhauer's respective approaches to the problems of moral philosophy, and next to pinpoint the profound roots of this difference, with the aim of arriving at a conclusion about the lasting value of both. This will be the purpose of our next chapter.

---

* "Le platonisme est vaincu grâce aux moyens mêmes qu'a fournis la pensée universelle issue de Platon, cette décriée civilisation occidentale et qui a su comprendre les cultures particulières, lesquelles n'ont jamais rien compris à elles-mêmes." Lévinas' disciple Alain Finkielkraut criticized structuralism out of the same concern for universal values, and protested against a cultural relativism which feels obliged to show our respect for all peoples, even if their cultural tradition involves practices we find abhorrent (e.g., female circumcision). He therefore takes great pains to distinguish between Lévinas' philosophy of 'the Other' and the apology of 'difference' one finds in structuralism. "In spite of the similarity in vocabulary, Lévinas is not to be ranged in this line of thought. True, he corrects the adage "*homo homini lupus*" in pointing out that hate of the *other* man is meant; but he rejects at the same time the confusion between 'otherness' and 'difference' which lies at the bottom of the topic of ethnocentrism. It is not his specific qualities or his peculiar habits that constitute the Other's otherness, but the nakedness of his countenance: it is this which appeals to my responsibility and which refuses to be immersed even into its own exoticism." ("Malgré la similitude du vocabulaire, Lévinas ne s'inscrit pas dans cette ligne de pensée. Il corrige bien en haine de l'autre homme l'adage "*homo homini lupus*", mais il recuse simultanément la confusion de l'altérité et de la différence qui fonde la thématique de l'ethnocentrisme. Ce ne sont pas ses qualités spécifiques ou ses coutumes singulières qui font l'altérité de l'Autre, c'est la nudité de son visage: appèl à ma responsabilité, et refus de se laisser englober même dans son propre exotisme." This does not mean, he insists, that all ethnic groups would have to assimilate into one hegemonic culture, but only that they must accept a common system of values that, in the spirit of the Enlightenment, favours the rights of the individual above those of (ethnic or other) collectivities. (Cf. Finkielkraut's remarks on Herder in Chapter I, footnote on p. 30).

VII

Ethics And Metaphysics

a. Kant's and Schopenhauer's moral philosophies: a critical comparison

"To preach morals is easy; to ground morals is difficult" ("Moral predigen ist leicht, Moral begründen schwer"), Schopenhauer says in the epigraph to his treatise on the two basic problems of ethics (thus quoting himself from *On Will in Nature*). But if this was meant as a sly dig against Kant, it was beside the point. Kant wished no more than Schopenhauer to 'preach' some new ethics; like Schopenhauer (though in a different way, to be sure) he sought a solid foundation for any moral *philosophy*. Now it is of course true that the subject matter of this department of philosophy makes an aloof, 'objective' treatment somewhat difficult. Even Kant (of all people) managed to become lyrical about (of all things) such a sober and prosaic thing as 'duty'.*[367] And how about Schopenhauer in this regard? After all, he prided himself on his strictly empirical method, with the core

* Here, of course, we have in mind the famous passage in the *Critique of Practical Reason*: "*Duty*! thou sublime and great name, that contains nothing popular and charming, but demands subjection; but that neither threatens with anything naturally repulsive and abhorrent to the mind in order to move the will; but that only formulates a law which of itself finds access into the mind and yet involuntarily gains worship (though not always obedience), and before which all inclinations fall silent (though they secretly work against it) — what is the origin, worthy of you, and where does one find the root of thy noble descent which proudly declines any kinship with inclinations and the descent from which is the indispensable condition of that value that human beings can only themselves bestow upon themselves?" ("*Pflicht*! du erhabener großer Name, der du nichts Beliebtes, was Einschmeichelung bei sich führt, in dir fassest, sondern Unterwerfung verlangst, doch auch nichts drohest, was natürliche Abneigung im Gemüthe erregte und schreckte, um den Willen zu bewegen, sondern bloß ein Gesetz aufstellst, welches von selbst im Gemüthe Eingang findet, und doch sich selbst wider Willen Verehrung (wenn gleich nicht immer Befolgung) erwirbt, vor dem alle Neigungen verstummen, wenn sie gleich in Geheim ihm entgegen wirken, welches ist der deiner würdige Ursprung, und wo findet man die Wurzel deiner edlen Abkunft, welche alle Verwandtschaft mit Neigungen stolz ausschlägt, und von welcher Wurzel abzustammen, die unnachlaßliche Bedingung desjenigen Werths ist, den sich Menschen allein selbst geben können?"). The answer, as is to be expected, is that this origin lies in human *personality*, i.e., in the freedom and independence from the mechanisms of nature that distinguishes man insofar as he belongs to the intelligible world.

of his argument against Kant residing in the claim that his own moral philosophy is more scientific than Kant's in that the latter proceeds from "aprioristic sand castles" whereas he himself remains firmly on empirical ground by offering an analysis of actual human conduct. Still, moral passion, indignation about cruelty (especially against animals!), and eloquent scorn for false prophets is what drips from every page. Indeed, if Schopenhauer wished to play the role of an impartial observer of human conduct, he played it badly. And how could it be otherwise? A man like Hegel, who was able to discern the "cunning of reason" behind all apparent misery, can afford, in speaking about ethics, to be purely 'descriptive' rather than 'prescriptive'. Someone like Schopenhauer cannot.

As a matter of fact, Schopenhauer in this connection meant, not so much empirical knowledge in the ordinary sense (which is always mediated and which would indeed yield 'ethology' rather than 'ethics'), but rather the immediate experience we have of our own inner states (such as distress), which enables us to infer similar inner states in other people, and thus makes the phenomenon of 'compassion' possible. We must recall here that Schopenhauer contested the appropriateness of the concept of 'law' in moral philosophy. He did so for two reasons: it is, at most, a secondary phenomenon, originating in a pre-existing, moral *sentiment*; moreover, no imperative can be other than 'hypothetical' — it is never 'categorical'.

For a start, we note that Kant did not deny the existence of such a thing as moral sentiment. Only, he was prepared to admit it solely as a *subjective* effect of the law upon our will, since its *objective* grounds are demonstrated by Reason only. And indeed, if we pose the crucial question of the *validity* of ethics, Schopenhauer offers us nothing but the bald assertion that an examination of actual human conduct shows there to be actions that we immediately recognize as ethically valuable and others that provoke our spontaneous indignation. Thus his *Mitleidsmoral* (a "morals of compassion", to use Nietzsche's scornful expression) is a sentimental ethics, rooted in a capacity for sympathetic understanding. But this, as we have seen already in connection with Rorty, is a poor legitimation of ethics indeed. Schopenhauer must have felt, himself, that an appeal to what we spontaneously feel does not suffice here. This is why, upon reaching the end of his discourse on the foundations of morals, he informs the reader that, although the question posed by the Royal Danish Society has now been answered and his task has really been accomplished, he will, by way of some kind of supplement, provide an outline of his metaphysics for a better understanding of his ethics.[*368] Thus in the end, the Hindu doctrine of *tat tvam asi* (cf. Chapter VI, p. 153) appears as the ultimate foundation of Schopenhauer's moral philosophy ... though in a peculiarly modified way, as we shall see.

But what about Kant in this respect? He wished ethics to be rational, i.e., a matter of universal laws, since only along this path could he hope to find a

---

[*] He compares moral conduct with Goethe's *Urphänomen* which stands at the end of any scientific investigation while leaving open the question of its ultimate foundation.

foundation as certain as that of science. Nay, *more* certain; for he was of course well aware that necessity in the domain of science is always hypothetical (with even mathematics starting from axioms); moral law, however, should be categorical, which implies that (other than in science) any appeal to experience must be banished completely (whereas Schopenhauer insisted not only that ethics, like any science, should start from experience, but also that it may contain no *a priori* whatsoever). This, then, is the justification for that particular formality of arguing so often held against Kant. He did not ask so much what specific, concrete actions are good, but rather by what general, abstract maxim they must be guided in order to be considered as truly moral actions. Here, too, his investigation was a *transcendental* one; and it seems hardly refutable as such. For, first of all, *provided that there can be such a thing as a well-founded ethics in the first place*, it must, *pace* Schopenhauer, of necessity be imperative in form. Even Schopenhauer, for all his rejection of "slave-morals" (an expression later to be adopted by Nietzsche), could not avoid clothing his own fundamental principle in imperative terms (*neminem laede, imo omnes quantum potes iuva*), as he himself admitted, although trying to play down its significance (as we have seen in the previous chapter, p. 152). He even went so far as to write down (in the context of his discussion of compassion as the motive force of ethical behaviour) a passage such as the following: "The first degree of the effect of compassion is therefore that it hampers the sufferings to be inflicted upon others as a consequence of the antimoral potencies that rest inside me; that it calls out to me: 'stop!', and places itself before the other like a protecting wall which keeps him from that injury towards which my egoism or malice would otherwise drive me."[*369] A commandment if ever there was one! Schopenhauer, with that straightforward frame of mind of his, could imagine a commandment only as a thundering voice accompanied by the bang of a fist on the table; he was unable to recognize (although in fact he says as much in the passage quoted) that defencelessness can also compel.[†370] We remember from the previous chapter (p. 150) his invocation of

* "Der erste Grad der Wirkung des Mitleids ist also, daß es den von mir selbst, in Folge der mir einwohnenden antimoralischen Potenzen, Andern zu verursachenden Leiden, hemmend entgegentritt, mir "Halt!" zuruft und sich als eine Schutzwehr vor den Andern stellt, die ihn vor der Verletzung bewahrt, zu welcher außerdem mein Egoismus, oder Bosheit, mich treiben würde."

† This is a recurrent topic in the work of a contemporary thinker like Lévinas who (to quote one instance), in speaking about the "countenance" (*le visage*), calls it "inviolable", and continues thus: "these absolutely unprotected eyes, the most naked part of the human body, nevertheless offer an absolute resistance to possession, an absolute resistance, to which corresponds the temptation of an absolute negation. The Other is the one being that one may be tempted to kill. This temptation to murder and this impossibility of murdering constitute together the very aspect of the countenance. To see a countenance means already to hear: 'thou shalt not kill'." ("ces yeux absolument sans protection, partie la plus nue du corps humain, offrent cependant une résistance absolue à la possession, résistance absolue où s'inscrit la tentation d'une négation absolue. Autrui est le seul être qu'on peut être tenté

Christian morals, with love put above everything else, against the 'apotheosis of lovelessness' of Kant's duty-based ethics; and an orthodox Schopenhauerian might perhaps be tempted, therefore, to regard Schopenhauer's moral philosophy as the New Testament replacing (or at least bringing to fulfillment) Kant's severe Old Testament! But first of all, 'love' is very well presented as a commandment in the New Testament; moreover, in spite of Schopenhauer's suggestion to the contrary, Kant does not seek a 'slave-morals' any more than Schopenhauer does. For to him the moral law is not only universal and categorical, *but also self-given.* Kant believed — rightly so, in our view — that this is where his true originality resided, in that, according to him, all previous systems of ethics were in one or another way *heteronomous*, whereas his own is *autonomous.* That is why, *pace* Heine, the *Critique of Practical Reason* is at least as revolutionary a book as the *Critique of Pure Reason.* And indeed, such a combination of an unconditionally binding law with human autonomy (suggested to Kant, no doubt, by the phenomenon of science, where by means of free, individual investigation we arrive at universally valid laws) is bound to exert a powerful attraction upon modern man, while at the same time its claim to being founded on the universal and solid human faculty of *reason* suggests the equally tempting possibility of a common platform against the gloomy perspective of a 'clash of cultures', such as evoked by prophets like Huntington.[371] I therefore wish I could believe it, but ...

Let us consider, once again, the first formulation of the Categorical Imperative. According to Kant, my freely given maxim (e.g., not to lie) must be universal if it is not to be self-defeating. But what compelling reason is there that the maxim of my conduct should be such as to be able to serve as a universal maxim? There just is no *logical* reason why I must impose, so to speak, my own rules on the whole of mankind. As Leszek Kolakowski has shrewdly observed, "I am not inconsistent at all if I prefer other people to follow rules which I do not want to abide by. If I lie whenever it pleases me but I wish everyone else to remain invariably truthful, I am perfectly consistent. I may always, without contradicting myself, dismiss the arguments of people who try to convert me or prompt me to mend my ways by asking me 'what if everybody did the same?', because I can consistently reply either that I do not care about other people's conduct or that I positively wish them

---

de tuer. Cette tentation du meurtre et cette impossibilité du meurtre constituent la vision même du visage. Voir un visage, c'est déjà entendre: 'Tu ne tueras point'"). This is a moral, not a real impossibility, to be sure, for in reality murder is of course possible, since we are not concerned here with the experience of an obstacle which is too great for our power; but it is possible only when you have not looked the other in the face. Schopenhauer could not have objected to this statement, which indeed is very much like his own. Still, he would surely not have failed (probably muttering something about *foetor judaicus*) to object to a commandment from the Mosaic Decalogue being introduced into moral *philosophy* (especially when Lévinas adds: "And all that I can learn from God and understand about God, who is invisible, must have come to me by the same, unique voice." ("Et tout ce que je peux entendre de Dieu, et à Dieu, qui est invisible, doit m'être venu par la même voix, unique.")), and to add indignantly that a dog has a countenance and eyes, too ...

to obey norms which I refuse to observe."[372] Indeed, no *logical* contradictions are involved in pursuing maxims unfit to be universalized; there are only *social* objections. And here the problem rests. A law requires a lawgiver who, in Kant's view, can be no one else but the autonomous subject. Man as a rational being is free, i.e., the law for his actions must be given to him by himself, says Kant; but why, then, should not 'authenticity' be proclaimed the highest value, and why should the law not simply say 'Be yourself!' (or, in Nietzsche's formulation, "Become what you are!": "Werde was Du bist!")?

Against this Kant would of course argue that to give to oneself a law which could not serve as a general law would mean to give to oneself a law that is not a law, which would be a contradiction unworthy of a rational being. For, as in science (though of course in a different way) it is the very concept of law which guarantees universal validity; hence Kant's allowing for at least one sentiment, namely, 'respect for the law'.* Now it is no doubt true that, once more from a *transcendental* point of view, the capacity to be universalized is the criterion *par excellence* for a moral law worthy of that name ... *always provided, once again, that there is indeed such a law*!

Now the reader will have noticed that we believe Schopenhauer to have involuntarily demonstrated the existence of such a law (unless one assumes, as Schopenhauer seems to do, that 'stop!' is not a commandment); nevertheless, it seems to us that Iris Murdoch is right in saying that Kant with *his* demonstration allowed himself a *petitio principii* (as a matter of fact, we recall from the previous chapter, p. 146, that Kant himself felt compelled to admit that he had demonstrated the existence of the moral law in a circular way). For "Kant assumed freedom and rational will in view of the existence of the moral law, and then went on to prove the reality of this moral law through the concept of freedom."[373] This is what made him introduce that interesting man who, although by nature indifferent to human suffering, still (as a rational being) renders benefits to mankind out of a mere sense of duty, i.e. out of respect for the law! Why, that man does not exist. A man acting thus acts, not out of respect for the law, but out of fear, as Schopenhauer rightly suspected.† Indeed, Schopenhauer hits the nail on the head when he says that the demand for universability is bound up with the consideration that, in an action, I am not always the *active* part, so that the true formulation of the first version of the

---

* cf. Chapter VI, footnote on p. 143.

† Cf. Chapter VI, p. 152. Let it be granted: human nature is always able to surprise one by the contradictions that can be found in one and the same specimen of the genus. Even "that man" might therefore exist after all (though until further notice we persist in our lack of belief). And there *are* examples of the opposite case, as we know now. Living in the backward 19th century Schopenhauer was mercifully spared the acquaintance with the type, exemplified by a man like Himmler who (probably rightly) claimed to be not cruel-hearted or blood-thirsty by nature at all, but to be firmly dedicated to fulfilling the hard and bitter duty of exterminating the eternal foe of Aryan culture. Evidently, such a sense of 'duty' has nothing whatever to do with the Categorical Imperative (though Himmler may have imagined it did).

Categorical Imperative should rather run: 'Act always according to such maxims that you could wish were reciprocal'; also when (previous chapter, p. 152) he adds that this might be useful as a foundation for the principle of political order, but not for the principle of morals.

Schopenhauer proceeds here once more by means of a *genetic* approach;* not, this time, as a complement to the transcendental approach but rather as its complete substitute, since for him there is no such thing as 'Practical Reason'. In this, as in other things, he stands at the cross-roads between Kant and Nietzsche. When we compare Nietzsche's "genealogy of morals", and its devastating critique of all traditional moral values, with Schopenhauer's investigation (which rather resulted — to his satisfaction — in a philosophical justification of these very values)†[374] we see how relatively 'unmodern' Schopenhauer still was. Anyhow, proceeding along his genetic line, Schopenhauer believes he had found the real motive of moral behaviour (which, it is to be repeated, he accepts as such without any questioning) in 'compassion' provoked by the experience of our being confronted with our fellow man (and also our fellow animal). Characteristically, 'cruelty' is for him the principal sin, whereas for Kant it is rather 'lying'. Probably most people will instinctively agree with Schopenhauer (as we just saw Rorty do, who mentions Schopenhauer nowhere), especially so when learning of Kant's argument against telling lies in his 'Metaphysics of the doctrine of law' ('Metaphysik der Rechtslehre'), which is that society could not exist if lying became a universal maxim. For not only does Kant involuntarily demonstrate here the truth of Schopenhauer's and Kolakowski's diagnosis of the Categorical Imperative (at least in its first formulation) as being of an essentially *social* nature; he also learned, to his dismay, that his very argument against telling lies could be turned against him, when he was confronted with Benjamin Constant's remark that society could not exist *if one ought to tell the truth under any circumstances whatsoever*!‡[375] One is inclined, therefore, to suspect that Kant would have sided

---

* We have seen him doing so before in Chapter IV, p. 111.

† "I venture to say, therefore, that my ethics is wholly orthodox with regard to the Upanishads of the holy Veda's and to Buddha's world religion, nay more, that it does not contradict old, genuine Christianity." ("Ich getröste mich demnach, daß meine Ethik, in Beziehung auf den Upanischad der heiligen Veden, wie auch auf die Weltreligion Buddhas, völlig orthodox ist, ja, selbst mit dem alten, ächten Christenthum nicht im Widerspruch steht.") This must have been reassuring to Schopenhauer, says Thomas Mann, going on with lovely irony to quote Gretchen: "ungefähr sagt das der Pfarrer auch, nur mit ein bißchen andern Worten." ("that is by and large what the priest also says, only in somewhat different words").

‡ As is to be expected, and in spite of "love of truth being his principal, perhaps his only, virtue" ("Wahrheitsliebe ist seine größte, vielleicht seine einzige Tugend", as Johanna once said — and who would be a better judge?), Schopenhauer, too, held that lying is under certain circumstances rightly allowed by any normal man, even though, generally speaking, it is of course reprehensible. However, this reprehensibility is made to follow by him, not so much from its undermining effect on any legal system, as Kant thought, but rather from

with that truth-fanatic out to expose everybody else's self-deception about his own life even at the price of plunging them into despair and destroying their lives (Gregers Werle in Ibsen's *Wild Duck*)[*376] ... yet the matter still does not end here. For we believe that those very objections directed by Schopenhauer and Kolakowski against the first formulation of the Categorical Imperative turn into a recommendation once it is presented, not as the first, original one (as Kant did), but as the second, derivative one.

To illustrate this, at first sight, somewhat paradoxical statement, let us recall Max Weber's famous (or perhaps notorious) distinction between *Gesinnungsethik* (an ethics of conviction which, in view of the absoluteness of its demands, does not ask about consequences) and *Verantwortungsethik* (an ethics of responsibility). The distinction may certainly have been brought into play too often and it may also lend itself too easily to political abuse, but, still, it remains a fruitful one.[†377] 'The

---

its connection with *violence*, since lying is an instrument of *cunning*, i.e., of coercion by means of motivation. In order to protect myself from violence I am therefore allowed under certain circumstances to lie. Similarly, to protect my privacy ('ask me no questions and I'll tell you no lies'). Constant went still further — he held against Kant that under certain circumstances (for example, when someone with the intention of killing a person who happens to be in my house asks me about the whereabouts of his intended victim) to lie is not only allowed but even obligatory. Constant's argument was countered by Kant by means of what in our opinion comes down to a miserable sophism quite unworthy of its author.

[*] That for Kant lying is the principal sin has, of course, everything to do with the rational foundation of his ethics, which Schopenhauer rejected; and indeed the truth commandment can be seen as the link between 'theoretical' and 'practical' reason, as we argued in Chapter VI, footnote on p. 138. But the difference is in our opinion that in the first case it is *the* virtue, whereas in the second case it remains a virtue, of course, yet one that must be weighed against other virtues (e.g., 'mercy', as in the case of Gregers Werle). Kant's pathos of truth calls to mind Chesterton's acute remark: "the modern world is not evil: in some ways the modern world is far too good ... The modern world is full of the old Christian virtues gone mad. The virtues have gone mad because they have been isolated from each other and are wandering alone." And already Pascal confessed that an excessive degree of one virtue is for him admirable only when it is accompanied by a no less excessive degree of the opposite virtue ("Je n'admire point l'excès d'une vertu comme de la valeur si je ne vois en même temps l'excès de la vertu opposée.").

[†] According to Weber, the *Gesinnungsethiker*, out of the absolutism of his demands, cannot bear the ethical irrationality of the world; his brand of humaneness is therefore called by Weber "a-cosmistic". This kind of ethics (of which the Sermon on the Mount may be regarded as the classic example) is becoming to a saint, not a politician. For, whereas from a-cosmistic love-ethics it follows that you should not resist evil, the politician is obliged to resist evil since otherwise he is responsible for its expansion. Moreover, the attainment of an ethically good end is often dependent on the use of ethically doubtful or dangerous means; therefore, although logically the *Gesinnungsethiker* has no other possibility than to reject any action which makes use of ethically dangerous means, in reality we see again and again that the *Gesinnungsethiker* suddenly turns into a chiliastic prophet, who, after having first preached 'love against violence', summons in the next moment to violence — namely,

road to hell is paved with good intentions,' popular wisdom says; Chesterton, always eager to be contrary, commented: "that is exactly the one thing it cannot be paved with."[378] Chesterton was not exactly what one would call a Kantian, but on this point Kant would have whole-heartedly agreed. And we are all Kantians nowadays insofar as everybody will probably agree that it is the *intention* which is decisive for the moral value of an action. Yet the proverb quoted does contain a limited truth. For not only wickedness, but also boundless and thoughtless goodness is bound to create havoc, and is therefore unfit to be universalized. On the 'theoretical' level, Kant has taught us, Reason can only be called truly 'rational' if it recognizes its own boundaries (*Grenzen*), that is, if it is self-critical in that it accepts that the very tools of its trade lead only to sterile phantasies if applied beyond its legitimate domain. Just so, we would add, on the 'practical' level Reason can only be called truly 'rational' if it recognizes, this time not *its own* boundaries, but those of 'good will' (which, *pace* Kant, in the last resort, has its origins not in the 'head' but in the 'heart'!) in that it combines the latter with that critical reflection which takes the *consequences* of human actions (or of human passivity) into account. This is why, unlike Schopenhauer, we do not object to Kant's using the expression 'Practical Reason'; but we take it in the sense of that wisdom which leaves room for the badly needed *moderation* of absolute ethical demands, not in the sense of the *source* of any such demand. In Kant's failure to make proper distinctions here, we spot the source of such confusion as is amply demonstrated by his statements on lying. These evidently arise from ethical absolutism (*Gesinnungsethik*), yet they are presented by Kant as motivated by the demands of society (i.e., using an argument which belongs to the ethics of responsibility (*Verantwortungsethik*)), which is how a man like Constant (who *really* thought in terms of an ethics of responsibility) could so easily deflate the claim.

All this leads to the possibly surprising conclusion that the first formulation of the Categorical Imperative seems appropriately to serve as the ultimate foundation for what Kant called *Rechtslehre* (the doctrine of law), and the second rather as the ultimate foundation of what Kant called *Tugendlehre* (the doctrine of virtue). Such a distinction fairly corresponds to Schopenhauer's distinction between the 'negative' virtue of *justice* and the 'positive' value of *humaneness*, with the former, according to him, being more specifically the domain of man, and the latter of woman.*[379]

---

to the ultimate violence that will bring about the destruction of all violence. Indeed, the experience of the 20th century has amply confirmed this observation by Weber.

* Women, according to Schopenhauer, are less rational than men, i.e., they are to a lesser extent gifted with the faculty of distinguishing general principles; therefore, in the virtue of justice they generally rate below men. But they exceed men in the virtue of human sympathy — here the motive is as a rule more directly tangible so that it exerts a more straightforward appeal to compassion, to which women are certainly more susceptible. "The idea of seeing women holding the office of judge provokes laughter; but the sisters of charity exceed even the brothers of charity." ("Der Gedanke, Weiber das Richteramt

Now does this not sound a bit too much like a gentlemen's agreement? Schopenhauer himself would no doubt have protested against his own maxim '*Neminem laede, imo omnes, quantum potes, juva*' being lumped together with Kant's second formulation of the Categorical Imperative.* More was involved here for him than his overall stance that compassion (the source motive of all ethical behaviour) has nothing to do with any 'imperative' whatsoever — a view whose shortcomings we have already sought to reveal. He also (as we recall from the previous chapter, p. 152) raised some more specific objections. For example, the Kantian Imperative fails to apply to a criminal about to be executed, who is very well — and rightly so — used as a means (namely, of deterrence). Unfortunately, this argument can with equal justice be directed against his own formulation: ought we to abstain from hurting that criminal? Ought we, on the contrary, to help him? If so, help him with what? Perhaps with escaping from prison? More important, surely, is Schopenhauer's argument that his own formulation is superior because it includes the animals as well. This argument deserves somewhat closer scrutiny, since we are confronted here with a serious and, in our own times, very pertinent problem.

Long before Schopenhauer did so, the founder of utilitarian ethics, Jeremy Bentham (who is not mentioned by Schopenhauer, although he praises the "sensitive" English nation for its striking compassion with animals enabling it — in spite of its otherwise unworthy "cold superstition" — to correct by its legislation the gap left by her Bible-based morals),†[380] had in anticipation criticised what was soon to appear as Kant's position on the subject. To Bentham the question of whether animals could reason, and also whether they could speak, was irrelevant; what mattered to him was "that they can suffer". Just as the French Revolution, by abolishing slavery in the colonies, proclaimed that blackness of skin is not a sufficient reason for abandoning a sensitive being to the caprice of a torturer, so, he says, the day may come when "the number of legs, the villosity of the skin or the termination of the *os sacrum*" will be deemed equally insufficient reasons for such treatment.[381] Now it is certainly true that Kant did not grant "rights" to animals. However unpleasant such a refusal may appear to us, does it

---

verwalten zu sehen, erregt Lachen; aber die barmherzigen Schwestern übertreffen sogar die barmherzigen Brüder.") *Mirabile dictu* our old misogynist is here to a large extent in agreement with a present-day feminist author, namely, Carol Gilligan, in her book *In a different voice*.

* Kant actually offers several variations, but for all practical purposes we may confine ourselves to the distinction between the 'formal' and the 'material' version.

† "Diese feinfühlende englische Nation sehen wir, vor allen andern, durch ein hervorstechendes Mitleid mit Thieren ausgezeichnet welches sich bei jeder Gelegenheit kund giebt und die Macht gehabt hat, dieselbe dem sie übrigens degradirenden 'kalten Aberglauben' zum Trotz, dahin zu bewegen, daß sie die in der Moral von der Religion gelassene Lücke durch die Gesetzgebung ausfüllte."

lack justification if considered from Kant's (and also, really, from Schopenhauer's) point of view? If not 'happiness' (as in utilitarianism), but 'freedom' (as Kant asserted, and Schopenhauer agreed) is the ultimate object of ethics, then 'having interests' (as animals no doubt do) is not indeed a sufficient reason for recognition as a legal person, as the ability to be a partner in a freely entered contract then forms the criterion. We therefore join Luc Ferry in holding that, since animals evidently lack that ability, one cannot speak of animal *rights* in any strict sense.[*382] Still one *can*, and should, speak of human *duties* towards animals (and towards nature in general).

Now it is certainly true that in Kant's 'Metaphysics of the doctrine of virtue' we find expressed the thesis, so indignantly rejected by Schopenhauer, that cruelty against animals should be avoided out of duty, not towards them, but towards man himself. For this Kant gives as his reason that compassion towards human suffering — surely recognized by him as a virtue! — tends to be blunted by callousness towards animals, with the consequence that a natural disposition very useful for morality is thus weakened and gradually extirpated.[383] Still, Kant has a little more to say on the topic (just a little, to be sure, and only in a footnote) in the *Critique of Judgement*. Speaking about the constructions built by beavers, he remarks that, although such artisanal capacities (*Kunstvermögen*) are rightly attributed, not to reason, but to instinct, they nevertheless show an *analogy* between man and animal in the sense that animals, too, act according to *representations* [†] and that they are *not* therefore machines as Descartes had wrongly claimed.[384] This is certainly not a sufficient admission for us, since Kant failed to infer from this insight that we have duties towards animals. Yet it seems that with this concept of analogy Kant has pointed here, however imperfectly, towards precisely the right middle road that runs between Cartesianism and some extremist forms of contemporary ecological movement ('deep ecology'). For the analogy to human freedom displayed not only by the animal's capacity to act in a purposeful way (as shown by Kant), but also by their capacity for suffering (their 'vulnerability', as stressed by Bentham; a capacity which likewise shows that, *pace* Descartes, they are not just machines), ought to make us regard them as our less

[*] One might object that this goes for children, too; but Ferry would no doubt retort that children, although for the moment still lacking the ability to be a partner in a freely entered contract, as human beings in principle do have it.

[†] This seems at odds with our own opinion on animal representations as given in Chapter IV, p 107. Indeed we believe that in the case of the beaver (Kant's example) one cannot strictly speak of 'representations' since its activity is wholly programmed by its instinct. It is otherwise with the behaviour of some apes who seem to be able to act with discretion; anyhow, in the case of the beaver, too, we are not concerned with a 'machine' in the sense of Descartes! For indeed, as the utilitarianists rightly stressed, it is the animal's capacity to *suffer* (a capacity oddly denied it by Descartes) which makes it analogous to man, even though they failed to see the link between this capacity and the *freedom* rightly stressed by Kant.

gifted or infant brothers, and, to the extent that that is what they are, to arouse the solidarity with them which Schopenhauer demanded.

Still, the argument for this should properly run 'from above to below', so to speak, since human freedom (on which we have more to say in what follows) ought to be its starting-point, whereas in Schopenhauer it is rather the other way round, biological life, which in man becomes conscious of itself, forming his own point of departure. This is more than a merely formal distinction, as it raises the question of *value* in nature. It is indeed a logical mistake (however well-intentioned) of many contemporary ecologists to say that extra-human nature 'has value in itself'. Only man can give 'value' to something, and at least in this regard 'anthropocentrism' is inevitable.[385] *How* inevitable it is appears with striking clarity from the writings of our, in theory, so anti-anthropocentric Schopenhauer. For what else is he doing all the time but making value judgements about nature — a nature which, to him, is by no means the 'big Mother' that it is in Romantic *Naturphilosophie*, but is most strikingly characterized instead as a *bellum omnium contra omnes*? By way of one more example, we need only think of how he retells a story by Junghuhn, in his book on Java, of a squirrel hypnotized by a snake staring at it and thus compelled to run, wailing with terror, into the snake's wide-open jaws.[*386] He shudders, and we shudder with him. But in all fairness *tat tvam asi* ought to go for the snake, too, who cannot help being a carnivore. (It is true that the squirrel has a far larger 'caressibility factor' ... to use a term coined by a Dutch essayist and lover of cats and pigs, in particular).

In short, the biosphere gives life, not only to baby seals and forests and brooks but also to the HIV virus, pestilence, and cholera, as Ferry rightly notes,[†387] and therefore we cannot avoid treating different natural phenomena differently, thereby applying *human* values. This is why Kant's concept of 'analogy' seems to us to provide the best clue for our dealings with non-human nature. Only, it certainly stands in need of further elaboration, in the particular sense that Kant's "Realm of Ends" ought to be 'democratized', so to speak, in order to include our lesser brothers as well!

The "Realm of Ends"! Here, finally, we are at the metaphysics underlying Kant's ethics. Feeling (though not pronouncing) that 'duty' alone would not do *to provide a motive* for acting morally, he postulated a general "Realm of Ends in themselves" as a "glorious ideal" (*herrliches Ideal*), apt to awaken in us a lively *interest* in the moral law.[388] Similarly, Schopenhauer, feeling on his part that 'inclination' alone (in this case, compassion and empathy) would not suffice *to*

---

[*] He quotes Aristotle: "What a terrible nature is it to which we belong."("Was für eine entsetzliche Natur ist dies die wir angehören.").

[†] We would add that, on the other hand, when people talk about nature, unconsciously they often have parks and gardens in mind, that is to say, places in which, thanks to human cultivation, beautiful and rare species are enabled to survive which in real nature might have been selected out in the struggle for life.

legitimate morals,* fell back for the latter on the ancient Indian wisdom of *tat tvam asi*. In these different metaphysical backgrounds, then, resides the fundamental difference between our two thinkers in the ethical domain. We shall therefore take a closer look at these, especially with regard to that fundamental problem (or 'pseudo-problem', as logical positivists will say) — the problem of freedom.

But is not this very strange? After all, Schopenhauer believed that precisely here lay his fundamental *agreement* with Kant, namely, in their shared belief in the ideal nature of space and time and in the consequent distinction between the empirical and the intelligible character, which is the very distinction that makes the coexistence of freedom and necessity possible. But in fact Schopenhauer, as a consequence of his metaphysical background, made something quite different out of this. We surely need a new section to explain what all this really amounts to.

b. The metaphysical background to Kant's and Schopenhauer's moral philosophies, and its consequences for the problem of freedom

Is one justified in speaking of Kant's 'metaphysical background' in the first place? Schopenhauer, who, for all his strident quarrels with German Idealism, was not content with "severe and sober", purely critical philosophy, either, denied it. According to him, Kant held any metaphysics to be impossible. And what made him do so was —Schopenhauer thought — Kant's definition of metaphysics as "the science of that which lies beyond the possibility of any experience", together with the conclusion drawn by Kant from this premise "that such knowledge can never be found using principles derived from experience in the first place, but that only what we know *before* (i.e., independently *from* any experience) can transcend possible experience."† For, in view of the fact that all *a priori* knowledge that we possess (i.e., knowledge from pure reason) consists merely of forms of our intellect, which are valid only for our apprehension of the world and cannot therefore transcend the possibility of experience, nothing remains for us on this score but resignation.[389] Schopenhauer then draws the distinction between himself and Kant as he sees it, by granting that all this is quite true for earlier, dogmatic metaphysical efforts, but that complete resignation would follow as its logical

---

* Certainly Schopenhauer who, although a metaphysician, was a decided atheist, would vehemently have rejected Ivan Karamasov's claim that, if there is no God, anything is permitted. And we, for our part, do not in the least intend to join those who think either a 'theist' or 'atheist' (like Buddhism) metaphysical foundation *a priori* necessary for genuinely moral conduct. But we do think it *a posteriori* necessary, so to speak, since as reasonable beings we cannot be content with the simple assertion that either our biological instinct or the historically determined norms of our particular civilization happen to incite us to moral conduct. Cf. for this also the final section of Chapter VIII.

† "Metaphysik ist Wissenschaft von Demjenigen, was jenseits der Möglichkeit aller Erfahrung liegt." "Ein solches kann nimmermehr gefunden werden nach Grundsätzen, die selbst erst aus der Erfahrung geschöpft sind, sondern nur Das, was wir *vor*, also unabhängig *von* aller Erfahrung wissen, kann weiter reichen als mögliche Erfahrung."

consequence only if it had indeed been proven that the material for the solution to the world-riddle cannot rest in the empirical world itself, but can only be sought outside it. As long as no such proof has been forthcoming, no sufficient ground for resignation presents itself. More than that, Schopenhauer thinks that the world-riddle can indeed be solved, as a result, precisely, of our understanding of this empirical world of ours. It is not up to metaphysics to pass over that experience whereby the world presents itself to us, but rather to understand it from the very foundation, since outer and inner experience *together* provide the source of all knowledge. He believes, therefore, that by connecting outer experience at the right point with inner experience, and thus bringing about a synthesis of these two apparently heterogeneous sources of knowledge, the world-riddle can indeed be solved ... even if only within certain limits that are inseparable from our finite nature.[390] The Thing-in-itself therefore has nothing to do with the 'Absolute'; it should be sought, not in cuckooland, but in the things of this world, i.e., "in the table at which you are writing, in the chair under your bottom", as Schopenhauer impatiently wrote to his (at times somewhat dense) disciple Frauenstädt.[*391]

Well then, in investigating the world as it lies before us, Schopenhauer believed he had found the key to the world-riddle in the very thing of which we have the most intimate knowledge, which is in our own "*Will*". In the next chapter we shall take ample time to tackle all that is problematic about this particular derivation; for the moment, we confine ourselves to Schopenhauer's suggestion that, deep down and tucked away almost in secrecy, to Kant, too, the Will was the 'Thing-in-itself'.[392] And indeed, in a sense it may be argued that for Kant, 'Will' was the 'Thing-in-itself' ... as far as Man is concerned! For it is the "good will" through which man as a moral being demonstrates his citizenship of a noumenal realm, though at the same time as a natural being he continues to belong to the phenomenal realm. And this makes it immediately clear that Kant's 'Will' is something rather different from Schopenhauer's 'Will'. Not by chance does *The World as Will and Representation* carry by way of an epigraph Goethe's words "Ob nicht Natur zuletzt sich doch ergründe?" ("Will Nature not after all fathom her own secret?")[393] Schopenhauer, although taking his starting point in man's inner experience, proclaims "Will" as the "Thing-in-itself" by considering Nature as a whole, of which man is only a part (albeit the most highly developed part); it is, therefore, not in man as a *moral* (which for Kant is synonymous with a 'rational') being, but precisely in man as a *natural* being that the Thing-in-itself reveals itself!

Still, his very conviction that man with a part of his being belongs to a 'noumenal' realm shows that Kant, in spite of what Schopenhauer suggested, did have a metaphysics, too. Certainly, not only did it have nothing to do with a "Will"-metaphysics like Schopenhauer's (not even secretly, as Schopenhauer believed), but of course nothing at all either with the kind of rational knowledge of trans-empirical reality he had shown to be impossible in the *Critique of Pure*

---

* " ... im Tisch, daran Sie schreiben, im Stuhl unter Ihrem Werthesten ..."

Reason. But it was precisely this "critique" that was seen by Kant as the necessary preliminary to a metaphysics of a novel kind. After having accomplished the necessary demolition of the old kind of metaphysics (an accomplishment which earned him the name *Alleszermalmer*, in the wording of Moses Mendelssohn, but which had a 'positive' aspect, too, in that it was at the same time a kind of meta-science which sought to lay down the *transcendental foundations* of any possible science), he would proceed to the constructive task of sketching at least the outline of a new kind of metaphysics. This would distinguish itself from all earlier ones precisely by its having gone through the purifying fire of the "critique". Unfortunately, Kant was already well into his forties when Hume awoke him from his "dogmatic slumbers"; it then took him twelve years to find the proper tools for the 'negative' part of what he saw as his life's task; thus, when he had finally written down the three monumental *Critiques* he was an old man who no longer had the full mental powers required to accomplish the 'positive' (i.e., to him the most important) part. In this respect there can be no greater difference than between him and Schopenhauer. Had the latter died at the age of thirty, i.e., after the completion of the first volume of *The World as Will and Representation*, the world would certainly have missed some interesting pieces of writing like its second volume, or *On Will in Nature*, or (last, but not least) the exercise in moral philosophy which is the topic of our present chapter; but really the essence of the latter two works is already to be found in the second and the fourth books of the main work, respectively, whereas the former is in fact a series of footnotes (expanded into short essays) to this main work. Thus, whereas Schopenhauer had essentially completed his life's work at a young age, Kant was in old age no longer able to complete his own: what we possess are fragments, from which, however, it is possible to get an inkling of what the completed edifice would have looked like.[394] In any case, it is an indisputable fact that Kant recognized a dimension of reality that transcends experience (and thus in this sense was in fact *more* of a metaphysician than Schopenhauer, whose "Will" transcends only *outer* experience) — a dimension, however, which (because of its lying beyond the scope of theoretical knowledge) can reveal itself solely in *moral practice*. At bottom, therefore, *Kant's metaphysics coincides with his ethics*.[*395]

Kant has often been compared to Plato, and with some justice; however, the parallel should not be pushed too far. To Plato the phenomenal world, as it is also to Schopenhauer, is more or less like the Veil of Maya, whereas to Kant it is essentially the world of — successful — science and quite 'real' as such. Moreover, the Platonic distinction between δοξα *(doxa)* and ´επιστημη *(epistèmè)* suggests the human ability to attain knowledge of a higher world of Ideas, whereas to Kant any theoretical knowledge of the noumenal world is impossible. After all, Kant's 'Ideas' are only 'regulative', not 'constitutive', for any kind of science; if

[*] It is not therefore by chance that Kant's 'positive' period opens with the '*Metaphysics of Morals*' (*Metaphysik der Sitten*), consisting of a 'Metaphysik der Tugendlehre' and a 'Metaphysik der Rechtslehre' (see above), as Raymund Schmidt has rightly pointed out.

used in the latter sense they can give rise only to the pseudo-sciences of rational psychology, rational cosmology and rational theology. This already suggests that Kant's 'Ideas' are very different from Plato's, which observation led Schopenhauer to reproach Kant for his improper usage of the term.*[396] Finally, although it is true that to Plato the Idea of the "Good" is supreme, so that he, like Kant, may be regarded as an essentially ethical thinker, one should keep in mind that Plato's "Good" means 'perfection' in *any* domain (hence its overall status), rather than a specifically *ethical* category: it is the one Idea that turns all Ideas into Ideas, just as the Sun makes all things visible.†[397]

All these significant differences notwithstanding, it is certainly true (and to that extent the parallel with Plato is indeed valid) that the Categorical Imperative *transcends* the phenomenal realm; it is like the Idea of the Good "beyond Being" ('επεκεινα της 'ουσιας; *epekeina tès ousias*).‡ We agree therefore with Iris Murdoch (who otherwise prefers Plato, as need not overly surprise in a literary author) that Kant's definition of morality in terms of an *unconditional* demand, "demonstrates his vision of an absolute at the centre of human existence".[398] For all Schopenhauer's objections to the concept of an 'Unconditioned', we believe that it cannot be dispensed with, even if not in an ontological, then at least in a deontological sense. Though it may not exist as a 'Being', it exists at least as a force in us; to deny this amounts to a consistent scepticism which evidently refutes itself as one can deny 'Truth' only by rational argument, i.e., in the name of that very 'Truth'. Conceived thus, however, the 'Unconditioned' ought to be seen as an unending quest, or (to adopt the expression used by Kant for demonstrating the necessity of the postulate of Immortality) as an "infinite progress", which we are well aware will never arrive at a final stage, but in which we sense it to be our task to remain engaged. *Imaginable* is the 'Unconditioned' only in the 'horizontal' sense of a limit continually on the move; but our very awareness of the relativity of all results attained thereby demonstrates the connection between our intellect and an unimaginable, 'radically Other' which breaks 'vertically' into this historical

---

* We come back to this in Chapter IX, p. 240.

† It is in fact not otherwise with Aristotle's *Nicomachean Ethics*, as is to be expected from their common Greek background. Hence, in spite of all the differences between them as political thinkers, for both men a good society is a society where everything is in its right place: 'justice', e.g., is therefore quite compatible with slavery. It would of course be wholly unhistorical to reproach them for this; but it would be no less unhistorical to forget that it is a far cry from there to Kant's doctrine that every human being should be treated as an end in itself. Although he does not refer to Plato's "Idea of the Good", Kant in the *Groundwork* rejects "the ontological concept of *perfection*" as a foundation for morality. His reason is that, although it is to be preferred to the theological concept of a divine will on the one hand and to the empirical concept of happiness on the other, it has in common with these that it turns *heteronomy* of the will into the primary foundation of morality.

‡ 'Being', mark you, to be understood here not as empirical being but precisely as the *true* Being of the Ideas); the Idea of the Good is therefore of a fundamentally different nature from the other Ideas.

process as an inexorably critical voice (perhaps to be called, with Nietzsche, even if in not quite the same sense, 'intellectual conscience'). Every finite step in the endless process of human thinking, striving, and creating can only be *recognized as finite* thanks to this 'voice', just as the very same 'voice' always reminds us that the historical sphere remains forever at this side of, never beyond, the Unconditioned. The former insight is what positivism ignores, whereas the latter insight goes against speculative idealism *à la* Hegel, which, for all its emphasis on 'Becoming', in the final analysis presents us with the end product of all previous metaphysics — a new version of Parmenides' sphere, i.e., a closed circle instead of Kant's open, infinite line.

Once more, can an 'Unconditioned' thus conceived be said to 'exist'? The question calls to mind all those debates over the Ontological Proof (the one which holds that the idea of a most perfect being must entail the existence of that being, as it would otherwise lack one quality)! Kant, who in his 'Transcendental Dialectic' refuted, as we know, that proof like the other two traditional proofs of speculative theology (the 'cosmological' and the 'physicotheological' ones), argued that 'existence' is not a predicate since a hundred pounds (*Thaler*) that I merely imagine have all the same predicates as a hundred real pounds in my purse. Schopenhauer for his part considered a critique of reason, as Kant had given, wholly unnecessary for its refutation since it is quite easy to see that this ontological proof is nothing but a quibbling (*spitzfündig*) word-play without any power to convince.[399] Yet one might ask whether, with his objection to Lamarck's biological evolutionism, to wit, that the 'lower' cannot give rise to the 'higher' (an objection which forced him to introduce the Platonic Ideas into his metaphysical system, where they look at first sight rather like a *Fremdkörper*), he did not unintentionally present a new version of it?* Simone Weil would much later write in the same vein that the whole idea of Progress is the atheist idea *par excellence* since it implies that the mediocre may produce the better all by itself, which is a negation of what she calls "the experimental ontological proof".[†400] What she means by the latter she explains as follows: "I have in myself no principle of ascension. I cannot climb into the air up to heaven. Only when I direct my thought at something better than myself this something draws me towards the higher, and if I am really drawn this something is real; for an imaginary perfection would not be able to draw me even one millimetre upwards, since it would find itself necessarily at the same level where I, who imagine it, find myself".[‡401] And Iris Murdoch, who

---

* We come back to this in the final section of Chapter VIII.

† "L'idée-athée par excellence est l'idée de progrès, qui est la negation de la preuve ontologique expérimentale, car elle implique que le médiocre peut de lui-même produire le meilleur."

‡ "Je n'ai pas en moi de principe d'ascension. Je ne puis grimper dans l'air jusqu'au ciel. C'est seulement en orientant ma pensée vers quelque chose de meilleur que moi, que ce quelque chose me tire vers le haut. Si je suis réellement tiré, ce quelque chose est réel. Aucune perfection imaginaire ne peut me tirer en haut, même d'un millimètre. Car une

was strongly influenced by Simone Weil, has in her turn devoted (in that, in its best moments, wonderful book of hers) a highly instructive chapter on the subject of Ontological Proof — in particular, to Anselm's way of explaining it. His early critic and fellow Benedictine monk Gaunilo had objected, in essence, that 'if one is going to argue from perfect essence to real existence then could one not argue anything into existence from the imagined idea of a single perfect instance (e.g., a perfect island)?'; whereupon Anselm retorted that only in the case of God is it such that, if you can conceive of this entity, you are *ipso facto* certain that what you are thinking of is real. "You often repeat that I assert that what is greater than all other beings is in the understanding, and if it is in the understanding it exists also in reality for otherwise the being which is greater than all would not be greater than all. Nowhere in my writings is such a demonstration found,"[402] he writes, thus anticipating Kant's rejection of the proof. But he continues that one should distinguish between the real existence of a being which is said to be greater than all other beings (i.e., a being, which *happens to be* greater than all other beings) and the real existence of one that is said to be a being than which a greater one cannot be *conceived*. Only in the latter case are we concerned with a being whose *non-existence is impossible*. This he illustrates with Plato's Idea of the "Good". We recognize and identify goodness and *degrees* of good, and are thus able to have the idea of a greater conceivable good. In Murdoch's paraphrasing: "We 'see' God through the morally good things of the world, through our ability to distinguish good and evil, and through our just God-fearing understanding of what is not good ... God exists, he exists necessarily, we conceive of him by noticing *degrees of goodness*, which we see in ourselves and in all the world which is a shadow of God."[403] And this second version of the Ontological Proof is, according to Murdoch, not affected by Kant's critique (it is in fact not unlike his own, *ethical*, demonstration which removes the issue from the domain of Theoretical Reason to that of Practical Reason); she adds, however, that, in this guise, the Proof works like a boomerang (the metaphor is not hers) in that, from a modern viewpoint, it entails not that God exists, but that he *cannot exist*. She quotes J.M. Findlay: "If God is to satisfy religious claims He must be a being in every way inescapable, and it is absurd (even ungrammatical) to attribute existence to such a Being."[404]

The staunch atheist Schopenhauer would no doubt have whole-heartedly agreed, but Findlay assures us that all this needs cause no grief to the truly religious mind, since it discredits the idolatry of "merely existent" things and brings out clearly what many religious people actually understand, namely, that any adequate religious object is necessarily non-existent. As Murdoch puts it: Any existing God would be less than God. An existing God would be an idol or a demon.[405] But what led us to conceive of him *does* exist. "Morality and demythologized religion are concerned with what is absolute, with unconditioned

---

perfection imaginaire se trouve automatiquement au niveau de moi qui l'imagine, ni plus haut ni plus bas."

structure, with what cannot be 'thought away' out of human life, what Plato expressed in the concept of the Ideal Form of the Good, and Kant in the Categorical Imperative. What is in question here is something unique, of which the traditional idea of God was an image or metaphor and to which it has certainly been an effective pointer."[406]

Perhaps we may sum up all this somewhat freely in the following way: Monotheism, rightly understood, may be interpreted as a kind of atheism, insofar as rightly understood atheism may be interpreted as the radical consequence of the Second Commandment![*407] God is the Eternal and the Infinite, and existence means finitude in time and space; it would therefore be blasphemy to say that God exists! Simone Weil distinguishes, therefore, between two kinds of atheism of which one is in fact a purification of the notion of God: according to her, religion insofar as it is a source of consolation is an obstacle to genuine faith, and in this sense atheism is a useful exercise in purification.[†408] Whether all this suffices to satisfy true believers, we do not venture to decide; but, if at the very least we join Schopenhauer in believing in the existence of an indestructible "metaphysical need" in man, we may say that we are presented here (to vary Lévinas) with a metaphysics for grown-ups.[‡409] For this is a journey towards the unknown which, precisely because of what Dame Iris very appropriately calls its "iconoclastic pilgrimage"[410] — that is, its ongoing destruction of false images — leaves intact a respectful *distance* from the 'Absolute', instead of whining for it like a child for the moon. And, although Kant did not explicitly arrive at so radical a conclusion (unless just possibly in the most secret recesses of his mind, to make use, for once, of Schopenhauer's insinuation ...), *it is fully implicit in his entire mode of thought* — one for which the difference between the 'phaenomenal' and the 'noumenal' worlds amounts, in the last analysis, to the difference between two grammatical forms: the world of the Indicative and the world of the Imperative. This is indeed the difference between 'is' and 'ought', but grounded here in a metaphysical foundation.

We have already spoken of the 'truth-commandment' as the link between 'Theoretical Reason' and 'Practical Reason', and of the resulting primacy of the latter. Indeed, the commandment in theoretical reason is hypothetical: you must accept certain rules if you wish to play the game of science. But why play the game in the first place? The decision taken thereby to submit to the commandment of truth is a moral decision, which results therefore from the very same Categorical Imperative of 'Practical Reason' which, as we saw, together with the Idea of the

---

[*] Curiously, a similar statement is made by Sartre: "The absence of God is no longer the closing but rather the opening of the Infinite. The absence of God is greater and more divine than God." ("L'absence de Dieu n'est plus la fermeture; c'est l'ouverture de l'Infini. L'absence de Dieu est plus grande, elle est plus divin que Dieu.").

[†] "La réligion en tant que source de consolation est un obstacle à la véritable foi: en ce sens l'athéisme est une purification."

[‡] Lévinas has called Judaism "une religion d'adultes".

Good was advanced by Iris Murdoch as the core concept of "demythologized religion" as she calls it (or of 'metaphysics for grown-ups', as, perhaps somewhat presumptuously, we chose to call it). But what, meanwhile, about the "traditional idea of God" which, according to Murdoch, served as "an image or metaphor" of that which "cannot be 'thought away' out of human life" (and for which the Idea of the Good, or the Categorical Imperative, provides the mature expression)?

Why, this "traditional idea of God" (against which we saw Simone Weil object that it is an obstacle to true faith in that it is a source of *consolation*) appears with Kant when he argues that from the existence of the moral law we may infer a supreme lawgiver who will introduce *happiness* into the *summum bonum* ('the highest good'). But Dame Iris rightly points out that, strictly speaking, this must be regarded as a slip, since, by Kant's own account, a search for happiness is *heteronomous*, a concession, that is, to selfish desires.[411] And we also recall from Chapter VI, p. 150, Schopenhauer's comment that Kant, in so doing, surreptitiously reintroduced into his moral philosophy, not only eudaemonism, but theological thinking as well. Had Heine got it right after all?

No, he had not, insofar as no contradiction exists between the Kant of Theoretical Reason and the Kant of Practical Reason. For we must reiterate that, in the *Critique of Practical Reason* no less than in the *Critique of Pure Reason*, Kant emphatically affirms that neither God nor Immortality can be demonstrated theoretically, *God's existence moreover not being necessary at all for the foundation of morals*;[*412] just as, on the other hand, in the *Critique of Pure Reason* no less than in the *Critique of Practical Reason* he had already pronounced God a necessary postulate.[†413] So Heine had been warned in advance!

Yet there *is* a contradiction *inside* the Kant of Practical Reason, insofar as it is true that (in the *Critique of Practical Reason* rather than in the *Groundwork* or the *Metaphysic of Morals*) God as a safeguard of *happiness* is offered as a kind of *bonus* to the autonomous moral will. In summary, Kant's pertinent argument runs as follows. Though virtue is the highest good, it is not yet the whole and complete good, since for that, happiness, too, is required. Hence, in the highest good, virtue

---

[*] "Here it should be noticed that this moral necessity is *subjective*, i.e., a need, and not *objective*, i.e., a duty itself; for there can be no duty whatever to accept the existence of a thing (since this regards only the theoretical uses of reason). Nor is meant by it that the acceptance of God's existence *as a foundation of any possible engagement* is necessary. For the latter, as has been sufficiently demonstrated, is based solely on the autonomy of reason itself." ("Hier ist nun wohl zu merken, daß diese moralische Notwendigkeit *subjektiv*, d.i. Bedürfniß, und nicht *objektiv*, d.i. selbst Pflicht sei; denn es kann gar keine Pflicht geben, die Existenz eines Dinges anzunehmen (weil dies bloß den theoretischen Gebrauch der Vernunft angeht). Auch wird hierunter nicht verstanden, daß die Annehmung des Daseins Gottes, *als eines Grundes aller Verbindlichkeit überhaupt*, notwendig sei. Denn dieser beruht, wie hinreichend bewiesen worden, lediglich auf der Autonomie der Vernunft selbst") (*Critique of Practical Reason*, Part I, Book II, Chapter II, section 'Das Dasein Gottes als ein Postulat der reinen praktischen Vernunft').

[†] He did so in the third chapter of the 'Transcendental Dialectic'.

and happiness are conceived of as necessarily connected. But this connection is neither of an analytical nor of a synthetic *a priori* nature. It is not analytical because evidently 'virtue' and 'happiness' are not identical. Nor is it synthetic *a priori*, because in that case there ought to be a cause–effect relationship between virtue and happiness; there is no such relationship, neither in the sense that desire for happiness is the motivating cause of the maxims of virtue (for in that case these maxims would not be moral at all), nor in the sense that the maxim of virtue is the efficient cause of happiness (since, alas, in this world of ours the relationship between causes and effects, even if these have their origin in human volitions, is determined, not by the moral quality of these volitions, but by the acting person's knowledge of the laws of nature and by his ability to use these for his own purposes). So once again we face here an antinomy, since, on the one hand, to further the highest good (in which virtue and happiness are necessarily connected) is an *a priori* necessary object of our will, because it is repugnant to our moral sense to see a rascal happy, whereas, on the other hand, this connection has been shown to be non-existent.

The antinomy, Kant goes on, can be resolved in the same way as the one between necessity and freedom, namely, by considering the acting person to belong at the same time to the phenomenal sphere and to the noumenal sphere. For the former synthetic statement ('a desire for happiness is the motivating cause of the maxims of virtue') is *intrinsically* impossible, whereas the latter ('the maxim of virtue is the efficient cause of happiness') is so only *conditionally*. This is so because it is not impossible for a moral disposition, even if not directly connected as a cause to happiness as its effect in the sensual world, nonetheless to be indirectly connected to it, namely, through the medium of an intelligent Originator (*Urheber*) of Nature. Therefore, the immortality of the soul and the existence of God are necessary postulates of Practical Reason. The former is required for the *completion* of virtue, the latter for guaranteeing its *connection to happiness*.

As to the former, a complete congruence between the will and the moral law is called *holiness*, which is a state of perfection no rational being in the sensual world can hope to attain in this life (at least according to the Protestant Kant ...) ; since, however, such a congruence is nevertheless necessarily required by Practical Reason, it can be found only in an *infinite progress* towards it.* As to the latter, since the acting rational being in this world is not at the same time its cause, he is unable to achieve by his own force that complete unanimity between his happiness and his moral principles which is nevertheless postulated as necessary by Practical Reason; therefore, the Highest Good is possible in the world only insofar as the existence is postulated of a cause of the whole of nature but which is, at the same time, separated from nature and which contains in itself the complete agreement between happiness and morality, that is to say, of a Being which, on the strength of

* Note that the Protestant Kant is actually postulating here something rather like Buddhist reincarnation or, at the very least, like Roman Catholic purgatory!

its Intellect and its Will, is the cause of Nature, that is to say, God.*[414]

We may very well admire Kant's ingenious way of thus safeguarding, by means of God, the prospect of happiness for people who might be repelled by the severity of Kantian ethics; yet it is not too difficult to sympathize a little with the mockery this magic formula was bound to provoke in our poet. It was bound, indeed, to raise the suspicion that Kant introduced God here with a view to the Lampes of this world. Still, more is at stake here. We saw Kant speak elsewhere of a "Realm of Ends", meant as a necessary postulate for arousing an *interest* in the Categorical Imperative, and we called it a tacit admission that sense of duty alone will not do. But, apart from this 'interest' (which once again gives rise to the suspicion that Kant was surreptitiously reintroducing eudaemonism here), we believe that the "Realm of Ends" is above all a necessary postulate for harmonizing Kant's claims of 'autonomy' and of 'universability'.

This is an intricate problem indeed. One can certainly agree with Kant that a certain *disinterestedness* in the sense of 'not being dictated by natural necessities', as is the case with ordinary selfishness, is an essential characteristic of the actions of an autonomous subject;†[415] but does that suffice? For what about the disinterestedness of, for example, Nietzsche's *Übermensch* (superman) or of the cheerful perpetrators of Gide's *acte gratuit* ('gratuitous act')?‡ Kant would perhaps deny that we may speak here of 'disinterestedness' in that these are 'inclinations', and evil ones to boot; yet the fact remains that this kind of inclination is to be

---

* Cf. Chapter VI, p. 136, on Kant's warning in the Preface to the *Critique of Practical Reason*.

† One may of course deny that such a thing as 'disinterestedness' exists. Sophisticated people who see through everything will certainly sneer at such naiveté (to please them, we might replace talk about 'disinterested' actions with the less unctuous term 'nonfunctional' actions). Schopenhauer for his part would certainly remind us, as he reminded Kant, that there can never be any action without a motive and thus (*pace* Gide, cf. our next footnote) neither a crime without a motive. Even so, we are concerned here at the very least with a *category of motives* that is different from those of natural egoism. And that is at the same time the answer to an objection Schopenhauer raised to the thesis of Kant's third antinomy. The example Kant had given of rising from one's chair may not have been very fortunate as a demonstration of an unconditioned beginning, and Schopenhauer is of course right when he says that it is as impossible for Kant to rise without a motive as it is for a ball to roll without a cause. For even if Kant only wished to demonstrate by his rising that it is possible to act without a motive, this very wish would be the motive here. But the point is that man can act from motives that are not dictated by natural necessity, and is as such 'free'.

‡ As to Nietzsche, we recall here (and we return to it in our next chapter) that in the preceding section we already put the question as to why (speaking consistently) Nietzsche's maxim: "Become what you are!" ("Werde was Du bist!") might not be regarded as an adequate formulation of a law given to himself by an autonomous being? As to Gide (who was strongly influenced by Nietzsche), we think here, e.g., of Lafcadio, from *Les caves du Vatican*, who lives from spontaneous impulse to spontaneous impulse, thus at one moment risking his life by saving children from a burning house, and at another moment pushing an elderly man from a train, just to get the kick of committing a 'crime without a motive'.

found only in a 'free', 'autonomous' being. Which is to say that freedom, although certainly the cornerstone of 'human dignity', is nevertheless ambiguous in nature. (True, the expression 'human dignity' was scorned by Schopenhauer for its readily acknowledged proneness to being turned into a mere phrase, yet it is an indispensable one because the autonomy of man as a rational being remains the foundation of *responsibility*, without which there can be no ethics whatsoever). Undoubtedly, *pace* Schopenhauer there *is* an intrinsic relationship between 'freedom' and 'law' — 'true freedom listens to the laws', as a Dutch proverb has it; and, paradoxically, it is precisely because the animal is not 'free' that it may do what it can do. Thus Kant's statement that precisely in the notion of *duty* does freedom reveal itself is not nearly so paradoxical as it seems at first sight; but it is still one step further to proclaim the virtual *identity* of freedom and duty, as Kant does. That he did so was in itself a logical consequence of his wish to combine autonomy with objectivity and humaneness. For accepting that a tension may exist between freedom and duty would imply that the moral law means a *limitation* of my autonomy (an autonomy which, it must be repeated, remains a necessary condition for any ethical possibility at all). And why not, one would ask? Certainly heteronomy deriving from fear of divine punishment or from social conformity is to be rejected as an unworthy foundation for ethics, but there seems nothing unworthy in the kind of heteronomy so eloquently described by Schopenhauer in what we dared to call his own 'commandment', or, for that matter, by Kant in the second formulation of his Categorical Imperative. But that *is* not heteronomy, both Kant and Schopenhauer would protest — in Schopenhauer's case because the Other and I are in fact One, and in Kant's case because both are reasonable beings who participate as such in a systematic community through commonly held, objective laws. But, confining ourselves for the moment, to Kant, these commonly held laws are, according to him, at the same time self-given; hence it seems that harmony can arise from such a community only *when all reasonable beings want the same*. That, Kant would answer, is precisely what they do, insofar as they are reasonable beings, that is, insofar as they belong to the noumenal realm. In fact, 'good will' is for Kant a pleonasm: fundamentally, man can never *will* evil, because of his being aware as a rational being (= as an *intelligible* I) of the moral law. Actual evil derives (to adopt Schopenhauer's distinction between 'willing' and 'wishing', not used as such by Kant)* from his *wishing* to allow for exceptions from the moral law as far as his own individuality (= his *empirical* I) is concerned, i.e. from his making the motives of self-love and its inclinations into a condition of his following the moral law instead of the other way round.[416]

Now this comes uncomfortably close to Rousseau's notorious *volonté générale* (General Will). That is not by accident. Well-known is the anecdote that Kant, when reading *Emile*, got so utterly absorbed as for once in his life-time to forget the hour for his daily stroll, which otherwise was used by the Königsberger citizenry to set their watches by. Indeed, for the domain of 'practical reason'

* Cf. Chapter VI, p. 154.

Kant's encounter with Rousseau has scarcely been less decisive than his encounter with Hume had been for the domain of 'theoretical reason'. The 'Profession de foi du Vicaire Savoyard' (to be found as an interlude in the fourth book of *Emile*) deals with the same problem which vexed Kant, namely, how the belief in the Ideas of God, Freedom, and Immortality can be justified in view of the fact that Reason cannot prove their existence. The great difference is, of course, that Rousseau solved the problem by declaring Reason incompetent in favour of sentiment and a 'return to nature', whereas Kant took refuge in an alternative ('practical') kind of Reason. Moreover, Kant's awareness of the "radically evil in human nature" seems a far cry from Rousseau's belief in the fundamental goodness of the latter.* Yet, just as the *actual* existence of evil was of course recognized by Rousseau (and attributed to the corrupting influence of civilization and current society, the original sin being the introduction of private property), so Kant's belief that man can *wish*, but not *will*, evil suggests that for him, too, man's innermost nature is 'good'. (This seems to be confirmed by his statement that the "capacities for the Good" (*Anlagen zum Guten*) in man are 'original' and the inclination towards evil 'accidental').

Only that for him it is as a *rational* being that man is good whereas it is his being partly a *natural* being which is responsible for his evil inclinations. At least that is how it appears at first sight; in fact, Kant's position is more subtle. He knew very well that the "tendency in man to be animal-like" ("die Anlage für die *Tierheit* im Menschen") was responsible only for some of his possible vices (in general, the coarser ones such as gluttony and voluptuousness), whereas it is precisely the "tendencies to be human-like" ("die Anlagen für die *Menschheit*") that are responsible for the more refined ones such as jealousy, ingratitude and gloating. As a matter of fact, both tendencies are considered by Kant as good in themselves since they necessarily belong to human nature; they are evil if improperly used since this improper use does *not* necessarily belong to human nature (that is what Kant meant when he called the capacity for the good 'original' and the inclination towards evil 'accidental'). It is otherwise with the *third* tendency: the "tendency towards being a personality" ("die Anlage für die *Persönlichkeit*"). For here we meet again that famous susceptibility to respect for the moral law in us that we already encountered! *This* tendency (in contrast to the other two) can*not*, according to Kant, be turned into evil; therefore it cannot, strictly speaking, be called a "*tendency* towards being a personality" since the Idea of the moral law together with the 'respect' that is inseparably connected with it is

* "On the radically evil in human nature" ('Über das radikale Böse in der menschlichen Natur') is the subtitle of the first part ('On the living together of the evil principle and the good one': 'Von der Einwohnung des bösen Prinzips neben dem guten') of Kant's book '*Religion within the Boundaries of Reason Alone*' (*Die Religion innerhalb der Grenzen der bloßen Vernunft*), first published in 1793 (the publication caused him problems with the Prussian censorship because it was judged to undermine biblical and ecclesiastical faith).

the very *essence* of 'personality'.[417] Here, therefore, we are concerned only with what is 'original' in man and not with what is 'accidental'.

To us the distinction between 'original' and 'accidental' seems somewhat questionable, together with that between "tendencies towards being human-like" and the "tendency towards being a personality". For one might object that the "improper use" of the "tendencies towards being human-like" *does* necessarily belong to human nature, precisely because man is man and not an angel (as one might believe from Kant's introducing the "tendency towards being a personality" as something *different* from "tendencies towards being human-like").* Thus, although Kant knew very well that freedom is the necessary condition of the 'accidental' deviation of the moral law since otherwise man could not be blamed for it,[418] i.e. that *freedom is ambiguous in nature* as we formulated it in the above, he in fact recoiled for this insight by proclaiming an abstract, purely rational man as 'true' man who as such belongs to that ideal community he called the "Realm of Ends". 'True' freedom is therefore indeed for Kant as well as for Rousseau synonymous with obeying the law. The difference is, of course, that Kant's "Realm of Ends" belongs strictly to the 'noumenal' sphere, i.e, it is really unhistorical in nature. It was to be otherwise later on with Hegel's Absolute Spirit and Marx' classless society, in which we discern the legitimate descendants of Rousseau's *volonté générale*. For Kant, history could never arrive at an end because of the "crooked timber" out of which human nature (i.e., the nature of really existing terrestrial man) was hewn, although mankind certainly has the duty as well as the capacity to *approximate* progressively the ideal of a perfectly moral society.† Therefore he saved the *complete* harmony of wills (which was for him no less a longed-for ideal than for Rousseau and later for Hegel and Marx) for the 'noumenal' sphere; fortunately so, but characteristically he does not question the *desirability* of that ideal of complete harmony. Moreover, his insistence on the goodness of man's 'original' nature leaves the origin of evil (unless one accepts the biblical doctrine of original sin, which Kant, however, emphatically rejected)[419] rather mysterious, as he himself confessed.‡[420] For Schopenhauer this problem did not of course arise, as *his* metaphysics rather raises the opposite question of how there can be any *good* in this world!

To this, Schopenhauer's answer is, in brief, by means of an insight into *tat tvam asi*. And thus we are back at *Schopenhauer's* metaphysics. It should be clear that this hoary, Indian doctrine has been transformed by Schopenhauer in that he believed he had identified *what* 'is you' as Will; and we shall have something to

---

* cf. Chapter VI, footnote on p. 150.

† In Kantian terms it is a regulative Idea', just like 'eternal peace'. Cf. also our exposition in Chapter V, p. 131.

‡ Indeed, one of his commentators (Günther Patzig) remarks in this context that one wonders how for Kant the morally bad man can have reality at all, in view of his teaching that man as a 'natural being' is in fact only how man as a 'rational being' *appears* to himself.

say about the questionable compatiblity of this metaphysics of his with his professed adherence to Kant's doctrine of intelligible freedom. But first consider the *tat tvam asi* as such.

Certainly Schopenhauer's doctrine going under that name is impressive, and there was a time when I also thought it convincing. True, I was already disquieted a little by good old obese Chesterton's protestation (directed not against Schopenhauer, to be sure, but against Annie Besant) that he wanted to love his neighbour, "not because he is I, but precisely because he is not I", and also that "if the world is full of real Selves, they can be really unselfish Selves; but upon Mrs. Besant's principle the cosmos is only one enormously selfish person".[421] But of course, there seemed to be no reason to take too seriously a notorious paradox-hunter who, over his entire life-time, used his considerable wit to defend popular prejudice. Many years later, however, I picked up from Lévinas that the religious category of God as the "radically Other" reveals its ethical dimension in the confrontation with the countenance (*le visage*) of the "other man" (*Autrui*), which forces itself upon me as an *external reality*, transcending my *ego*, which it is therefore impossible for me fully to grasp or to dominate.[*] Translated into Schopenhauerian terms, it seemed to suggest — and as such it sounded convincing — that, contrary to what Schopenhauer taught, it is the whole-hearted *affirmation* rather than the *abolition* of the *principium individuationis* which forms the proper root of the ethical attitude.

Perhaps the reader is inclined to dismiss this as just a battle over words, and to insist that we are all agreed in any case that *altruism* is the core of ethics; and that, since this beautiful sentiment is painfully absent from Kant's harsh 'duty'-ethics, Schopenhauer has a point here against Kant. However, Kant rejected altruism only as a *sentiment*;[†422] moreover, it has rightly been argued that it is a rather primitive idea that egoism is simply 'the bad' and altruism simply 'the good';[‡423] finally, one

---

[*] Cf. footnote about Lévinas on p. 167.

[†] That is, insofar as sentiment in general is banned by Kant from the domain of ethics. But of course this does not imply, as Schopenhauer suggests, that Kant's ethics is at bottom an egoist one, for, according to Kant, altruist actions may certainly be obligatory on rational grounds. In the 'Metaphysik der Tugendlehre' he phrases this as follows: "Love is not understood as *sentiment*, i.e., not as love from *pleasure* (for there can be no duty of having sentiments), but should be thought of as 'practical', i.e., as a maxim of *benevolence* resulting in doing well." ("Liebe wird nicht als *Gefühl*, d.i. nicht als Liebe des *Wohlgefallens* verstanden (denn Gefühle zu haben, dazu kann es keine Verpflichtung geben), sondern muß als Maxime des *Wohlwollens* (als praktisch) gedacht werden, welche das Wohltun zur Folge hat.").

[‡] Günther Patzig rightly says that not egoism but rather the incapacity of men to restrain their irrational strivings and impulses is what produces the greater part of man-made calamity in the world, and that much that was originally done *also* out of — enlightened — self-love deserves nonetheless to be highly appreciated from a moral point of view (he mentions the Marshall-plan as a good example).

may well ask whether Schopenhauer's moral philosophy can indeed be qualified as one of altruism?

As regards this question, we wish to recall young Nietzsche's description (in his *Birth of Tragedy*) of the original Dionysus-cult as a "horrible witches' brew of sensuality and cruelty" ("jener scheußliche Hexentrank aus Wollust und Grausamkeit"). Although the connection may not seem too obvious at first sight, it becomes more so if, with Julian Young,[424] we see here an implicit criticism of Schopenhauer's doctrine that the overcoming of plurality and the insight that all individual entities are nothing but drops in the ocean of metaphysical Will is *eo ipso* bound up with altruism; *and that therefore the attainment of such a state of mind is intrinsically blissful.*[*] Now it would surely be an insult to the great mystics of East and West (whom Schopenhauer professed to admire) to put their illumination on a par with the Dionysian ecstasy of orgiastic fertility cults (which Schopenhauer no doubt abhorred). Still, it seems obvious that both have a common root in the original, human trauma of the separation between 'I' and 'World'. The gradual development of the infant's consciousness chases it out of the paradise of non-duality (*Advaita* in Sanskrit) — a lost paradise which for ordinary man is only transitorily regained in sleep, but the regaining of which on a higher level may be said to constitute the essence of the mystic's striving.

'On a higher level', indeed; for, far from with a simple-minded regression to an infantile level of consciousness, we are concerned here, on the contrary, with a degree of concentration which precisely presupposes a highly developed consciousness. But, with all admiration due to such an effort, one might well ask whether the desire to overcome the *principium individuationis* is a truly *ethical* one. It is not by chance that Schopenhauer's doctrine of compassion was not his final word on the topic. His ethics really consists of three layers, or, to be more precise, his ethics in the strict sense of the word is the middle part of a triptych. At one side it is flanked by the worldly wisdom of his 'Aphorismen zur Lebensweisheit', in which he gave a range of hints as to how this cursed life of ours may be made relatively happy (or at least painless) after all. Almost needless to say, this is the 'ethics' according to which he lived himself. At the other side stands the ultimate ideal of the ascetic saint as evoked in the fourth book of *The World as Will and Representation.* "It no longer suffices to him [namely, to the man who, having seen through the *principium individuationis*, thus recognizes the essence of things and by it the Whole, PL] to love others as himself and to do as much for them as for himself; but there arises within him a horror of the nature of his own phenomenal existence — the Will to Live, the core and inner nature of a world recognized as being full of misery."[†425] In other words, Schopenhauer is an

---

[*] We come back to this in Chapter IX, p. 244.

[†] "Nämlich es genügt ihm nicht mehr, Andere sich selbst gleich zu lieben und für sie soviel zu thun, wie fur sich selbst; sondern es entsteht in ihm ein Abscheu vor dem Wesen dessen Ausdruck seine eigene Erscheinung ist, dem Willen zum Leben, dem Kern und Wesen jener als jammervoll erkannten Welt."

ethical thinker insofar as there is for him a difference between 'is' and 'ought'; the latter is rather an 'ought-not-to-be-at-all', thus turning existence itself into guilt! In this, Schopenhauer may of course be right; but in any case 'world-negation' (*Weltverneinung*) seems hardly to offer an incentive for trying to provide other people, too, with the opportunity of enjoying the good things of life which one enjoys oneself (and which, by the way, one can really 'enjoy' only because one knows from experience the contrast with lacking them, which means in a sense a positive appreciation not only of 'world' but even of 'suffering'!). 'World-affirmation' (*Weltbejahung*) — which is the very condition for any improvement of the world, however inevitably imperfect or even ambiguous always — seems to be intrinsically bound up, therefore, not only with vulgar hedonism, but with a truly ethical world-view as well.

Returning now to the middle part of the triptych, it is time for us to become fully explicit about what, contrary to Schopenhauer, we consider to be the essence of a truly ethical attitude towards life ... always with the proviso that the choice between an 'ethical' and a 'meta-ethical' world-view (like Schopenhauer's or the Buddha's) is indeed a *choice*, rooted in the last resort in temperament and not in rational argument. Phrased negatively, it is a refusal to join in that lamenting of man's 'alienation'; phrased positively, it is a striving after a kind of unity which leaves intact the difference between Self and Other,* because it does not regard plurality as an evil to be combated or as a Fall from the Absolute, but rather as the very condition for practising goodness (which in a certain sense implies again, paradoxical as it may sound, a positive attitude toward 'suffering', since its existence is also a necessary condition for such practice). Certainly, Schopenhauer is right when (contrary to Kant) he holds compassion and empathy to be the only motor drives towards bestowing benefits upon others (though he was too much of a realist to join Rousseau and many philanthropists in believing in an innate kindness of the human heart, the natural inclination of which he rather believed to be egoism). It is also true that these sentiments result from a partial identification with the other (even though, unfortunately, the same is true of 'malice' — Schopenhauer's third basic motive for human action —, for how could, for example, a sadist enjoy inflicting pain on others without being aware that they are, like himself, beings capable of suffering?). Indeed — one might agree with Schopenhauer — the other is 'like me', but (one might go on to object to him in the same breath) he is so in the sense that he, too, is a *person* in the Kantian sense and as such at the same time *different* from me (an 'I' which is not an 'I myself').

* 'Correlation' instead of 'identity' once again! We have already met this fundamental difference in Chapter II, p. 44, and Chapter IV, p. 93. Would it be an all too crude simplification to say that the heart of the problem presented by Schopenhauer's philosophy consists of his failure to stick firmly to the 'correlation' between subject and object (which he recognized as such), lapsing instead into subjective idealism of Berkeley's kind in his epistemology, and into a mystical philosophy of identity in his ethics? (cf. our remarks on what we consider to be Schopenhauer's πρωτον ψευδος (*prôton pseudos;* 'primal fallacy*'*) in the final section of our Chapter 8).

To a lesser extent (the 'analogy' mentioned on p. 175!) this goes for animals as well (or at least for those higher animals to whom Schopenhauer's — as almost everybody's — love was actually confined) in that most definitely one cat is not any other. The latter consideration was also the ground for our statement in the preceding section, namely, that Kant's "Realm of Ends" at least had to be democratized in order to be fully acceptable to us; after the more fundamental objections we have raised in the present section against this Kantian concept we are all the more happy to illustrate our position with Kant's own beautiful description of friendship in the 'Metaphysik der Tugendlehre'. He defines it as a state of mind in which 'love' and 'respect' are combined like attraction and repulsion in mechanics, with the former aiming at ever growing intimacy and the latter never ceasing to impose the required distance.[*426] Leaving aside the comparison with mechanics, would it be too far-fetched an assertion to see in this ambiguity the core, not only of 'friendship', but also, much more generally, of an 'ethics for grown-ups' in the sense that the element of distance implicit in 'respect' is not only necessary for accepting the other as 'other' but also for dealing with him or her on equal terms (pure 'compassion' lending itself all too easily to a certain condescendension towards the other, who thus tends to be reduced to the pitiful status of a victim)? And since for a long time already our patient reader may have painfully missed Goethe, we may also invoke here a characteristic statement in which he illustrated his respectful attitude towards Nature by comparing her with a girl that attracts one but escapes one at the very moment one believes one has grasped her ...[†427]

But we will leave Goethe for the moment (in our final chapter we shall have to say something more on the frequently ignored, ethical dimension of his mode of thought) and pursue, a little further, our analysis of the ethical value of Schopenhauer's metaphysics. After having spoken of its general principle of *tat tvam asi*, it is time now to say something about Schopenhauer's particular interpretation of the universal being underlying all individual differentiation as "Will"! Leaving to our next chapter the questionable side of Schopenhauer's derivation of the Will as Thing-in-itself from inner experience, we here declare ourselves quite ready to admit that this particular approach distinguishes Schopenhauer's metaphysics favourably from both pre-Kantian dogmatism and post-Kantian German Idealism, in that it did not indulge in abstractions like 'Spirit' (or, for that matter, 'Matter'), but led the reader straightforwardly into what a few decades ago it was fashionable to call the 'existential' sphere. In this

---

[*] The combination of mutual love (*Wechselliebe*) and respect (*Achtung*) is held by Kant to be normative for the relations between reasonable beings in general; but in friendship these two elements are "linked together in the most intimate way" ("am innigsten verbunden").

[†] "Die Natur ... ergiebt sich nicht einem Jeden. Sie erweist sich vielmehr gegen Viele wie ein neckisches junges Mädchen, das uns durch tausend Reize anlockt, aber in dem Augenblick, wo wir es zu fassen und zu besitzen glauben, unsern Armen entschlüpft."

context, the reader might wish to recall Rosenzweig's statement, quoted before in Chapter V, p. 129, when we were discussing Hegel's philosophy of history. But we must add, now, that Rosenzweig said something more: "True, Schopenhauer incorporated Kant's doctrine of the intelligible character into his own doctrine of Will, but (in a manner opposite to that of the great idealists) he took away its worth. By turning Will into the essence of the world, he did not so much fuse Will into the world as fuse the world into Will, thus destroying the very distinction of which he had such a lively awareness — the one between the Being of man and the Being of the world."*[428]

This is certainly true — Schopenhauer's Will is a Cosmic Will. Given this much, we may well ask how Kant's distinction between the empirical and the intelligible character (a distinction so abundantly praised by Schopenhauer since it is the warrant of human freedom) fits into Schopenhauer's metaphysical vision? The Cosmic Will is 'free'; but since it is one-and-the-same Will that underlies all phenomena, it is difficult to see why the human individual should in any sense be more 'free' than any other part of nature. This means not only (as Cassirer pointed out) that the intelligible essence is sought in a 'mythical past' rather than in the 'ethical future',†[429] but also that there is really no place for a realm of freedom in Schopenhauer's metaphysics!

This goes without saying for the 'empirical character', as Schopenhauer, in his 'On the Freedom of Will', had proven himself to be a perhaps even more thorough-going determinist than Kant. But it is not really otherwise with the 'intelligible character' as presented by Schopenhauer. Consistency would require one to speak here of only *one* 'intelligible character' for the whole world, since plurality presupposes space and time and thus belongs to the phenomenal world only. In fact the 'intelligible character' is in Schopenhauer's ethics as much of a *Fremdkörper* as the Platonic Ideas are in his aesthetics. Both are a sort of hybrid. They are not 'Thing-in-Itself', since both are plural. But they are not 'appearance', either. Why not? The Platonic Ideas, as pre-existing ideal types, are nonetheless 'one' in relation to the many individual samples of each. By the same token, man's 'intelligible character' as conceived by Schopenhauer may be said to represent in

---

* "Schopenhauer hat zwar Kants Lehre vom intelligibelen Charakter in seiner Lehre vom Willen aufgenommen, aber ihren Wert, und zwar in entgegengesetzter Richtung wie die großen Idealisten, entwertet. Indem er den Willen zum Wesen der Welt machte, ließ er zwar nicht den Willen in die Welt, aber die Welt in den Willen aufgehen und vernichtete so die in ihm selbst lebendige Unterscheidung zwischen dem Sein des Menschen und dem Sein der Welt."

† One may doubt whether 'future' as such is more 'ethical' than 'past'; at least the condition should be made that the future remains *open*, as quite enough atrocities have been committed in the name of a future, into the secrets of which the perpetrators believed themselves initiated. In any case, we agree with Cassirer if he meant to say that in an ethical interpretation the intelligible essence appears, not as *gegeben*, but as *aufgegeben*. Yet, as we shall see later on (Chapter IX, p. 258), we believe that respect for the 'given' is also an essential element of true ethics.

Reinkultur (in its pure form) one specific kind of 'willing' which is characteristic of his personality but which, in his 'empirical character', appears invariably mixed up with other drives — "will" vs. "wishes"! In the last resort the human individual is to Schopenhauer only an ephemeral manifestation of the one, universal Will. To him, as to Hegel (though in a different way), individuals come and go whereas the One remains — the latter being a favourite topic not only in Eastern thought but also (though less predominantly) in Western thought from Parmenides onward, now presenting itself in the guise of what Schopenhauer chose to call "Will".

Now in a sense this is true even for Kant's "Realm of Ends" (as we have just observed), insofar as tensions, due to the differences among men, are not wholeheartedly accepted there.[*430] In any case, Kant's position is at least consistent since (1) the Realm of Ends is of a *noumenal* character and thus *beyond* Nature, and (2) Reason is after all the same in all men. In contrast, Schopenhauer's universal "Will" is rather the innermost *core* of Nature ("Ob nicht Natur zuletzt sich doch ergründe?"), whereas at the same time "Will" in Schopenhauer's sense (about which we shall have something more to say in our next chapter) is precisely that, one might say, which constitutes man as an individual being — an individual being, moreover, which (not only when his name is Faust!) has "two (or more) souls in his breast". The *noumenal* "Thing-in-itself" of Kant is really transposed by Schopenhauer into the *phenomenal* realm, albeit into its deeper, subconscious layer, and it is difficult to see how it can then reveal itself in the human individual as his 'intelligible' character. And this has consequences for the problem of freedom. First, Schopenhauer's doctrine that freedom is to be sought not in the *operari* (which is in *time*, and therefore subject to the principle of sufficient

---

[*] Kant's 'loveless' metaphysics of respect for the law, i.e., for the abstract universal reason which is the same in all men rather than for concrete man in his endless variety, and Schopenhauer's 'lawless' metaphysics of universal love which tends towards an ecstatic absorption of the 'Self' into the 'Other' seem thus, paradoxically, to suffer from the same weakness. And this coincidence may well point to a fundamental tendency in human psychology. Without wishing in the least to connect our two thinkers with 20th-century totalitarianism (which would in their case be still more unjust as well as unhistorical than in the cases of Rousseau and Hegel), one may well ask whether the secret of the totalitarian temptation does not lie in its harsh and severe policy of repression clothed in the idyllic vision of universal comradeship? And this in its turn gives rise to the question whether (unlike the customary, superficial mode of rejecting totalitarianism, which stresses the mendacious discrepancy between its reality and its ideal) one should not rather dispute the beauty of the ideal itself? We think here of how Milan Kundera made his libertine painter Sabina comment upon Soviet propaganda films: "The conventional interpretation of these films is that they showed the communist ideal, whereas communist reality was worse. Sabina always rebelled against this interpretation. Whenever she imagined the world of Soviet kitsch becoming a reality, she felt a shiver run down her spine. She would unhesitatingly prefer life in a real communist regime with all its persecution and meat queues. Life in the real communist world is still liveable. In the world of the communist ideal made real, in that world of grinning idiots, she would have nothing to say, she would die of horror within a week."

reason, appearing here as *motivation*) but in the *esse* (which is nothing but the particular form in which Will, before and beyond all time, chose to manifest itself) is not unlike the doctrine of predestination with which he indeed compared it himself,*[431] whereas Kant considers obeying the law in principle possible for any reasonable being. Second, although Kant's freedom, precisely because of its 'noumenal' nature, is a somewhat unearthy freedom in that it hovers above the realm of necessity rather than realizing itself (however imperfectly) in the confrontation with it (and thus exposes itself to the reproach, voiced by Charles Péguy, that Kant has clean hands but has no hands)†[432] it is nevertheless genuine freedom in the sense in which we for our part would define it: namely, as the human capacity to detach oneself as a rational being (even if necessarily up to a point only) from being conditioned by *nature*, and also (once again: up to a certain point) from being conditioned by one's *particular culture*. Certainly for Schopenhauer, too, 'Reason' is what differentiates man from animal, but it is not man's kernel. Therefore reason appears for him only as 'instrumental reason' in Horkheimer's terminology; consequently, the only kind of freedom Schopenhauer in fact knows of is the relative one we have seen him describe in his treatise on the freedom of Will (our Chapter VI, p. 155) and (Chapter II, p. 48) in the *Fourfold Root*.

Or perhaps not the only kind after all? Certainly, 'Understanding' is, for Schopenhauer, a faculty we share with the animals, and 'Reason', although specifically human, has for him a much narrower scope than for Kant and *a fortiori* for German Idealism. 'Intellect' in general is secondary to 'Will', in the service of which it stands. Yet we see young Schopenhauer in the secrecy of his notebooks introducing a faculty which he calls "better consciousness" ("besseres Bewußtseyn") over and above the 'normal' intellect.‡[433] The expression itself is not used in his published writings, but in the third book of *The World as Will and Representation* we learn that in exceptional cases cognition tears itself loose from its subservience to will so that the subject ceases to be a mere individual (=willing)

---

* The Augustinian doctrine (reappearing in Luther's *De servo arbitrio*) that will is not free but subject to the inclination towards evil is considered by Schopenhauer as the original and truly evangelical Christian doctrine, whereas modern rationalism is nothing but a new edition of the 'vulgarities of Pelagianism'. The contradictions within this Augustinian theology (e.g., a God who created the world, yet is not judged responsible for the state of things in it, God's goodness and the misery of the world, freedom of will and God's foreknowledge), which have provoked the Pelagian reaction, are, according to him, due to the Jewish fundamental dogma that man is the product of another's instead of his own will; remove this dogma and all becomes clear. But alas modern rationalism in Schopenhauer's opinion clings precisely to this Jewish legacy and rejects instead those profound dogmas which are characteristic and essential for Christianity (which was interpreted by Schopenhauer as akin to Indian thought).

† "Kant qui a les mains pures mais qui n'a pas de mains ..."

‡ It appears for the first time in notes of the young student in Berlin of the year 1812 and from then on often reappears in his notes.

subject and becomes a "pure subject of cognition", a "clear world eye" ("klares Weltauge"),[434] which no longer restlessly pursues relations on the basis of the principle of sufficient reason but is absorbed in quiet contemplation of the object. This is the *aesthetic* way of looking at the world; however, it provides only temporary release. Only in the ascetic saint is the bondage of Will definitely overcome. And this is crucial: what fundamentally distinguishes Schopenhauer from 'vitalist' worshippers of 'Life-Force' is precisely his 'pessimism', which always prevented him from idealizing "Will" into anything like the healthy alternative to abstract 'Reason'. Thus, although Will is the Thing-in-itself, there is a way out: Intellect, though not in its customary guise, can play tricks upon its master ("kann ihm ein Schnippchen schlagen", to use Thomas Mann's expression).[435] Thus there is genuine freedom after all, but only in the one sense of Will arriving at full self-consciousness and consequently at self-negation in some people; and this illumination cannot be intentionally compelled, but seizes one as an effect of 'grace' (in the Christian terminology to which Schopenhauer refers here once more).[436] But unfortunately we believe that, in raising this possibility, Schopenhauer became ensnared in an antinomy comparable to the one that we met previously in his epistemology. The analysis of this antinomy and of the metaphysical paradox arising from it forms the subject of our next chapter.

VIII

THE PARADOX: CAN GOD COMMIT SUICIDE?

a. Schopenhauer's great doctrine: Will as Thing-in-itself

Here is how Schopenhauer reasoned his way towards identifying the Thing-in-Itself as Will.

The willing subject, so he had argued in *The Fourfold Root*, is in fact an *object* for the knowing subject (namely, an object of the fourth class of the objects for the subject). Nevertheless, there remains the *identity* of the willing subject with the knowing subject, by virtue of which the word 'I' necessarily includes and defines both. This identity he called "just incomprehensible" ("schlechthin unbegreiflich") in the first version,[437] and "the world-knot and therefore inexplicable" ("der Weltknoten und daher unerklärlich") in the second;[438] and in both versions he comes to the conclusion that he "who fully realizes the incomprehensible of this identity will call it with me the miracle κάτ' 'εξοχην (*kat' exochèn)*", i.e., the pre-eminent miracle, the miracle-if-ever-there-was-one.[*439] In the main work, however, Schopenhauer somewhat inconsistently conceded that this book was an attempt to explain the miracle after all;[440] and his solution amounts to the following.

Insofar as I recognize my will as an object, I recognize it as my *body*, and the latter belongs to the objects of the *first* class of the objects for the subject in which the principle of sufficient reason appears as *causality*. But this first class of the objects for the subject in its turn finds its true explanation in the *fourth* class, and there the principle of sufficient reason appears as *motivation* — defined by Schopenhauer (as we saw in Chapter II) as "causality seen from within". For, although we observe how in the outer world mechanical, physical, and chemical effects promptly follow upon their respective causes, it remains a mystery how it is *possible* in the first place for *this* cause necessarily to produce *that* effect. Similarly mysterious would be the movements and actions of animals and of human beings if we did not know from *inner* experience that we are concerned here with a *volition* provoking the *motive*. The way in which the motive works is therefore known by us, not only indirectly, from without, but also directly, from within (cf. Chapter II, p. 47). And in the same way one might say that one's own

[*] "Wer aber das Unbegreifliche dieser Identität sich recht vergegenwärtigt, wird sie mit mir das Wunder κάτ' 'εξοχην nennen."

body, although it is an object of the first class, is distinguished from all other objects in that class in that it presents itself to consciousness in still another, *toto genere* different way, namely, as *Will*. Precisely of our own body we possess this kind of dual knowledge — it informs us *both* about its own actions and movements (guided by motives) *and* about what it suffers owing to external causes — knowledge that we do not possess for any other object of the first class. Our own body, therefore, serves in Schopenhauer's philosophy as the one hinge that links the phenomenal world ("World as Representation") to the Thing-in-itself ("World as Will"). Indeed, we perceive ourselves as material bodies, that is, as a thing among things. But, since all changes of our body (accompanied as they are by pain or pleasure) present themselves to us at the same time as changes in what we feel, the body is a very special kind of object, just as motivation is a very special kind of causality. To us it is the one *direct* object, for the knowledge we have of it is not gained solely *by way of representation*, as is the case with all other objects. The body is the same thing that we sense from inside as *will*, but now cast into the form of *outer* intuition, i.e., of *representation*. Hence, between body and will there is, not a causal relationship, but pure identity (cf. Chapter IV, p. 100).

Now this means, according to Schopenhauer — and this is a very important point — that in the fourth class of the objects for the subject *we are actually no longer dealing with an object for a subject*. True, it must be granted that what I recognize in inner experience as 'will' is still a phenomenon, for the very reason that it is experienced. Nonetheless, Schopenhauer thinks, it distinguishes itself from other phenomena in that it is not draped in three veils (space, time, and causality), but in one only (time).[441] For it is in space only insofar as it is at the same time my body (in which capacity it belongs to the first class); furthermore, although motivation is "causality seen from within", motives in Schopenhauer's view never determine more than what I will at *this* time, at *this* place, and under *these* circumstances — not *that* I will in the first place, nor *what* I will.[442] Thus, using a military metaphor, it may be said that the object of the fourth class of the objects for the subject shows us the secret underground corridor which leads us, as if by treason, into the fortification of the Thing-in-itself.[443] The knowing subject (which as such is impersonal) is an *individual* (i.e., a willing subject) owing to its special relation to this one body which, beyond this special relation, is just a representation like all others.

We are now ready for the crucial point as Schopenhauer sees it. The individual thus considered may suppose either of two things. *Either* he supposes that the peculiarity of this one representation (a peculiarity, to repeat, which consists of the fact that only of *this* visible object has he *dual* knowledge) is to be explained, not by a difference of this object from all other objects but only by a difference of his *knowledge* of this one object from his knowledge of all others. *Or* he may suppose that this one *object* is fundamentally different from all others, in that it is at the same time Will and Representation, whereas the others are only Representation, i.e. mere phantoms. The latter point of view is what Schopenhauer calls "theoretical egoism" (we would nowadays say 'solipsism'), which, although it can never be refuted by proofs, as a sincere conviction is to be encountered only in the lunatic asylum.[444] We must decide, therefore, in favour of the former option and suppose that *the inner essence of all objects that are not our own body is also Will.* Seen in that way, the willing subject is nothing but an individuation of the Cosmic Will which at this stage of its development has arrived at consciousness of itself, thus at the same time becoming a *knowing* subject. Or, to rephrase in Schopenhauer's own words, "The Will, on having so far followed its own urge quite securely and without fail through the dark, at this level kindled a light to itself, as a means which became necessary to get rid of the drawbacks which, precisely for its most perfect phenomena, would arise from their crowding and from their complex constitution."[*445]

b. An unrecognized antinomy, and the paradox that results

This is surely a grandiose vision. It may seem pedantic, superfluous even (since hardly any orthodox Schopenhauerians will be left today), to say that, besides some other possible objections which will occupy us in Section d., unfortunately it leads to a little difficulty. Yet, to say so is indispensable for an understanding of the precarious position of Schopenhauer's philosophy in the history of Western thought. In order to make this clear we would ask the reader to call to mind, once more, our treatment of the 'antinomy' in Chapter IV (p. 94). Its origin, as we saw there, lay in the circumstance that, in spite of Schopenhauer's own explicit prohibition, a *causal relationship* was postulated between subject and object. 'The knowing subject in fact creates (and thus 'causes') the object — an object which is at the same time said to have caused previously the sensation in the sense-organ,' so we rephrased the antinomy in our own words. Now we believe that Schopenhauer's attempted solution of the "miracle κατ΄ ΄εξοχηv" leads to a new antinomy, only, this time Schopenhauer did not recognize it as such. It might run thus: 'At a certain moment in *time* the Will creates (and thus '*causes*') the Intellect — the very Intellect about which it is simultaneously being said that causality and time (and also, less to the point here, space) are nothing but its *a priori* forms.'[†446]

[*] "Der Wille, der bis hieher im Dunklen höchst sicher und unfehlbar seinen Trieb verfolgte, hat sich auf dieser Stufe ein Licht angezündet, als ein Mittel, welches notwendig wurde zur Aufhebung des Nachteils, der aus dem Gedränge und der komplizierten Beschaffenheit seiner Erscheinungen eben den vollendetesten erwachsen würde."

[†] It is reassuring to find ourselves in agreement here with such a distinguished thinker as Cassirer. He pointed out that, on the one hand, according to Schopenhauer the Intellect with its *a priori* forms of space, time, and causality is only a secondary, nay even a tertiary, phenomenon, since it is only a late product of Will, appearing at the final stage of its successive objectivations ("Far from being simply the first (as, e.g., *Fichte* taught), it is really tertiary, since it presupposes the organism, and this in its turn the will" ["Weit entfernt, das schlechthin Erste zu seyn (wie z.B. *Fichte* lehrte), ist es im Grunde tertiär, indem es den Organismus voraussetzt, dieser aber den Willen"], Schopenhauer says, whereas, on the other hand, it follows logically from Schopenhauer's own doctrine that all

Will as 'Thing-in-itself' is One and Changeless; only to the Intellect, i.e., in the "world as representation", does it appear as Plurality and Becoming. This applies even to the (Platonic) Ideas, introduced by Schopenhauer in the third book of *The World as Will and Representation*, which are an object not of workaday empirical experience but of Sunday aesthetic experience, so to speak. For, although they are timeless and unchanging, they *do* form a plurality; hence, Schopenhauer does not consider them to be the ultimate metaphysical reality. Here he sees himself as differing from Plato;[*] yet as a metaphysician he remained basically a Platonist insofar as the timeless nature of ultimate reality is concerned.[†] In a metaphysical sense there can be no question, for Schopenhauer, of any 'history', since it is only by entering into the *a priori* forms of the intellect and thus becoming "world as representation" that "world as will" manifests itself as manifold and developing. In our opinion this involves (or at least ought to involve) that not only "Will" (being the ultimate reality, i.e., the '*transcendent*' — though in the sense of the 'deepest', not of the 'highest'!), but also 'Intellect' (being the very thing that makes history possible in the first place, i.e., the '*transcendental*'), is beyond history. Had Schopenhauer really been the consistent thinker he prided himself on being,[‡] he ought to have arrived at some kind of dualist metaphysics, such as that which Max Scheler was to suggest much later [§447] ... or, better still, he ought to have accepted

---

these 'objectivations' can be there only *for*, not *before*, the Intellect (for the very reason that he holds space, time, and causality to be *a priori* forms, i.e., nothing but the necessary ways in which the Intellect grasps the world), so that in this respect Intellect ought to be considered as primary. Metaphysically, therefore, Will and Intellect would be wholly on a par.

[*] This opinion was not of course quite correct, insofar as the Idea of the Good (which is comparable to Parmenides' 'One', albeit with the important difference that it is 'beyond Being' rather than 'true Being') stands above and is fundamentally different from the other Ideas (cf. Chapter VII, p. 179). So in this respect there is no difference between Plato and Schopenhauer (not counting, to be sure, the trifle that for Schopenhauer the 'One', far from being the 'Idea of the Good', is wicked "Will"). It may however be doubted (as in the present chapter we seek to show) whether this is really Schopenhauer's last word on the matter.

[†] This also explains his attitude towards history (cf. Chapter V, p. 127).

[‡] Cf. Introduction, p. 5.

[§] Scheler distinguishes between what he calls the "classical" and the "negative" theory of Man. The former comprises all those theories which, in one form or another, grant "Spirit" (*Geist*), not only autonomy and original being, but also force and activity, nay even supreme power and force. The latter comprises all those which in one form or another (Schopenhauerianism being one of them) hold that Spirit comes into being only as a result of the specific human capacity to say 'no' to reality, whether in acts of asceticism, of repression, or of sublimation. Rejecting both theories, Scheler presents his own view that Spirit, although possessing being and laws of its own, is in its 'pure', original form without any power, force, and activity. It is only through man's 'negative' activity that it is endowed with energy and thus with the possibility to manifest itself. Thus Being has two that Intellect is at the same time 'primary' from a *transcendental* point of view and 'secondary' (or even 'tertiary') from a *genetic* point of view. That way, he might well have recognized that we are facing here at the metaphysical level the very same, inevitable, *methodological* antinomy we met before (Chapter IV, p. 111) at the epistemological level.

There are, of course, other theoretical possibilities. If, for example, out of a desire to avoid any dualism and yet to satisfy his 'metaphysical needs' Schopenhauer had felt obliged to reject both solutions suggested in the above, he might, Hegel-like, have proclaimed Intellect (in the shape of the 'Absolute Idea') to be the ultimate metaphysical reality, while at the same time postulating *evolution* (in dialectical form) to be its very essence, since (along the detour of its self-externalization in Nature) it finally arrives at self-consciousness as Absolute Spirit. Or he might, Nietzsche-like, have stuck to the primacy of 'Will',* going on for the sake of consistency to drop any 'transcendental' claim to *a priori* validity for the forms and categories of the intellect. Such a juxtaposition of Hegel and Nietzsche may seem rather arbitrary at first sight — hardly a greater difference can be imagined than between these two philosophers, in that Nietzsche, as a (however heretical) disciple of Schopenhauer, held the foundation of the world to be irrational. Still, both may be said to represent some kind of historicism in philosophy. Hegel wished to cling to a divine, intelligible world, and to that end he found it necessary to introduce a historical dimension into his notion of truth. Nietzsche, in spite of his instinctive and frequently outspoken opposition to Hegelianism, nonetheless spoke of "Hegel's amazing find" ("den erstaunlichen Griff Hegels") meaning the latter's daring doctrine, against all logical convention, that concepts develop one out of the other, thus, as a precursor of Darwinism, introducing the idea of *development* into science: "no Darwin without Hegel" ("kein Darwin ohne Hegel").[448] In his own philosophy, Nietzsche proceeded in a much more radical way. The notion of a transcendental subject is dropped, and Kant's question 'how are synthetic judgements *a priori* possible?' is replaced with the question 'why is the belief in such judgments necessary?' They may be inevitable, subjective conditions of any possible interpretation of reality, Nietzsche concedes, but they are not 'transcendental' in the sense of having any *a priori* validity. The constraint of holding them to be true does not derive from their being 'true' in the transcendental sense; rather, it is the other way round, i.e., it is from the constraint of 'holding certain things to be true' that the notion of truth has been derived. But whence this constraint? Nietzsche answers this by saying that hidden

---

attributes: "Spirit" (*Geist*) and "Urge" (*Drang*), the former needing the latter to realize itself in the world-process, culminating in Man.

* Certainly to Nietzsche this 'Will' is no longer a metaphysical being beyond the veil woven by the *principium individuationis*, but rather the will-to-power of the creative individual. Or at least this is so for the later Nietzsche, as in this respect the youngster who wrote *The Birth of Tragedy* was still a more or less orthodox Schopenhauerian ('more or less' indeed, as we shall see in the next chapter).

behind this epistemological constraint lies the biological constraint of the reproduction of life. Our belief in the truth of a certain amount of 'synthetic *a priori* judgements' and in *these* rather than in others is in the last resort a matter of how we value them; we prefer always those which fit best into the task of securing our existence. Synthetic *a priori* judgements are thus in the end value-judgements — which in their turn are only a consequence of the fact that all life seeks to expand its force: life is nothing but Will-to-Power.[*449]

These alternatives are indeed theoretical insofar as both would of course have been unacceptable to Schopenhauer, being the man he was; as a matter of fact, both only at first sight seem to be less inconsistent than Schopenhauer's own position.[†] Their unacceptability from Schopenhauer's point of view is already apparent from the fact that, in a sense, both are a relapse from Kant, whose only legitimate heir Schopenhauer insisted he was. This is so, insofar as both may be regarded as new versions of those two philosophical positions (classic rationalism and empiricism, respectively) which Kant's critical philosophy had sought to overcome. Hegel, who greatly admired Spinoza, presented indeed a (to some extent) novel version of Spinozist pantheism (albeit in a more dynamic, 'evolutionist', and at the same time more 'idealist' way, in that of the two equal attributes of Spinoza's one 'Substance' — 'Thought' and 'Extension' — only the first remains, which moreover must realize itself in history) and thus of rationalist 'dogmatism'. Similarly, Nietzsche's rejection of transcendentalism might in one sense be regarded as a new version (a much grimmer and more 'biologist' one) of Humean 'scepticism'.

Kant and, in his footsteps, Schopenhauer, who here appears indeed as Kant's true heir, chose the 'narrow path' (to use a Gospel term) between the two abysses. A difficult enterprise, this, no doubt, and precisely what made us speak in the above of the 'precarious position' of Schopenhauer's philosophy. What Schopenhauer thought of Hegelianism is known well enough; but Nietzscheanism,

---

[*] Up to this point Nietzsche's "will-to-power" may be said to be universal in nature after all, yet not in the monistic sense of a metaphysical world behind the 'veil of Maya', but rather in the pluralistic sense of an interplay between an infinity of forces. Man (i.e., each individual man) can be seen as the intersection of these forces; and certainly man always brings some kind of order and hierarchy to those forces. To that extent, then, man is for Nietzsche what he was for Kant and, in however contradictory a manner, for Schopenhauer too — a free and autonomous being. But unlike Kant, Nietzsche does not ground this autonomy in Reason, and therefore it is not the source of universally valid, moral values (cf. Chapter VII, footnote on p. 185).

[†] 'Only at first sight'; for the problem with Nietzsche, as with all critique-of-ideology (e.g., Marx's, too) is that he cannot possibly avoid the position of that famous Cretan who must have lied in any case (cf. also footnote on p. 226). On the other hand Hegel's position can be kept consistent only by dint of a dialectics which enables him to incorporate the irrational into the rational (cf. Chapter V, p. 132). Insofar as it is up to us to choose, we therefore prefer Schopenhauer's inconsistency, while recognizing it *as being* an inconsistency.

too (had he lived to know it) would surely have been repugnant to his pessimism-inspired *Mitleidsmoral*. Think only of Nietzsche's devastating critique of what he chose to call "the fable of intelligible freedom" ("die Fabel von der intelligiblen Freiheit")!*[450] This, as we have seen in the two preceding chapters, was the very piece of Kant's legacy which Schopenhauer sought to preserve in his own fashion, as it was precisely what he needed in order to leave open the possibility of an escape from the tyranny of Will.

It is difficult to see how such a *tour de force* could be brought about without actually adopting some kind of what we might call 'amphitheism'. And indeed, Schopenhauer tacitly recognized as much. Here is the famous closing passage of *The World as Will and Representation*: "We rather make a frank confession: what remains after Will has been fully cancelled is, to all those who are still will-ridden, nothing. But conversely, to those in whom Will has reversed and negated itself, this so utterly real world of ours with all its suns and galaxies is ... Nothing."†[451]

What, one might ask here, is now the *true* 'Thing-in-itself'? Is it "Will", or is it 'Nirvana'? Even Schopenhauer's faithful 'arch-evangelist' Julius Frauenstädt, heedless of his master's warnings, already believed he had detected certain inconsistencies in his system.‡[452] In a letter to the venerated master he tried,

---

* According to Nietzsche's exposition in *Human, All too Human*, man was at first made responsible for his *effects* (because particular actions were called 'good' or 'evil' without consideration of their motives but only with regard to their useful or harmful consequences); then for his *actions* (by imagining that the quality 'good' or 'evil' is inherent in an action in the same way as — equally wrongly — a stone was believed to be hard in itself or a tree green in itself); later still for his *motives* (which made the action as such morally equivocal); and finally for his *essence* (from which the good or evil motives were held to arise). This was still the position of Schopenhauer, who inferred the existence of moral responsibility from the phenomenon of our *sense of guilt* (for which there would be no reason if *both* all human *action* occurs with necessity — which he acknowledged — *and* man's whole *essence* is acquired with the same necessity — which he denied). According to Nietzsche, Schopenhauer is subject here to the fallacy that the *fact* of our sense of guilt leads him to inferring its *being justified*; "and from this wrong conclusion Schopenhauer arrives at his phantastic consequence of so-called intelligible freedom" ("und von jenem Fehlschluß aus kommt Schopenhauer zu seiner phantastischen Konsequenz der sogenannten intelligiblen Freiheit"). That is how Schopenhauer failed, in Nietzsche's view, to make the truly final step, which is to stop making man responsible for anything at all.

† "Wir bekennen es vielmehr frei: was nach gänzlicher Aufhebung des Willens übrig bleibt, ist für alle Die, welche noch des Willens voll sind, allerdings Nichts. Aber auch umgekehrt ist Denen, in welchen der Wille sich gewendet und verneint hat, diese unsere so sehr reale Welt mit all ihren Sonnen und Milchstraßen — Nichts."

‡ Schopenhauer's remark in his letter of 31 March 1854 to Johann August Becker about the complete futility of seeking contradictions in his work (cf. Introduction, p. 1) referred to an article by Frauenstädt, about whom he goes on as follows: "But it seems to me that my good Frauenstädt, in his desire to show up his sharp wit, has brought home only quibbling, petty criticisms. However, I should forgive him much once I realize that only through his ongoing, steadfast striving for eight years now, is my philosophy finally gaining an

however timidly, to point to a paradox which inevitably arises from the antinomy we discussed in the above and which neither he nor, consequently, Schopenhauer himself recognized as such. That paradox rests in the Intellect being nothing but the servant and instrument of omnipotent Will while apparently being able, in principle, to annihilate Will. As Frauenstädt (with his knowledge of Schopenhauer as a person) might have foreseen, the result of his innocent query was an explosion.*[453] We do not know whether Frauenstädt was persuaded by the reply he received; to us, at any rate, the problem remains, and it comes down precisely to what is not our joke but Schopenhauer's own irony in putting his disciple right — 'can God commit suicide?' For it may be all very well that, as Schopenhauer impatiently pointed out, the "Will" is not to be located in cuckooland ("where the Jews' God sits enthroned"; "da wo der Judengott sitzt"), and that it is wholly immanent in nature. Still, "Will" in Schopenhauer's system forms the imperishable core of the visible world, which remains the same under all the world's innumerable phenomenal forms as these come and go, and, hence, in spite of Schopenhauer's protestations to the contrary Will serves as a kind of 'Absolute' (even if not in Hegel's sense). At the same time Schopenhauer tells us that, in aesthetic intuition, Intellect can temporarily appease Will, and in the asceticism of the saint it can even vanquish it for good. But if that is so, one must *either* see the

---

audience, and the professors' intrigue is frustrated." ("Aber mir scheint, daß der gute Frauenstädt hat seinen Scharfsinn zeigen wollen und nur spitzfindige Kritzeleien zu Markte gebracht hat. Ich muß ihm aber vieles zu Gute halten, wenn ich bedenke, daß allein durch sein jetzt achtjähriges unausgesetztes, standhaftes Bemühen meine Philosophie jetzt endlich ins Publikum dringt und die Kabale der Professoren vereitelt wird.") But how difficult it was for Schopenhauer at times to let his gratitude prevail over his annoyance with his slow-witted disciple becomes clear in a letter of 21 August 1852 from which we have already quoted a passage in our preceding chapter, and which opens with the words: "I must call to mind, my dear friend, all your many and great merits for the preaching of my philosophy in order not to lose all patience and countenance when reading your last letter." ("Ich muß, mein werther Freund, mir alle Ihre vielen und großen Verdienste um die Verkündigung meiner Philosophie vergegenwärtigen, um nur nicht außer aller Geduld und Fassung zu gerathen, bei Ihrem letzten Briefe.").

* The explosion forms the main content of the letter of which we quoted the first sentence in the preceding note. It was none other than Schopenhauer himself who on this occasion ironically coined the phrase we have chosen as the most adequate description of the Schopenhauerian paradox (a paradox, to be sure, of which he denied that his metaphysics leads to it). What provoked Schopenhauer to his explosion was the manner in which Frauenstädt spoke of the Thing-in-itself, namely, as "the eternal, uncreated, and imperishable primal being" ("das ewige, unentstandene, und unvergängliche Urwesen"). This conception of the Thing-in-itself, the master sneered, has evidently been taken from the synagogue. He then went on: "The meaning of your long digression is, in short, that the good God cannot commit suicide. Quite so! for how should he? and, for that matter, why would he? when he warbles *panta kala lian* ('all is very fine')." ("Und der langen Rede kurzer Sinn ist, daß der liebe Gott keinen Selbstmord begehen kann. Richtig! wie sollte er auch? und wie möchte er? wenn er παντα καλα λιαν trillert.").

world in Manichean fashion as the battlefield of two 'Absolutes' (as we suggested in coining the term 'amphitheism'), *or* accept indeed the paradox that the 'Absolute' (not, that much may be granted, God exactly, but rather the Devil) can 'commit suicide'.

To this Schopenhauer might still retort, as in fact he did, that the redemption of the individual is not yet the redemption of all. Hence, in the ascetic saint only one particular individuation of Will is killed, not Will itself.[*454] However, not only for the (in Schopenhauer's kind opinion) somewhat dense (*begriffsstützige*) 'evangelist' Frauenstädt, but also for the 'apostle' Adam von Doss (really appreciated much more by the master, but unfortunately unable to bring himself to become an 'evangelist' because of his reluctance to write down his thoughts),[†455] the question was precisely *how* this can be so, in view of the metaphysical oneness of the Will. A saint would be able to annihilate the world! To this Schopenhauer replied that the question cannot be answered until we know "how deep the root of individuality goes" ("wie tief die Wurzel der Individualität geht"), and this we shall never know — as a transcendent problem, we must needs leave it unresolved.[‡456]

Yet at the end of *The World as Will and Representation* we read: "Should the maxim [of denial of will] become general, then the human race would become

---

[*] Mark you: in the ascesis of the saint, not in (individual) suicide! Bertrand Russell complains that Schopenhauer never explains clearly why the saint's purpose to come as near as possible to non-existence cannot be achieved by suicide. But in fact Schopenhauer (who actually rejected *individual* suicide!) does explain it and even clearly (whether one is wholly convinced by his argument is of course another question), when he argues that suicide is not a denial but on the contrary an expression of strong affirmation of will, since a person who commits suicide *does* will life and is only dissatisfied with the particular conditions under which it is given to him; and especially when he adds that the essence of denial of will does not lie in an abhorrence of the sufferings but precisely of the enjoyments of life ("Der Selbstmörder will das Leben und ist bloß mit den Bedingungen unzufrieden. Weit entfernt Verneinung des Willens zu seyn, ist dieser ein Phänomen starker Bejahung des Willens. Denn die Verneinung hat ihr Wesen nicht darin, daß man die Leiden, sondern daß man die Genüsse des Lebens verabscheut.") It is particularly this rejection of the good things of life which (in Chapter VII, p. 191) made us call Schopenhauer's attitude 'meta-ethical' rather than 'ethical'.

[†] Angelika Hübscher calls Schopenhauer's regret over this reluctance ("My best apostles do not write"; "Meine besten Aposteln schreiben nicht") 'the only shadow ever to fall over a relation filled with mutual sympathy'. Quoting Doss ("Now that he is dead I feel a bitter regret over my irresolution and hesitation that was rebuked by himself"; "Nun er todt ist, fühle ich bittere Reue über meine von ihm selbst gerügte Unschlüssigkeit und Zaghaftigkeit") she adds that Doss' subsequent illness, which compelled him to go into retreat already at the age of forty-four, unfortunately prevented him from ever becoming Schopenhauer's 'writing evangelist'.

[‡] Cf. the letter to Frauenstädt already quoted and Schopenhauer's letter to Adam von Doss of 22 July 1852. We are reminded here of the final line of the *Tractatus logico-philosophicus* (Wittgenstein, as is well known, underwent Schopenhauer's influence).

extinct; and I believe I may safely assume that with the highest phenomenon of Will its weaker reflection, the animal kingdom, would fade away as well, just as half-shadows disappear together with full brightness. With the complete abolition of existence, the remaining world would automatically fade into nothingness; for without a subject there can be no object."*[457] He quotes the *Veda*: "Just as in this world hungry children crowd around their mother, so all creatures long for the holy sacrifice" [i.e., human self-sacrifice].†[458] He quotes Angelus Silesius: "Man! All love thee; there is much crowding around thee / All run towards thee to reach God."‡[459] He quotes Meister Eckhardt's comment upon John 12:32 ("And I shall draw all men to myself, when I am lifted up from the earth"): "Thus the good man shall carry all things upward to God, their first origin. The great teachers guarantee us as truth that all creatures have been made for the sake of mankind. You will find in all creatures that one creature uses the other: oxen the grass, fish the water, birds the air, animals the forest. In the same way all creatures are of use to the good man: one creature upon another is carried by a good man to God."§[460] And finally he quotes the Buddha upon saddling his horse for the last time when leaving his father's palace: "For a long time already thou art there in life as well as in death; but from now on thou shalt cease to bear and to drag. Only for one more time carry

---

* "Wenn die Maxime allgemein würde, würde das Menschengeschlecht aussterben und ich glaube annehmen zu können, daß mit der höchsten Willenserscheinung auch der schwächere Widerschein derselben, die Thierheit, wegfallen würde; wie mit dem vollen Lichte auch die Halbschatten verschwinden. Mit gänzlicher Aufhebung der Existenz schwände denn auch von selbst die übrige Welt in Nichts; denn ohne Subjekt kein Objekt."

† "Wie in dieser Welt hungerige Kinder sich um ihre Mutter drängen, so harren alle Wesen des heiligen Opfers." He might also (so we would add) have quoted St. Paul ("All creature longs for the revelation of the children God's"; *Epistle to the Romans*, 8:19).

‡ "Mensch! Alles liebet dich; um dich ist sehr Gedränge: / Es läuft dir Alles zu, daß es zu Gott gelange."

§ "So soll der gute Mensch alle Dinge hinauftragen zu Gott, in ihren ersten Ursprung. Dies bewähren uns die Meister, daß alle Kreaturen sind gemacht um des Menschen Willen. Dies prüfet an allen Kreaturen, daß eine Kreatur die andere nützt: das Rind das Gras, der Fisch das Wasser, der Vogel die Luft, das Thier den Wald. So kommen alle Kreaturen dem guten Menschen zu Nutz: eine Kreatur in der andern trägt ein guter Mensch zu Gott." This passage from Meister Eckhardt may at first sight appear ambiguous, as it seems to subscribe to the notorious idea that 'animals are there for man, who may therefore do with them as he pleases'. But Schopenhauer is certainly right in interpreting it in the sense that animals help to lead man to his destiny, namely, to redeem himself and thus the animals as well. It was this vision which made Thomas Mann speak of Schopenhauer's "humanism" (cf. Chapter IX, p. 255).

me from here, o Kantakana; and when I have attained the law [i.e., when I have found Illumination] I shall not forget thee."*[461]

c. Schopenhauer and Buddhism

All this is very fine; yet precisely Schopenhauer's venerated Buddha shows that 'to attain the law' need not entail the Schopenhauerian paradox. Buddhism (in in its original guise, which, so at least its adherents claim, lives forth in its present 'Southern' version)†[462] is not a 'εν και παν (*hen kai pan)* ("One and All") doctrine. It is so neither in Spinoza's sense of true pan*theism* nor in Schopenhauer's sense of what might better be called a pan*daemonism*.‡[463] Certainly, Buddhism, too, sees in 'desire' (which may be somewhat freely translated as 'Will') the source of all suffering. But these desires have no cosmic quality at all; they are not the essence of the 'World', i.e., they are not a trans-individual 'Thing-in-itself' but precisely *individual* desires that *separate* man from the Ground of Being.§ Strictly speaking, no such 'Ground of Being' exists at all (as it does in Brahmanism); there is nothing permanent, neither a World-Soul (*Brahman*) nor an individual soul (*Atman*, which according to Brahmanism is identical with the World-Soul). Only man's thirst for life makes him cling to the illusion of a permanent Self; when Illumination is reached man can enter the state of *Nirvana*, which the Buddha consistently describes in negative terms only. Therefore the doctrine of abolishment of suffering by insight is logically much more consistent in Buddhism than it is with Schopenhauer. Contrary to what Schopenhauer repeatedly claimed,**[464] Buddhism

---

* "Schon lange Zeit bist du im Leben und im Tode da; jetzt aber sollst du aufhören zu tragen und zu schleppen. Nur dies Mal noch, o Kantakana, trage mich von hinnen, und wann ich werde das Gesetz erlangt haben (Buddha geworden seyn), werde ich deiner nicht vergessen."

† That is, the 'Small Vehicle' (*Hinayana*), as distinguished from the 'Northern' version of the 'Great Vehicle' (*Mahayana*). It is actually more correct to use for the former the name by which it designates itself, namely, *Theravada* ('pure doctrine'), since 'Hinayana' is really a term of abuse coined by the Mahayanists. We added the proviso 'so at least its adherents claim', because it has been suggested from the side of an expert (Richard F. Gombrich) that the Buddha preached a religion of love (like Mahayana), whereas in the systematized Theravada-doctrine kindness and compassion, though recommendable, are not sufficient to reach Nirvana.

‡ Schopenhauer quotes with approval Aristotle: 'η φυσις δαιμονια, 'αλλ' 'ου θεια 'εστι (*hè fusis daimonia, all' ou theia esti*: "nature is demonic, but not divine"). For other remarks by Schopenhauer on pantheism cf. Chapter IX, p. 250 and 253.

§ Although we can accept only with much reservation Heidegger's view of voluntarism as the secret Western metaphysics (cf. Chapter I, footnote on p. 14) we do believe that only a Westerner could ever arrive at proclaiming 'Will' to be the essence of the world (even if this 'Will' is such a peculiar one as Schopenhauer's).

** According to Schopenhauer, the 'great, fundamental truth' of the need for redemption from an existence bound up with suffering and death, and of the attainment of redemption in its original guise is actually *less* mythological than his own Will-metaphysics. From a metaphysical point of view it has much more in common with what later was to be called 'logical atomism'.[465]

Schopenhauer is certainly not to be blamed for the incompleteness of his knowledge of Indian thought, which only in his life-time gradually came to be known in the Western world. Still, what he did know might have been sufficient for him to see at least some fundamental differences. One such difference obtains between the *Vedanta*-doctrines (a monist philosophy of Being, of which, as noted, he had made the acquaintance as a youngster by reading the *Oupnekhat* (*Upanishads*), and which contributed decisively to the making of his philosophy)* and Buddhism (a pluralist philosophy of Becoming, which at the time when he wrote his *magnum opus* he knew only through some essays from the earliest volumes of *Asiatic Researches*,[466] but which at a later time, when Buddhism had become a little better known in Europe, he enthusiastically welcomed as a confirmation of his, by then, completed philosophy). Another, even more significant difference that forever failed to strike him is between Indian thought in general and his own mode of thought. An expert in Indian religion and philosophy, Glasenapp, a few decades ago compiled an inventory of the similarities and differences between Schopenhauer's philosophy and the *Vedanta* doctrines on the one hand, and between Schopenhauer's philosophy and Buddhism on the other.[467] As to the first, both teach All-Unity and Veil of Maya. But Schopenhauer's "Will" as Thing-in-itself is incompatible with an eternal All-Spirit as the ground of this All-Unity, since the latter is by definition Being, Thought, and Bliss and therefore 'good', not, like Schopenhauer's "Will", 'bad'. As to the second, both teach atheism and the doctrine that the actions of any being in its present existence lead to its palingenesis (reappearance) in a new form and also that such palingenesis takes place without a permanent soul and is not therefore a 'transmigration', although it may be called a 'reincarnation'. But in Buddhism thirst for life and will to exist are only two of the numerous *dharma*'s that arise in functional dependence upon others, arising and perishing in their turn;† hence, there is no eternal, all-

---

through negation of will, runs counter to the natural inclination of mankind and is hard to fathom in its true foundations; it stands in need, therefore, of a mythical vehicle. A doctrine which remains inaccessible to the masses *sensu proprio* must be taught to them *sensu allegorico*. Hence, the real value of a religion depends upon how great or small a truth-content it carries under the veil of allegory, and on how clearly this becomes visible through the veil. Considered thus, Buddhism is to Schopenhauer the most perfect religion. Philosophy, unlike religion, must always speak *sensu proprio* and without a vehicle, since it adresses itself to the very few who are able to think. Needless to say, Schopenhauer has his own philosophy in mind here more than anyone else's.

* To be more precise, *Vedanta* philosophy is what Shankara around 800 made out of the likewise monist philosophy of the *Uphanishads*.

† Here it must be stressed that this, just as the absence of a Divine Redeemer (see below), goes only for Theravada-Buddhism. For at the popular level the 'Arhat' has given way in Mahayana-Buddhism to the 'Bodhisatva', by means of which concept divine, redeeming

embracing world-principle such as Schopenhauer's "Will". Still, Glasenapp concedes, there is some agreement between Schopenhauer's and the Buddha's ideas of redemption, in that redemption is not granted by a Divine Redeemer but won by self-acquired insight.*[468] Glasenapp also quite rightly observes that, even if Schopenhauer could have possessed a more precise knowledge of these matters, he was in any case far too subjective not to take from other systems what he could use;[469] anyhow, according to Glasenapp, Schopenhauer has the great merit of having been the first European philosopher who (quite unlike Hegel, for example) allowed himself to be genuinely inspired by Indian thought.[470]

So it is indeed; and this might suggest that Schopenhauer's philosophy is also of some topical interest in view of the contemporary flirtation with all kinds of Far-Eastern mysticism. All the more reason to make distinctions and to keep a cool head. It cannot be denied that the Buddhist 'eightfold path' to salvation is more rationally consistent than the Christian expectation and its secularized counterparts, in that it goes to the root and thus aims at the extinction of the flame of life; for a life in which suffering, death and sin are overcome is not a 'better' or an 'eternal' life but no life at all. However, can this seriously be practised anywhere else but in a monastery? And the contemporary Western *vogue* is concerned, more than with the serious world-negation of the monk, with the search by members of the Western consumer society for a new kick (just as the soberness of a simple glass of water can be an unexpected luxury after too much alcohol) or for a kind of psychotherapy against the neurosis created by contemporary society. But even of Buddhism in its purest form, which has little in common with the cult of the same name even in Theravada countries, it might be asked (apart from the 'ethical' vs. 'meta-ethical' question we glanced at in the preceding chapter) whether an 'I' which strives for the annihilation of the illusion of an 'I' and thus for the annihilation of the 'I' as well, does not somewhat resemble Baron von Münchhausen, who drew himself, by his own wig, from the swamp? But this may of course be a Western prejudice.

Be all this as it may; the fundamental point in our context is that Buddhism, unlike Schopenhauer, knows nothing of a Cosmic World-Will. Redemption is therefore a personal matter, not a cosmic drama which should result in the World-

---

beings are reintroduced; and at the philosophical level Mahayana-Buddhism's greatest representative, Nagarjuna, turned *Nirvana* into a cosmic principle (the 'Void') not unlike Vedantic *Brahman*, with which one can enter into a *unio mystica*. This meant really a reintegration of Buddhism into the great mystical tradition, whereas original Buddhism strikes us by its rational method (even if, other than in Western rationalism, the method is inner-directed).

* We add that Schopenhauer himself indicated still another point of agreement between his philosophy and Buddhism, which Glasenapp failed to mention. This is the rejection, in Buddhism, of *exaggerated* asceticism — the self-torture characteristic of Brahmanism. The Buddhist 'middle road' confines (*sic*! PL) asceticism to celibacy, voluntary poverty, humility and obedience of the monks, and abstinence from animal food and in general of all worldliness.

Will's self-annihilation ('God — or rather, the Devil — committing suicide'). True, Buddhism may be said to recognize, as Schopenhauer does, a 'pure subject of knowing', which is all that remains once the 'desires' have been overcome. And these desires may be said to constitute the 'willing' subject or the 'empirical subject', which, for the Buddha, was just as little a 'substance' as it was for Hume or for Kant, and was rather a stream of 'contents of consciousness' (in this respect he was even more radical, since he plainly denied the existence of a simple 'Self', on which Hume took only an agnostic position). What remains after the desires have been annihilated may be said to be the 'knowing subject' or the 'transcendental subject', which is of course 'real' in the same sense that it was 'real' for Kant and for Schopenhauer, to wit, as the 'synthetic unity of apperception' (Kant), or as the "extensionless centre of the sphere of all our representations, towards which the radii of the latter converge" (Schopenhauer).[*471] Without it no illumination would have been possible; but since its reality is of a 'functional' nature,[†472] it vanishes *pari passu* with the whole empirical world into Nothingness. This was Schopenhauer's last word, too, as we remember; still, his penultimate word was Will as ultimate being. And that is why our question 'What, then, is the *true* Thing-in-itself: "Will" or 'Nirvana'?' appears justified in his case, whereas it would be a senseless question in the case of the Buddha.

d. Schopenhauer as a precursor of Einstein ...

This leads us to a much-needed reconsideration of Schopenhauer's arguments for proclaiming "Will" to be the 'Thing-in-itself'. The possible objections are so obvious that they did not, as a matter of fact, escape Schopenhauer himself.

Take, first, the matter of terminology: Schopenhauer's "Will", in view of its irrationality, had better be called *Drang*, or *Trieb* ('urge', or, better still, 'vital

---

[*] " ... das ausdehnungslose Centrum der Sphäre aller unserer Vorstellungen [ist], deren Radien zu ihm konvergieren ..." Making this point in his 'Critique of Kantian Philosophy', Schopenhauer also refers to Chapter 22 of the second volume of *The World as Will and Representation*, where we read: "This focus of the whole activity of the brain is what Kant called the synthetic unity of apperception; only by means of it does Will become aware of itself, since this focus of the activity of the brain, or of cognition, conceives itself as identical with its own basis upon which it has arisen; and thus the I comes into being." ("Dieser Brennpunkt der gesamten Gehirntätigkeit ist Das, was Kant die synthetische Einheit der Apperception nannte; erst mittelst desselben wird der Wille sich seiner selbst bewußt, indem dieser Fokus der Gehirntätigkeit, oder des Erkennens, sich mit seinem eigenen Basis, daraus er entsprungen ist, dem Wollenden, als identisch auffaßt und so das Ich entsteht.").

[†] Recall here Wittgenstein's well-known remark on the eye which cannot see itself and the visual field from which one cannot gather that it is seen by an eye! Yet it might be objected that *as such* it is real, hence our remark above about Münchhausen.

impulse').*[473] Indeed, if Schopenhauer accused Kant of improperly using the word 'reason',† then the same, nay more, goes for his own usage of the word 'will'. For 'according to the linguistic usage of all peoples and times' (to adopt Schopenhauer's own wording) will is connected with consciousness, as Schopenhauer himself felt compelled to admit. 'Will' implies *hierarchy*. In fact he admitted that much by saying that one can *wish* many different things, but *will* only one,‡ but without drawing the consequence that his ascetic saint, far from 'killing' his will (although he may have 'killed', or at least repressed, sexual or other 'wishes') gives an impressive demonstration of a triumph of the will! The various, often contradictory 'wishes', 'desires' and 'urges' are subordinated to one overriding *purpose* or 'ideal'. Such a purpose or 'ideal' need not be lofty as it is in the saint (it may even be morally reprehensible as in the ruthless will-to-power of the tyrant and the conqueror), yet it always presupposes the faculty of distancing oneself from immediate needs. But in order to be able to turn 'Will' into a cosmic principle Schopenhauer was of course in need of an unorthodox usage of the word (what *made* him do so in the first place we shall see presently). This brings us to a second objection. Even if we were to grant that man's innermost nature is "Will", does this justify its being extended into a cosmic principle?§ And, thirdly and finally, is it really the case that an object of the fourth class (the 'willing subject') is actually not an object for a (knowing) subject, as Schopenhauer claims?

Let us set aside the first objection by pragmatically sticking to Schopenhauer's term "Will" while surrounding it with quotation marks when referring to his own mode of using it, and move to the second. It must certainly be admitted that Schopenhauer always stressed the metaphorical character of his cosmically extended "Will" ("The Will of which we speak is not the proper Will", he might have said, varying Lao-Tzu), and that therefore (other than with the *a priori* forms of empirical knowledge) there is no apodictic, but only approximate certainty here.**[474] Yet it is also true that he was not always consistent in this respect,

---

* Cf. Chapter IV, p. 100. In a letter to Frauenstädt of 6 June 1856 Schopenhauer refers to a book by Fichte (no Christian name is mentioned, but Immanuel Fichte must be meant) intended to refute him while tacitly adopting his concept of "Will" — a theft veiled by calling it urge (*Trieb*). Besides the matter of plagiarism, Schopenhauer objects to the term because we are concerned here with the Thing-in-itself, whereas the concept of *Trieb* has its origin in "the phenomenon, the representation, the consciousness of other things" ("in der Erscheinung, der Vorstellung, den Bewußtseyn von andern Dingen"), since it comes from *treiben*, which indicates a *vis a tergo*, visible only for him who looks in from the outside, in contrast to *Wille*, which reveals itself only in *self-consciousness* and is the whole content of the latter. To us this seems a rather banal piece of word-fetishism: as if the metaphor of a driven herd might not come to mind in the process of introspection as well!

† Cf. Chapter II, p. 48, and Chapter VI, p. 152.

‡ Cf. Chapter VI, p. 154.

§ Recall here Rosenzweig's remarks as quoted in the preceding chapter (p. 193).

** For example, in the letter to Frauenstädt of 21 August 1852, where we read: "["Will"] is Thing-in-itself only in a *relative* sense, i.e., in relation to phenomena: and phenomena are especially not in one of his weakest works, '*On Will in Nature*' (*Vom Willen in der Natur*). It is somewhat strange, therefore, to see Bryan Magee in his otherwise excellent book on Schopenhauer concerned to save precisely this aspect of Schopenhauer's doctrine. He rightly observes that "in choosing the word 'Will' to denote a noumenon whose essential nature is not conative at all but is through and through non-human and impersonal, without consciousness, without inner sense, without aim and — perhaps the most important point of all — without life, he has brought it about that all his formulations about it carry, hidden somewhere on board, something counter-intuitive in their cargo."[475] We can no longer agree, however, when Magee goes on to say that all this would have been avoided if only Schopenhauer had chosen another term: 'force' or, better still, 'energy'. The more so when, in his desire to present Schopenhauer's philosophy as a kind of unacknowledged energeticism, he refers in turn to two wholly different aspects of the doctrine, namely, its account of causality as given in the theory of knowledge, and its Will-metaphysics. Now this is an intricate point well in need of some further clarification.

Schopenhauer indeed sometimes used the word 'energy', or rather its Greek form ʹενεργεια *(energeia)*. We read for example: "As soon as we watch something real (*ein Wirkliches*), it presents itself *eo ipso* as material, just as, the other way round, something material necessarily presents itself as active (*wirksam*); these are indeed interchangeable concepts. Therefore the word 'real' is used as synonymous for 'material'; also the Greek κατʹ ʹενεργειαν *(kat' energeian)*, as opposed to κατα δυναμιν *(kata dynamin)*, reveals the same origin, since ʹενεργεια means 'working' in general."*[476] Indeed, it is a recurrent theme in Schopenhauer that "matter is causality through and through, its being is its working", as Magee rightly points out. That is also why (cf. Chapter IV, footnote on p. 95) Schopenhauer prefers German *Wirklichkeit* over the 'reality' of Latin-derived languages. But here speaks, not the metaphysician of "Will", but rather the nominalist out to reveal 'substance' (identified by him with 'matter') as a *flatus vocis* (an empty sound). True, besides 'causality' (which, as we have seen, was the only category preserved

---

phenomena only in relation to the Thing-in-itself ... I have never said what the Thing-in-itself is *beyond* this relation, since I don't know that; but *within* it, it's Will-to-Life." ("Dieses [i.e."Will"] ist Ding an sich bloß *relativ*, d.h. in seinem Verhältnis zur Erscheinung: und diese ist Erscheinung bloß in ihrer Relation zum Ding an sich ... Was das Ding an sich *außerhalb* jener Relation sei, habe ich nie gesagt, weil ich's nicht weiß: *in* derselben aber ist's Wille zum Leben."

* "Sobald wir ein Wirkliches anschauen, stellt es sich *eo ipso* als materiell da, wie auch umgekehrt ein Materielles notwendig als wirksam: es sind in der Tat Wechselbegriffe. Daher wird das Wort 'wirklich' als Synonym von 'materiell' gebraucht: auch das Griechische κατʹ ʹενεργειαν, im Gegensatz von κατα δυναμιν, beurkundet denselben Ursprung, da ʹενεργεια das Wirken überhaupt bedeutet." Thus, ʹενεργεια (*energeia*) means nothing but *Wirklichkeit*, for which 'actuality' would in Schopenhauer's view be a better English equivalent than 'reality' (as Greek κατʹ ʹενεργειαν (*kat' energeian*) corresponds to Latin *actu*, just as κατα δυναμιν (*kata dynamin*) corresponds with *potentiâ*).

by Schopenhauer from Kant's complicated collection, and which moreover he demonstrated in a very unorthodox way),* 'substance' was also in a sense recognized by him as a valid 'category', albeit one reduced to the status of a corollary to causality. For, according to Schopenhauer, causes and effects are events, i.e., *changes* of situations, and this suggests of course the existence of a permanent, unchangeable bearer of these changes. This bearer might be called 'substance', or 'matter',[477] but (and this goes against materialism) it should be distinguished carefully from 'stuff' (*Stoff*). The latter has already entered the veils of forms, whereas 'matter' is what remains of bodies once these have been stripped of all specific properties, and which is therefore the same in all bodies.[478] Now the forms and qualities thus set aside are nothing but the particular and specific kinds of *working* of the diverse bodies and, when we abstract from them, pure 'working' remains as such, i.e., nothing but the principle of causality: "matter is causality through and through, its Being is its Working."†[479]

From an epistemological point of view this sounds quite 'modern' if, with Cassirer and Dijksterhuis, we consider the shift of the concept of 'substance' (*Substanzbegriff*) to the concept of 'function' (*Funktionsbegriff*) as essential in the development of modern science.[480] But Magee goes a considerable step further than this when, from Schopenhauer's pertinent statements, he concludes: "Perhaps the most astonishing of all the many Kantian–Schopenhauerian anticipations of modern science lies not in the latter's forceful and concise exposition of the central core of Freudianism, nor in his sharp though uncoordinated glimpses of a theory of biological evolution ... but in the former's very specific announcement of one of the central doctrines of Einstein's theory of relativity more than a century before Einstein", going on somewhat naively to remark that "it is an extraordinarily impressive fact that this true, unobvious and amazing conclusion was reached purely by epistemological analysis."[481] 'Amazing', indeed — especially so when, in the same breath, Einstein's theory of the equivalence of mass and energy finds itself rephrased by Magee in Schopenhauerian terms: "force and substance are inseparable because at bottom they are one."[482]

Indeed, Schopenhauer maintains that also, insofar as, to him, the forces of nature are identical to the *Will* in us, and matter (substance) is the mere *visibility* of the Will and, to that extent, identical to it. Still, Schopenhauer would have vehemently protested against thus lumping together two quite different lines of argument in his philosophy. "'Force'," he wrote, "is that which, *although wholly different from the cause* [our italics, PL], provides every cause with its causality, i.e., with its possibility of working."‡[483] And this is the crux of the matter. Whereas 'causality' (and, hence, 'energy' in the way he uses the word) was for him the *proprium* of the *phenomenal* world, 'force' was in his view a word used by the

* Cf. our exposition in Chapter IV, sections b. and c.

† "Die Materie ist Kausalität durch und durch, ihr Seyn ist ihr Wirken."

‡ "'Kraft' ist das was, von der Ursache völlig verschieden, jedoch jeder Ursache ihre Kausalität, d.h. die Möglichkeit zu wirken, erteilt."

science of his day to denote what all scientific explanation runs into in the end without thereby becoming less mysterious. That is, 'force' in Schopenhauer's view serves as a pointer to the *noumenal* world, to the *Thing-in-itself* (which, although strictly speaking it is not of a 'noumenal' nature — being 'deeper' rather than 'higher' than the 'phenomenal' world — is for Schopenhauer no less sharply distinguished from that phenomenal world than for Kant). And that is why Schopenhauer was indeed willing to recognize the theoretical possibility of using 'force' (*not* 'energy'!) instead of 'will' as a *chiffre* for the Thing-in-itself, going on at once to reject the option with the argument that in that case what is better-known (namely, from direct inner experience) would be subsumed under the lesser-known. In other words, the word richer in empirical content, and therefore more meaningful (namely, 'will') would be dropped in favour of a less meaningful word (namely, 'force'). He prefered therefore to proceed in the opposite direction and, rather than regarding Will as a specific kind of Force, to regard the diversity of forces in nature as so many forms of "Will".

One may join Magee in regretting the decision; still, a certain logic for it cannot be denied. Moreover, even if Schopenhauer had opted for 'force' this would still have been something purely *qualitative*, and as such closer to romantic *Naturphilosophie* than to the energy concept of modern science. When, therefore, Magee, in discussing Schopenhauer's Will-metaphysics, tells us once again that 20th-century physics provides the most powerful confirmation imaginable, on the plane of science, for Schopenhauer's central vision of the world, since "in the fullest scientific sense we now *know* that matter and energy are equivalent; that at the subatomic level the concept of matter dissolves completely into the concept of energy; that every material object is, in its inner constitution, a concatenation of forces and nothing else; and that it is theoretically possible to transform every material object without remainder into the energy that constitutes it,"[484] this sounds to our ear hardly less anachronistic than those frequent proclamations of Democritus as a forerunner of modern atomic theory.*[485]

Moreover, if one insists at any cost on having a science-based metaphysics, it might still be questioned whether 'energy' alone would do. After all, it is hard to see how 'energy' alone could bring forms of life into existence. Does not, in accordance with the second law of thermodynamics, entropy tend to increase, so that from a physicist's point of view increasing disorder is much more probable than the increasing complexity (i.e., increasing order) suggested by the biologist's concept of evolution? Would it not be necessary, therefore, to introduce, let us say, 'information' as a second metaphysical principle? This would in any case be in surprising conformity with the very conception of a dual Deity which we just recommended the late Dr Schopenhauer to adopt... But in earnest, what would be the good of it? In the heyday of classical mechanics philosophers tended to look at the world as one large clockwork mechanism; 19th-century philosophers naturally were inclined to see it as a steam-engine writ large, and for contemporary philosophers the temptation has proved irresistible to interpret it as one big computer. Indeed, the desire for a science-based metaphysics seems to be still as wide-spread in our time as it used to be in the days of 'energeticism' and 'monism', as such a New Age-prophet as Fritjof Capra eloquently testifies!

The difference is of course that previous philosophers bent on mobilizing science for their own purposes usually adhered to some form of scientism in the sense of a belief in the omnipotence and unqualified beneficiality of science, whereas nowadays we are concerned mostly with people who, somewhat inconsistently, mobilize contemporary science (the no longer 'mechanistic' nature of which they like to stress) to demonstrate the very shortcomings and drawbacks of science-and-technology-based, modern civilization.[486] It may be questioned whether this latter-day scientism *à rebours* has improved matters. Not only has it been rightly and wittily remarked (by Kierkegaard, if I am not mistaken) that a

---

* Concerning 'atomic theory' it is interesting to notice that in Schopenhauer's opinion the atom was nothing but a methodological fiction: "the atom is without reality" ("das Atom ist ohne Realität"). That view is typical for his 'positivist' or 'phenomenalist' conception of science (cf. our exposition in Chapter V); it was later defended against Boltzmann by Mach, and also (until owning defeat in 1908) by Ostwald. The latter figure is not mentioned by Magee, which is somewhat surprising as Ostwald's 'energeticism' ought to have pleased him to the point of regarding it as an improved version of Schopenhauerianism. Ostwald himself recognized a kindred spirit in Schopenhauer's philosophy. In his lectures on natural philosophy he praised the latter's insight into the tremendous importance of Will, going on (not too unexpectedly) to correct it by confining Will to those processes in which living beings give out Energy, thus depriving it of the central and unique position accorded to it by Schopenhauer. The basic difference in philosophical attitude between the two men seems to be that Schopenhauer was anxious to separate the scientist's phenomenalist proceedings from the philosopher's task of satisfying man's metaphysical needs, whereas in Ostwald's 'scientism' one rather meets a curious blend of both.

faith that marries contemporary science is likely to be a widow tomorrow.* But also, a scientific concept, precisely because that is what it is, can never be the 'Thing-in-itself'. The latter must be (using Schopenhauer's phrase) *toto genere* different from the phenomenal world, to which not only daily experience but also the world of science belongs. On all this we once again prefer, by far, Iris Murdoch's view of things. Rightly observing that Christianity today (faced with a wide-spread loss of belief in the supernatural) may be tempted by various forms of a 'cosmic answer' as a substitute, she professes against this tendency her faith in the importance of the concept of a *transcendent* (if perhaps not 'supernatural') 'Good' as essential to both morality and science.†[487] And therefore it seems better not to go along with scientism in either guise, nor to seek to rescue Schopenhauer by its means.

Moreover, any attempt to save Schopenhauer's metaphysics by rebaptizing its ultimate principle, and in the same act turning it upside down by making 'force' or 'energy' no longer a special kind of Will, but on the contrary Will a special kind of Energy, ignores Schopenhauer's explicit assertion that we should learn to understand Nature by starting from ourselves, not the other way round.[488] For good reason, the supplements to the second book of the *World as Will and Representation* to be found in the second volume carry, by way of an epigraph, Goethe's remark: "Lies not nature's core in man's heart?".‡[489] Indeed, in the section 'Epiphilosophy' of the second volume of his *magnum opus* Schopenhauer, just after acknowledging that he stands in a 'εν και παν ('One and All') tradition, points to two original contributions he has made to that tradition. He has been the first to solve the problem of what this inner essence *is* which is one and the same in all things; also, whereas his predecessors had described Man as *microcosm*, he, conversely, has revealed the World to be *macranthropos* ('man writ large').§[490] An absurdity perhaps, but nevertheless the *proprium* of Schopenhauer's philosophy, without which it would be, not an 'improved' Schopenhauerianism, but no Schopenhauerianism at all!

Magee's proposal, if accepted, would amount therefore to far more than just the cosmetic surgery he believes he is performing upon Schopenhauer's metaphysics of "Will". This becomes clear, for example, in the latter's belief that "Will" is able to steal away from natural laws (and thus from the domain of science), and in table-rapping, clairvoyance, and healing by animal magnetism to reveal itself

* This is well illustrated in the anti-atomism of Ostwald's energeticism, which (as mentioned in the previous footnote) he later felt compelled to drop.

† Certainly a 'cosmic answer' need not in itself be a 'science-based answer', but in people like Capra that is what it turns into in fact.

‡ "Ihr folget falscher Spur, / Denkt nicht, wir scherzen! / Ist nicht der Kern der Natur / Menschen im Herzen?" ("You follow the wrong track, / do not think that we are joking! / Lies not nature's core / in man's heart?").

§ For his other objections against traditional 'εν και παν (*hen kai pan)* doctrines cf. footnote on p. 207.

unmediated. This, *pace* Magee, points to a truly "Will"-like character of his cosmic principle! "In table-rapping, *Will* shows itself in its original omnipotence: guiding the movement it is also the mover," Schopenhauer went so far as to write in a letter to Ernst Otto Lindner.*[491] In a letter on the same topic to Adam von Doss he scorns the refusal of most scientists to believe in paranormal phenomena: "People who know nothing but their retorts, galvanic batteries, and frogs' legs, dare explain by these means the world and man;" also, "among physicists a real fear of table-rapping manifests itself."†[492] He similarly gave vent to his anger with Frauenstädt when this generally faithful disciple had dared, likewise, to express some scepticism about the performance of a certain medium called Regazzoni.[493] But, in fairness to Schopenhauer, it must be recognized even by those who share Frauenstädt's scepticism about these matters that Schopenhauer's firm belief in paranormal phenomena was an obvious consequence of his insistence on the need to seek the Thing-in-itself (or at least its best possible approximation) in introspection, where it reveals itself as "Will". Furthermore, this at first sight somewhat bizarre insistence was in its turn an obvious consequence of his general stance that expressions are meaningful only if their meaning can be elucidated in terms of sensual experience, be that experience external or internal (which is why he preferred 'Will' to 'Force', as we have just seen). And it is this very attitude which distinguishes him, favourably in our opinion, from his German Idealist contemporaries. Moreover, his interest in *inner* experience did enable him to do some pioneering work in the psychological domain, as we shall see now.

e. ... or of Freud?

If, then, it must be deemed absolutely necessary to regard Schopenhauer as a 'forerunner' of this or that (but how unhistorical as well as unjust a concept this is!),‡ a much better case can be made for psychoanalysis. The more so since Freud explicitly recognized the filiation, as Magee mentions himself,§[494] whereas no such comparable statement is known to us from Einstein. We read, for example: "You may perhaps shrug your shoulders and say: That is not science, that is

---

* "Im Tischrücken zeigt sich der Wille in seiner ursprünglichen Allmacht: lenkt er die Bewegung, so ist er auch der Beweger."

† "Menschen, die nichts als ihre Retorten, galvanischen Batterien und Froschkeulen kennen, unternehmen damit die Welt und den Menschen zu erklären"; "unter den Physikern manifestiert sich ein wirklicher Angst vor dem Tischrücken."

‡ 'Unhistorical', in that all too often it entails projecting present-day tendencies or problem situations back onto the past; 'unjust', in that it suggests that the character concerned is not interesting for his own sake but only deserves an approving (also patronizing) nod for having enabled later people to stand upon his shoulders.

§ He refers here to Freud's acknowledgement that the theory of repression appears already in Schopenhauer, although of course Freud insisted that he had made this discovery independently (more on this below).

Schopenhauerian philosophy. But why, ladies and gentlemen, might not a bold thinker have guessed something which is then confirmed by sober, painstaking, and detailed research?"*[495] Or (against the well-known reproach of pansexualism): "For the philosopher Arthur Schopenhauer has already long ago impressed upon people the extent to which their acting and striving is determined by sexual urges in the customary sense of the word, and a world of readers ought not to have been able fully to banish from their minds such a telling admonition!"†[496]

Indeed, the filiation can easily and abundantly be illustrated. Thus, when Schopenhauer in an interesting essay 'On Madness' ('Über den Wahnsinn') writes that memory, although forming representations, is not a representational faculty itself but a faculty of Will, and that some events or details are thus hidden from intellect, "because Will cannot suffer the sight of them",‡[497] then he is describing rather exactly the phenomenon later to be called 'repression' (*Verdrängung*).[498] Will seeks to prevent events repulsive to it from being illuminated by the Intellect; yet mental health is possible only if the painful operation of their assimilation by our consciousness succeeds. If not, the thread of memory is broken, with the resulting gap being filled by arbitrary representations out of place there; thus arises madness.§[499] And (now that against our own warning we have embarked upon a search for parallels ...) it is interesting to note that our Schopenhauerian paradox, too, reappears in Freud. On the one hand, psychoanalysis denies the supremacy of Reason — human beings are moved by libidinal urges and they are not as a rule very susceptible to reason. But on the other hand it declares that "we have no

* "Sie werden vielleicht achselzuckend sagen: Das ist nicht Naturwissenschaft, das ist Schopenhauersche Philosophie. Aber warum, meine Damen und Herren, sollte nicht ein kühner Denker erraten haben, was dann nüchterne und mühselige Detailforschung bestätigt?"

† "Denn der Philosoph Arthur Schopenhauer hat bereits vor geraumer Zeit den Menschen vorgehalten, in welchem Maß ihr Tun und Trachten durch sexuelle Strebungen — im gewohnten Sinne des Wortes — bestimmt wird, und eine Welt von Lesern sollte doch unfähig gewesen sein, sich eine so packende Mahnung so völlig aus dem Sinn zu schlagen!"

‡ "... weil der Wille ihren Anblick nicht ertragen kann ..."

§ According to Schopenhauer the criterion proper for distinguishing sanity from madness rests in the ability to be quite as certain of a past event to which one has been a witness as of a present perception. As soon as I doubt whether an event remembered by me has really taken place I expose myself to the suspicion of madness; that is why suspicion of madness makes one unfit to act as a witness in court. For, although a madman may occasionally have witty ideas or single intelligent thoughts or even correct judgements, he is never a valid witness of a past event. It is interesting (although as far as we know this has never been confirmed statistically) that Schopenhauer considers actors to be particularly prone to madness because they strain their memory to premature exhaustion by studying a new role or brushing up an old one every day; moreover, their roles are often mutually contradictory, so the actor must be another person every evening.

means other than intelligence for controlling our being driven by urges."*[500] As a theorist, Freud acknowledged the overwhelming power of the Unconscious; yet as a medical man (which is what he always remained in the first place) he could not but believe in the possibility of healing neurotic patients, and thus proclaim Reason as the only road to salvation. (There is a difference, to be sure, as Freudianism is about individual, not cosmic 'salvation' — just as in original Buddhism, to which Freud is more akin in this respect than Schopenhauer himself).

Finally, now that the term 'the Unconscious' has made its entrance, this provides a fitting opportunity to address the *third* objection we have raised against Schopenhauer's proclaiming "Will" to be the Thing-in-itself. His consideration, quoted in the above, that we have to do here with a phenomenon "that is not draped in three veils, but only in one" would be more convincing if he had made a fundamental distinction between space and time, or at least between space and 'real' time (as was to be done later by Bergson, whose *durée* (duration) is to be strictly distinguished from the *temps* (time) of mathematical science which is really a dimension of space). Precisely in this respect, however, Schopenhauer remained an orthodox Kantian, to whom space and time were *both* 'forms of intuition' (whereas for Bergson only space is that) and thus had the same epistemological status.

Now here, too, we may safely assume that Schopenhauer was aware in advance of the possibility of such an objection, and that here, too, he might have said: 'The Will of which we speak is not the real Will'. For it is clear that, to Schopenhauer, *conscious* volitions are only the tip of the iceberg, so to speak. Is, then, the 'Id' the true Thing-in-itself? Indeed, of Schopenhauer, too, it may said (with Janaway) that to him as to Kant the Self has *three* layers.†[501] There is a little difference, to be sure. To Schopenhauer, as to Freud, the deepest layer is unconscious, which implies that, in the latter's wording, "'I' stands under a threefold servitude and thus is exposed to three dangers: the outer world, the *libido* of the Id, and the severity of the Super-Ego."‡[502] This seems rather far away from Kant's deepest layer (the intelligible self), which is precisely the proud castle of freedom!

---

* "Nun haben wir kein anderes Mittel zur Beherrschung unserer Triebhaftigkeit als unsere Intelligenz."

† Janaway writes: "Kant has three discernible conceptions of the self, which continually threaten to become conceptions of three selves, each distinct from the others. There is, first, the empirical self — the collection of mental states appearing in the inner sense. Secondly, there is the self as it is in itself — that *of* which those states are an appearance [i.e., the intelligible self, PL]. And thirdly, the self as pure subject — that *for* which those states are there, and which can think of them as its own" [i.e. the transcendental self, PL]. Cf. also our Chapter I, footnote on p. 20 and Chapter VI, p. 161 with appended footnote.

‡ "Aber andrerseits sehen wir dasselbe Ich als armes Ding, welches unter dreierlei Dienstbarkeit steht und demzufolge unter den Drohungen von dreierlei Gefahren leidet, von der Außenwelt her, von der Libido des Es und von der Strenge des Über-Ichs."

Here we recall that Schopenhauer at least believed that he, too, believed in 'intelligible freedom'. His big problem (the very one that forms the root of the Schopenhauerian paradox) was how to reconcile this with his belief in omnipotent "Will". It is interesting that Freud (in a context which of course has no direct connection to this problem) gave a hint when he wrote that "not only the deepest but also the highest of the I may be unconscious".*[503] For, whereas the 'Ego' (yes, Freud's Self, too, has three layers!) is essentially a representative of the outer world, of 'reality' (since it seeks to assert the influence of the outer world upon the Id and its intentions, or, in other words, to put the reality-principle in place of the lust-principle that reigns without limitation in the Id), the Super-Ego confronts the Ego as a solicitor of the *inner* world, i.e., of the Id. This appears from analysis, which teaches us that there are persons in whom self-criticism and conscience (achievements of the psyche that are commonly ranked very highly) are *unconscious* and produce unconsciously their most important effects. For example, the normal, conscious sense of guilt, i.e. conscience,†[504] is based on a tension between the Ego and the Ego-ideal; it is the expression of a condemnation of the Ego by its critical agent. But Freud mentions, in this context, the phenomena of compulsive neurosis (in which the sense of guilt speaks extremely loudly without being able to justify itself before the Ego) and melancholy (in which the Ego does not dare to make objections, confesses itself guilty, and submits to punishment). In both cases the Super-Ego has known more of the unconscious Id than the Ego has. Therefore Freud says: "The Id is wholly amoral, the Ego strives to be moral, the Super-Ego may become hypermoral and then as cruel as ever the Id,"‡[505] For the more man controls his agressions, the more the agressive impulse of his ideal towards his Ego increases, too. Against critics who complain that there must for Heaven's sake be some higher essence in man, Freud rejoins: "Certainly, and here is what that higher being is: it is the ideal of the Ego or the Super-Ego, the representative of our relationship to our parents. As children we have known, admired, and feared these higher beings; later we have assimilated them into ourselves."§[506]

---

* "Nicht nur das Tiefste, auch das Höchste am Ich kann unbewußt sein."

† This rather crude identification of 'conscience' and 'sense of guilt' is phrased more subtly elsewhere in the same essay 'Das Ich und das Es': "The tension between the claims of conscience and the actions of the I is experienced as a sense of guilt" ("Die Spannung zwischen den Ansprüchen des Gewissens und den Leistungen des Ichs wird als Schuldgefühl empfunden"); also in more elaborate fashion in chapters VII and VIII of his famous *'Civilization and its Discontents'* (*Das Unbehagen in der Kultur*).

‡ "Das Es ist ganz amoralisch, das Ich ist bemüht, moralisch zu sein, das Über-Ich kann hypermoralisch und dann so grausam werden wie nur das Es."

§ "Gewiß, und dies ist das höhere Wesen, das Ichideal oder Über-Ich, die Repräsentanz unserer Elternbeziehung. Als Kind haben wir diese höheren Wesen gekannt, bewundert, gefürchtet, später sie in uns selbst aufgenommen."

The 'Will' as Id, and the 'intelligible I' as Super-Ego! A tempting translation for contemporary readers, perhaps; but let us not forget that depth-psychology infers the existence of an unconscious from an analysis of various phenomena of consciousness, so that, to that extent, a primacy of consciousness remains. The unconscious can, in principle, be made conscious — the whole therapeutic method is based on this presupposition; in Freud's lapidary sentence, "Where Id was, I should appear" ("Wo Es war soll Ich werden").[507] The root problem that something which, after all, remains a *phenomenon* (and in this sense, *pace* Schopenhauer, 'an object for the subject') were to be elevated to the rank of Thing-in-itself would not therefore be solved by this 'translation'. Moreover, one can easily imagine Schopenhauer's response to Freudianism if he had lived to know it. Triumphantly proclaiming it to be striking proof of the irresistibility of his own ideas, he would in all likelihood have gone on to accuse Freud of plagiarism, and to snort at Freud's insistence upon independent discovery followed by pleased surprise at finding out that he had such an illustrious precursor. Nor is he likely to have failed to observe that, in the last resort, this Viennese psychiatrist remained a typically flat-headed *Medikaster* lacking all deeper metaphysical insight. Therefore, if we really wished to interpret Schopenhauerianism as a kind of Freudianism *avant la lettre*, we would expose ourselves to the same objection we raised against Magee, only the other way round. For, whereas Magee tries to save the universality of Schopenhauer's "Will" by reinterpreting it as really identical to the *physical* concept of 'energy', we in our turn would renounce universality by reinterpreting "Will" as being really identical to the *psychological* concept of 'Id', thus confining it not only to the world of living beings but even within the latter to the human domain. (This is so because only where there is self-consciousness can one speak of an 'unconscious'. Human beings, not only as a species but also as individuals, have a *history*, with the conflicts occurring in the individual being stored in the unconscious).[508] Even though we have shown the latter approach to be somewhat more justified (since it would do justice to what we at least find the most interesting aspect of Schopenhauer's philosophy), in this way, too, we would de-Schopenhauerize Schopenhauerianism, since it is precisely this *universalisation* of man's innermost being into a cosmic principle that constitutes Schopenhauer's *proprium*. Indeed, how could it be otherwise? Schopenhauer had to universalize man's innermost being *in order to avoid solipsism as he saw it*. Here lies the heart of the matter, for this problem is an inevitable outcome of Schopenhauer's subjective idealism. Drop subjective idealism, and the problem vanishes with it.* So one might leave Schopenhauer's metaphysics for what it is, while gratefully acknowledging that his obsession with "Will" has greatly contributed to our insight, not indeed into the 'noumenal' realm he himself aimed at, but rather into the deeper regions of the 'phenomenal' realm. But we think that there is something more to say.

---

* Cf. Chapter IV, p. 107.

f. Schopenhauer's broken world-view: his professed, and his hidden metaphysics

What, so we asked in the above, is the real Thing-in-itself: "Will" or 'Nirvana'? From any point of view that may rightly claim to stand in the tradition of the Buddha, Plato and Kant (which is indeed what Schopenhauer claimed) the answer, no doubt, must be: the latter. For in conceiving the idea of Nirvana and striving for it, the Buddha demonstrated perhaps the most radical example of the capacity of the human mind for detachment from nature ("Will" in the Schopenhauerian sense); in the most literal sense we have here the possibility of saying 'no'.[*509] And although Nirvana can be *attained* and, in this sense, is more than a Kantian 'regulative Idea', the very fact that it can only be conceived of in negative terms shows its radically transcendent nature: like Plato's Idea of the Good, it is in the most literal sense 'beyond Being'. In contrast, Schopenhauer's "Will" is rather the *essence* of Being. In an effort to resolve the dilemma, let us call to mind the architecture of *The World as Will and Representation*.[†]

There we notice a twofold subdivision. The first and most obvious is the one between 'World as Will' and 'World as Representation'. This corresponds with Books II ('The objectivation of Will') and IV ('Once self-knowledge has been attained: affirmation and negation of the Will-to-life'), and with Books I

[*] According to Scheler, man is the only living creature which, by force of its spirit, is able to take a principally 'ascetic' attitude towards life, i.e., abstain from immediately following its natural urges and impulses. "Compared with the animal, which always says 'yes' to reality — even when it abhors and flees it — *man* is the one *who is able to say 'no'*, the '*ascete of life*', the perennial Protestant against mere reality." ("Mit dem Tiere verglichen, das immer 'Ja' zum Wirklichsein sagt — auch da noch, wo es verabscheut und flieht — ist *der Mensch* der '*Nein-sagenkönner*', der '*Asket des Lebens*', der ewige Protestant gegen die bloße Wirklichkeit."). Indeed; and this is true not only for the ascete in the normal sense (although the Buddha presents indeed the most radical form of this tendency) but also for those who have a positive attitude towards real life, but who, precisely because of that, want to change and improve it. It is interesting to compare Scheler's description of man as the sole creature capable of saying 'no' with Sartre's conception of human existence as "pour soi" rather than just "en soi", which mode of existence he defines as the human propensity to say 'non'. Ultimately, therefore, human existence is "nothing" against the "Being" of all other creatures (*L'être et le néant* being the title of Sartre's first *magnum opus* in philosophy; as we shall see presently, in his *L'existentialisme est un humanisme* he also said of man that he 'is' nothing but can only 'become' something, namely, that into which he makes himself).

[†] The titles of the four 'books' this work consists of are: Book I 'Der Welt als Vorstellung erste Betrachtung: Die Vorstellung unterworfen dem Satze vom Grunde: das Objekt der Erfahrung und Wissenschaft'; Book II 'Der Welt als Wille erste Betrachtung: Die Objektivation des Willens'; Book III 'Der Welt als Vorstellung zweite Betrachtung: Die Vorstellung, unabhängig vom Satze vom Grunde: die Platonische Idee, das Objekt der Kunst'; Book IV 'Der Welt als Wille zweite Betrachtung: Bei erreichter Selbsterkenntniß, Bejahung und Verneinung des Willens zum Leben'.

('Representation subject to the principle of sufficient reason: the object of experience and science') and III ('Representation independent of the principle of sufficient reason: the Platonic Idea, the object of art'), respectively. But the titles themselves already suggest another subdivision: between Books I and II on the one hand and Books III and IV on the other. The first subdivision expresses Schopenhauer's *professed* metaphysics which is Kantian ('appearance' vs. 'Thing-in-itself') in a *formal* sense yet wholly un-Kantian in spirit, since the way in which 'appearance' is treated betrays a virtual lapse from transcendental into psychological idealism whereas the 'Thing-in-itself' is interpreted as the innermost *core* of Nature rather than as something *transcending* Nature. The second subdivision expresses Schopenhauer's *hidden* metaphysics, which presents no longer *immanent* "Will", but *transcendent* 'Nirvana' as the ultimate reality. Therefore, despite its 'Buddhist' clothing, the latter is in that regard more in the spirit of Kant than the former. Schopenhauer's system would have been much more coherent if he had made the following subdivision, which we vicariously offer here. Books I and II are about the *phenomenal* world; here Schopenhauer's originality consists of the definition of the body as "unmediated object" ("unmittelbares Objekt"), which enabled him, in the domain of 'outer' experience (Book I), to do much more justice than Kant to the part played by perceptual knowledge, and in the domain of 'inner' experience (Book II) to the part played by the Unconscious. Books III and IV are about the *noumenal* world. We get a glimpse of this in aesthetic experience (Book III) which, however, is bound to remain a play and a temporary palliative; rather, it is the ethical attitude of universal compassion, culminating in the world-negation of the ascetic saint which leads in the end to the great, mystical silence of Nirvana (Book IV). Instead, that which should have been treated as the deepest layer of the phenomenal world is raised by Schopenhauer in the second book to quasi-noumenal status by proclaiming it to be the 'Thing-in-itself'. In this he anticipated Nietzsche, who, substituting "will-to-power" for Schopenhauer's "will-to-life", proclaimed: "This world is the will-to-power ... and nothing else!"[*510] But Nietzsche drew the dual consequence: to deify "Will", and definitely to dismiss the 'true world' as a fable.[†511] That consequence it was impossible for Schopenhauer himself to draw

[*] "Diese Welt ist der Wille zur Macht — und nichts außerdem!"

[†] "Against the theory that the 'in-itself' of things must necessarily be good, blessed, true and one, Schopenhauer's interpretation of the 'in-itself' as will was an essential step; but he did not understand how to *deify* this will; he remained entangled in the moral-Christian ideal. To that extent Schopenhauer still stood under the dominion of Christian values that (now that the Thing-in-itself was for him no longer 'God') he could see it only as bad, stupid, and absolutely reprehensible." ("Daß ein 'An-sich der Dinge' notwendig gut, selig, wahr, eins sein müsse, dagegen war Schopenhauer's Interpretation des 'An-sich's'als Wille ein wesentlicher Schritt: nur verstand er nicht, diesen Willen zu *vergöttlichen*: er blieb im moralisch-christlichen Ideal hängen. Schopenhauer stand so weit noch unter der Herrschaft der christlichen Werte, daß er, nachdem ihm das Ding an sich nicht mehr 'Gott' war, es als schlecht, dumm, absolut verwerflich sehen mußte.").

because his negative attitude to "Will" forced him to stick to Kant's intelligible freedom in order (see above, p. 203) to safeguard the possibility of escaping from its tyranny. As a result, he tells us a 'grand narrative' (to use the postmodernist phrase) about a "Will" which attains insight into its own essence and can in the end therefore negate itself. And so much is clear that, if it could not attain insight into its own essence, Schopenhauer could not have written his book. But the difficulty is that if it can, it cannot be the Thing-in-itself. For to us (see above, p. 199) the big problem remains to understand how a *blind* "Will" can ever arrive at feeling the need to kindle a light to itself (to use Schopenhauer's own phrase)? One would then say that some elementary kind of consciousness must have been present from the very beginning; that is why we said that Schopenhauer, had he really been the consistent thinker he prided himself on being, ought either (if he wished at any cost to grant — because of its "not being draped in three veils, but only in one" — metaphysical status to his "Will")* to have arrived at some kind of dualist metaphysics, such as Max Scheler was to suggest much later † ... or (as after all becomes a miracle) to have left his "miracle κατ΄ ΄εξοχηv" a miracle. Had not his venerated master Goethe (here basically in agreement with Kant, the other venerated master) formulated this ultimate resignation in the maxim (cf. Chapter I, p. 25) that it is the finest fortune of thinking man to have investigated what can be investigated and quietly to revere what cannot?

Fantasizing a little on this theme, one might imagine Schopenhauer to have taken up Goethe's suggestion (cf. Chapter IV, p. 114) to complement Kant's critique of Reason by giving a critique of Understanding and of the senses. Starting from the general, abstract subject–object correlation ('no object without a subject, no subject without an object') which he rightly considered to be philosophically fundamental, he would then have shown how in concrete reality the intentional relationship between the organism and its surroundings becomes more complex in man so as to lead to a separation of (immediately experienced) inner world and (represented) outer world. He might moreover have said that the 'knowing' and 'willing' subject, which in animals are not only identical but also perfectly attuned to each other, are both raised in man to the second power so as to turn himself into an intrinsically divided being. Animal 'Understanding' is raised in man to the second power of 'Reason' (a faculty, *pace* Schopenhauer, really directed at the 'Unconditioned' in view of its capacity for ongoing self-criticism), just as animal urges and vital impulses ("Will" in his terminology) are raised in him (because he has a 'history') to the second power of the (individual as well as collective) 'Unconscious'. Indeed, Man is a *doubly* knowing subject, as we saw in Chapter IV (p. 107); and in this lies the root of his relative freedom. 'Relative' freedom not in the sense in which Schopenhauer used the term (Chapter VII, p. 195); yet *relative* freedom. For, though it is true (as Sartre said) that in man "existence precedes essence" so that man originally 'is' nothing, but can only 'become' something,

* See above, p. 198.

† See above, footnote on p. 200.

namely, that into which he makes himself,*[512] it is no less true that man remains also a 'natural' (and as such predetermined) being (like the animal), and also that he distinguishes himself from the animal not only by developing self-consciousness (i.e., 'freedom') but also by developing an unconscious life of the psyche (i.e., a new dimension of predetermination which is, however, at the same time a source of artistic creativity). Translating this into Kantian terms one might say that man is a creature stretched between two 'in-themselves', the 'higher', 'transcendent' (Kantian–Sartrean) one of the realm of freedom, and the 'deeper', 'immanent' (Schopenhauerian–Freudian) one of the Unconscious. A paradox, to be sure, but one which Schopenhauer might have been well advised to accept as an *Urphänomen* instead of mythologizing it into a Cosmic Will which in the end negates itself.

Indeed, we think that we are entitled to call this *conditio humana* in Goethean terms an *Urphänomen*, because neither 'materialist' nor 'spiritualist' explanations are able to account for it satisfactorily.†[513] Materialist theories (as we argued in Chapter VII, p. 180 when discussing the 'ontological proof') are not in the last resort able to explain how biological evolution can arrive at an insight into its own evolutionary process and thus at 'spirit' and 'freedom' (a statement which, let it be clear, is not in the least meant to question the right of natural science to proceed as far as it can on its road nor to despise its past, present, and possible future

---

* "L'exististentialisme athée ... déclare que si Dieu n'existe pas, il y a au moins un être chez qui l'existence précède à l'essence." "L'homme, tel que le conçoit l'existentialiste, s'il n'est pas définissable, c'est qu'il n'est d'abord rien. Il ne sera qu'ensuite et il sera tel qu'il se sera fait." Thus, Sartre's existentialism, with its 'situation-ethics', insists that, when facing a certain situation, I am compelled to adopt some attitude towards it. As a result, I am responsible for my choice not only towards myself but towards all men, because with my engagement I engage at the same time humanity as a whole — we cannot in fact engage in an action with which we 'create' the man we want to be without at the same time creating an image of man as he should be in our opinion. This may well be seen as an attempt to radicalize, on the one hand, Kant's idea of ethical autonomy by stripping it of its rational certainty, yet on the other hand to avoid Gide's theory of the *acte gratuit* (cf. Chapter VII, p. 185) since the latter means, according to Sartre, to act out of a mere whim (*caprice*). Whether the attempt is very convincing, and especially whether Sartre's political 'engagements' have actually been very fortunate, remains of course another question.

† On this topic the essential has long ago been said already by Pascal when, in the *Pensées,* he wrote: "Man is to himself the greatest prodigy in nature, for he cannot conceive what body is, and still less what mind is, and least of all how a body can be joined to a mind. This is his supreme difficulty, and yet it is his very being. *The way in which minds are attached to bodies is beyond man's understanding, and yet this is what man is.*" ("L'homme est à lui-même le plus prodigieux objet de la nature; car il ne peut concevoir ce que c'est que corps, et encore moins ce que c'est qu'esprit, et moins qu'aucune chose comme un corps peut être uni avec un esprit. C'est là le comble de ses difficultés, et cependant c'est son propre être: *Modus quo corporibus adhaerent spiritus comprehendi ab hominibus non potest, et hoc tamen homo est*"; the passage in Latin is from St Augustine).

accomplishments (sociobiology, gene-investigation)).*[514] For the fact remains that man can love not only life but also abstract ideas and risk his individual life for them, in short, act 'unfunctionally' as we called it;† nay, he can even negate life! On the other hand, in spiritualist theories the explanation has been smuggled in from the beginning and is therefore no explanation at all.‡

Alas, we shall have to acquiesce in the fact that Schopenhauer did not take our advice! This, in its turn, gives rise to two questions. One is: Why did he not? (or rather: Why could he not?); the other: what consequences would have ensued if he had? To the first question our answer is: because of what (to use his own favourite expression when criticising Kant's philosophy) we consider to be his πρωτον ψευδος ('primal fallacy'), §[515] namely, the idea that we are originally conscious only of events on our retina (more generally: 'underneath the skin') which is really

---

* Kant, in the second, teleological part of his *Critique of Judgement*, stresses in his sober way, on the one hand, the need for a non-mechanistic, teleological view of nature since we cannot hope that there will ever arise a Newton who could make plausible the coming into being of even a simple blade of grass by laws of nature that are not ordered by design or intention, but also the need, dictated by our reason, to go on as long as we can to regard nature's productive faculty in a purely mechanical way, since teleology, although a necessary postulate of reason, cannot scientifically explain anything without lapsing into tautologies. So according to Kant one could imagine a combination of mechanical and teleological laws in the products of nature without contaminating the principles of judgement or putting the one in the place of the other. Since it is impossible for our reason to determine how much mechanism nature introduces as a means to its ultimate end, we cannot know either how far mechanical ways of explanation may go; only that they are insufficient for things that we recognize as ends of nature, and that therefore we must subordinate mechanical reasons to a teleological principle. We are here back again at Kant's distinction between *Grenzen* and *Schranken* (cf. Chapter VI, p. 136).

† Chapter VII, footnote on p. 185.

‡ This goes not only for traditional religious explanations, but also for Hegel's objective idealism. Moreover, bereft of its 'idealist' clothing the latter leads to a *historical* reductionism which leaves as little place for human freedom as a *biological* one. In both cases we are concerned with the Cretan who must have lied in any case (cf. footnote on p. 202); for evidently the reductionist is convinced of the determinedness and, hence, the ultimate lack of truth of all statements ... except his own! For the rest, what we said in the above about the discipline of biology goes likewise for the discipline of history as such; indeed, it is a problem as legitimate as it is fascinating (and we have tried to contribute to its solution in the course of this study) why the *conditio humana* described in the above (while fundamentally universal) has assumed in the West that much more pointed guise which has resulted in a world which may be said to be broken by modern science.

§ This expression was actually used twice by Schopenhauer; in the 'Critique of Kantian Philosophy' as regards epistemology (Kant's statement that the object of intuition is *given* to us; cf. our exposition in Chapter IV, p. 96), and in 'On the Foundation of Morals' as regards ethics (Kant's statement that it is not what actually happens but what *ought* to happen, even if it never happens, that should be the starting-point of moral philosophy; cf. our exposition in Chapter VI, p. 149).

the very thing we are *not* conscious of.* But just as Kant's πρωτον ψευδος was for Schopenhauer only the other side of the coin of his immortal merits ("no lotus without its stem": "Kein Lotus ohne Stengel", so he quoted an old Indian proverb),[516] so we may say that in the case of Schopenhauer himself his πρωτον ψευδος was only the other side of his immortal merit of having been the first significant philosopher to philosophize *am Leitfaden des Leibes*,† i.e., to stress the central role of the body which he expressed in the formula that this own body of ours is the "immediate object" (*unmittelbares Objekt*) and as such is not causally linked to Will but is *is identical* to it. To go ahead and proclaim this "Will", i.e., the *object* of inner sense (the 'object of the fourth class of objects for the subject', in his wording, but in any case an object!) to be the Thing-in-itself was the only way for Schopenhauer to avoid the solipsism into which his πρωτον ψευδος threatened to lure him (and simultaneously to satisfy his pessimistic temperament by stressing the ultimate nothingness of the empirical world). This "Will", however, as we initially know it (with ourselves as willing subject), is, according to Schopenhauer, at the same time identical to the *knowing* subject; and Schopenhauer had to confess that the identity of the knowing and willing subject was a mystery. This mystery, far from being explained by the vision of a Cosmic Will which at a certain stage of its evolution arrives at consciousness, then — precisely as a result of this attempt — became on the contrary the source of an (unrecognized) antinomy which in turn gave rise to the metaphysical paradox analysed in the present chapter.

In response to our second question: Had he acknowledged the human condition to be a true *Urphänomen*, Schopenhauer would have been driven to the conclusion that any modern metaphysics (if such a thing is possible at all) can no longer be of a 'cosmological' but only an 'anthropological' nature. This was actually announced already by Kant, to whom (despite his insistence on the need to distinguish carefully between transcendental and anthropological method) the three questions: 'what can I know?'; 'what ought I do?'; and 'what may I hope?', culminate in the end in the question: 'what is man?'.‡ Hegel, in hypostasizing Kant's transcendental subject into a universal 'World Spirit', then lost sight of this question; but it reappeared at full strength with Feuerbach. Religion, according to him, had 'estranged' man from himself; and in making Reason autonomous,

---

* Cf. our exposition in Chapter IV, p. 92.

† Cf. our exposition in Chapter IV, p. 99.

‡ The last work Kant completed is the *Anthropologie* published in 1798. It is therefore a product of his old age and bears the signs of it; as (in agreement with Raymund Schmidt's view) we remarked in Chapter VII, p. 178, it was Kant's tragedy that he proved unable to accomplish the 'positive' part of his philosophy because he was already at an advanced age when he had completed the necessary preliminary 'critical' part. Kant's '*Anthropology*' surely offers many an interesting observation on human nature, but taken as a whole it reminds one of the worldly wisdom of, say, Schopenhauer's *Parerga und Paralipomena* rather than of an elaborated, philosophical anthropology.

speculative philosophy (culminating in Hegel's identity of Being and Thought) betrayed its still essentially theological character. Theology had to give way to anthropology, and the rehabilitation of the senses would put an end to human estrangement. Schopenhauer, whose attention had been drawn to Feuerbach by poor Frauenstädt, then wrote in a letter to his slow-witted evangelist that Feuerbach's philosophy was the "coarsest and most narrow-minded materialism" ("der plumpeste, bornirteste Materialismus"),[517] which is what one would expect. In fact, Feuerbach had more in common with Schopenhauer than the latter was prepared to admit. To him, too, the body was *unmittelbares Objekt*; moreover, Feuerbach's sensualist epistemology — as likewise Schopenhauer's, with the definite exception of 'happiness' — comprises indeed "facts like mental and physical suffering, pain, organic lust, passion and happiness, needs, wishes and urges, but also categories like resistance, matter, practice, intuition, imagination and — love", as Alfred Schmidt has rightly remarked.[*518] But it is evident that in unsubtle minds (and relatively unsubtle Feuerbach's must be called), 'anthropology' was likely (under the overwhelming impression of the successes of science in their days) to lapse into crude materialism ("man is what he eats" as the German pun puts it).[†] As a matter of fact, not only 'mechanicist' science, but paradoxically also its adversary, Romanticism, had contributed to the rise of materialism in that (cf. Chapter I, p. 33) the latter had denounced the pretensions of human reason in the name of a superhuman power, though finding this power no longer in the beaming heights of Heaven, but rather in the dark depths of Nature and the Unconscious, e.g. in the *Volksseele*. The new disciplines of sociology, biology and psychology which arose in the 19th century were to give all this a scientific (or pseudo-scientific) foundation. The fascinating aspect of Schopenhauer's philosophy lies in its partaking of this trend (physiological foundation of epistemology; elevation of the 'lower' region of 'Will' to quasi-metaphysical status), while at the same time trying to seek an escape in art and ascetic mysticism. And that is why Schopenhauer's system (the last of the German 'grand narratives', though very different from those of German Idealism!), despite its contradictions (or rather because of them) filled a need in an era that had outgrown traditional religion as well as philosophical idealism but which, in its best representatives, could not be satisfied with positivism and materialism either. And therefore, perhaps, in our own time, as well? Before we attempt a cautious

---

[*] "Zu Feuerbachs 'Sinnlichkeit' gehören Tatbestände wie seelisches und physisches Leid, Schmerz, Organlust, Leidenschaft und Glück, Bedürfnisse, Wünsche und Triebe, aber auch Kategorien wie Widerstand, Materie, Praxis, Anschauung, Phantasie, und — Liebe."

[†] Cf. our remarks in Chapter I, p. 38. In fact the same happened to Marx, whose originally anthropological approach (to be found especially in the Paris manuscripts) finally resulted in a philosophical materialism which pretended to give a fully fledged explication not only of the workings of society but also of those of nature (the latter perhaps more so in Engels' *Anti-Dühring* than in Marx himself).

answer to this question, we shall first provide a broad sketch of the history of the reception of Schopenhauer after the crucial year, 1848.

IX

IN PRAISE OF REALISM

a. Schopenhauer's reception after 1848

The Paris streetfights of February 1848 were to have repercussions in the whole of Europe. In Germany, the unrest began in Baden, spreading from there to other German states; on 13 March Vienna also joined in, and five days later armed conflicts took place in the Prussian capital, Berlin. In Schopenhauer's residency, Frankfurt on the Main, that famous Assembly gathered that took it upon itself to write a constitution for a united Germany. For all its peaceful rhetoric (which, as is well known, produced no result; still better known are Bismarck's devastating comment and the alternative path along which he was later to forge German unity), in September riots took place in Frankfurt, too. Schopenhauer's response to these events is likewise well known, especially the picturesque detail of his lending his *doppelten Operngucker* (opera-glasses) to an Austrian officer in command of the twenty soldiers who had occupied his study (and who, in spite of the unwelcome disturbance, were hailed by him as "good friends"), so as to be better able to shoot at what our philosopher chose to call "the sovereign rabble" ("die souveräne Canaille"). A few years later, in his will, he granted a significant sum to the "fund for the relief of the Prussian soldiers who ... had become disabled in their struggle for the maintenance of law and order in Germany, and of the next of kin of those who had fallen in this struggle".[*519]

Ready-made stuff, all this, for later commentators, especially, of course, for those of Marxist provenance. Not only did they denounce, in the allegedly so unworldly philosopher, his petty-bourgeois fear and trembling for his money. More significantly, they also believed that they had found here the clue to the riddle of the popularity that, after consistent neglect over the decades, suddenly befell Schopenhauer's philosophy in the 1850s. Thus Franz Mehring spoke of the "hang-over" that had come over the German bourgeoisie after the frustrated

[*] "Zu meinem Universalerben setze ich ein den in Berlin errichteten Fonds zur Unterstützung der in den Aufruhr- und Empörungskämpfen der Jahre 1848 & 1849 für Aufrechterhaltung und Herstellung der gesetzlichen Ordnung in Deutschland invalide gewordenen preußischen Soldaten, wie auch der Hinterbliebenen solcher, die in jenen Kämpfen gefallen sind."

revolution of 1848, for which Schopenhauer's doctrine now provided a welcome balm: "For a whole generation [Schopenhauer's philosophy] was of no concern to anyone in Germany. For as long as the German bourgeoisie was still animated by the militant spirit of her historical vocation she would indeed have been a complete fool if, from the richness of the historical world disclosed to her by Hegel, she had withdrawn into Schopenhauer's Buddhist Nothingness. Not until her revolution misfired and a very unhistorical hang-over took hold of her did Schopenhauer's philosophy become her favourite breviary for a life of wisdom."*[520] And Lukács, in a rehash and further extension of such superior insights, went ahead to call Schopenhauer, simply, the founder of modern irrationalism.[521]

It is certainly true that Schopenhauer's attitude in 1848 was somewhat bizarre, just as it may freely be admitted that fear for the security of his modest capital had something to do with it. After all, the latter provided the material basis for his philosophical independence, and this may serve as something of an excuse; just as on a more ideological plane it has been argued, not only by the anti-Marxist Popper,†[522] but also by the (admittedly rather heretical) Marxist Horkheimer [523] that Schopenhauer's fear of German nationalism had been a significant motive. We know already from the attitude he took in 1813 (Chapter I, p. 39) that he was far from a *Deutschtümler*; still, it would be undue flattery to put all the emphasis on this one aspect. Schopenhauer was against democracy in general, as he was against nationalism (two broadly parallel movements at the time). Only, in his case this had nothing to do with any reverence for traditional powers, about which he had no illusions at all;‡ precisely because he had no illusions about mankind generally. And that is really why he refused to believe in Progress and left no place in his system for a divinization of History.

This grim scepticism of his, then (and here we have the kernel of truth in the Marxists' point), fitted in with what Palmer & Colton, in their textbook *History of the Modern World*, have called the "new toughness of mind" that set in after 1848.

---

* "Ein Menschenalter hindurch kümmerte sich niemand in Deutschland darum. Solange die deutsche Bourgeoisie noch von der Kampfstimmung ihres historischen Berufs beseelt war, wäre sie ja auch eine vollkommene Närrin gewesen, wenn sie sich aus dem Reichtum der historischen Welt, die ihr Hegel erschloß, auf Schopenhauer's buddhistisches Nichts zurückgezogen hätte. Erst als sie nach dem Scheitern ihrer Revolution in den Zustand eines sehr unhistorischen Katzenjammers geriet, wurde Schopenhauer's Philosophie das Lieblingsbrevier ihrer Lebensweisheit."

† Popper's abhorrence of Marxism surely led him to overestimate the anti-nationalist component in Schopenhauer's attitude. In fact, Schopenhauer's utterances at the time (which even Popper himself acknowledged to be "rather revulsive") were rarely directed at the nationalist ideology of the revolutionary movement; so much the more do they betray "his apprehension that under mob rule he might lose his independence" which, of course, Popper, too, recognizes as having been an important component, but which serves him rather as an apology than as an indictment of Schopenhauer's conduct (witness his very choosing of the expression "mob rule").

‡ Recall here his remark to Frédéric Morin (Chapter V, p. 133).

For the rest, we may leave aside the worked-up, Marxist indignation over the cowardly treacherous role of the bourgeoisie and its failure to form a common front with 'the masses'. That indignation is as inconsistent (why indignation over an attitude that is historically determined?) as it is unhistorical and therefore unjust, in that it fails to take into account contemporary circumstances. We mention only that the need to attain both political freedom and national unity had saddled the German liberals with an almost impossible dual task, or that, had they followed the lead of the ultra-left (which was also Great-German in its outlook), a European war would have been the result, thus, in all likelihood, ushering in a period of much more severe and forceful reaction than was actually to take place. Safranski and others have quite rightly pointed out, therefore, that the new 'toughness' (to stick to that expression) did not carry any implications of pessimism becoming the predominant mood of the German bourgeosie. Belief in progress was now perhaps stronger than ever, only, it had become more 'realist' and more 'objective'. So it was even with Marx, himself. Putting an end to his habit of impatiently awaiting immediate proletarian insurrections, he now settled down to a patient (at least until the *Commune* of 1871) 'scientific analysis' of society and its core, 'capital'.[524]

Indeed, as already noted in Chapter I, philosophically the hour had struck for materialism and positivism (in tandem with the successes of science and the economic boom of the fifties and sixties); but also for at least the more popular part of Schopenhauer's philosophy, which was in tune with the sober sense of reality of the time and even with its materialism, insofar as Schopenhauer had expressed both a deep distrust of 'spirit' [*525] and, in the wake of Kant, an unconditional appreciation of empirical science, albeit with the proviso that the world it discloses to us is only 'representation'. Thus Schopenhauer's philosophy provided *both* many practical hints on how to make one's life as healthy and comfortable as one can *and* a justification for the current, optimistic belief in science, while *simultaneously* making up for an attendant, deep down, uncomfortable sense of loss of metaphysical profundity through its pointing to the relative nature of all this business. To that extent Schopenhauer's 'pessimism' served a felt need; but it could do so precisely because it was a brand of pessimism quite compatible with the worldly wisdom of his 'Aphorisms on wisdom about life'. Indeed, rather than of the more esoteric portions of his philosophy it was the

[*] For example: "Spirit? Who is that fellow? and how do you know him? might he not be just an arbitrary and convenient hypostasis which is not even defined, let alone deduced or proven by you?" ("Geist? Wer ist denn der Bursche? und woher kennt ihr ihn? ist er nicht etwan bloß eine beliebige und bequeme Hypostase, die ihr nicht ein Mal definirt, geschweige deducirt, oder beweist?"); or elsewhere: "This 'spirit', which roams about everywhere in present-day German literature, is anyhow a decidedly suspicious type, and one should therefore ask for his passport wherever he may be spotted" ("Überhaupt ist dieser 'Geist', der in jetziger deutscher Literatur sich überall umhertreibt, ein durchaus verdächtiger Geselle, den man daher, wo er sich betreffen läßt, nach seinem Paß fragen soll.").

fate of the *Parerga und Paralipomena. Kleine philosophische Schriften*, the first volume of which (published in 1856) ends with those 'Aphorismen zur Lebensweisheit',[526] to cause the first breakthrough, happily to be enjoyed by their author for the last four years of his life. According to Heidegger, this sudden interest in Schopenhauer did not at all prove therefore that he had vanquished German Idealism. It only meant that the Germans were no longer capable of living up to the heights of the latter; hence the success of what he calls Schopenhauer's 'commonplaces'.[527] Now in a sense that is what the aphorisms are; only, Heidegger did not appreciate the more esoteric portion of Schopenhauer's philosophy either, as we shall see. An interest in *this* portion had to wait until the 1880s; but even during his subsequent heyday Schopenhauer was to remain somewhat of an outsider to his foe of old, the world of academic philosophy, and this has in fact always remained so.[*528] Rather, his enthusiastic adherents were to be found, firstly, among what the Germans call *Bildungsbürger*; and then, more importantly and

* A typical example is provided, not only by Heidegger, but also by Husserl: "The difference between empirical and transcendental subjectivity remains inevitable; but their identity remains just as inevitable, albeit incomprehensible at the same time. I myself, as a transcendental I, 'constitute' the world; at the same time, as a soul I am a human I in the world. Understanding, which prescribes its law to the world, is my transcendental Understanding, and this shapes me in accordance to those very laws, even though it is at the same time a faculty of my own, i.e., of a philosopher's soul. The I of which Fichte speaks, i.e., the I which posits itself, can it be another I than *Fichte's*? If this is to be, not a veritable absurdity but a soluble paradox, how then could any other method clarify this but that of consulting our inner experience and of an analysis following in its wake?" ("Unvermeidlich bleibt die Differenz zwischen der empirischen und der transzendentalen Subjektivität, und doch auch unvermeidlich, aber auch unverständlich, ihre Identität. Ich selbst als transzendentales Ich 'konstituiere' die Welt und bin zugleich als Seele menschliches Ich in der Welt. Der Verstand, der der Welt sein Gesetz vorschreibt, ist mein transzendentaler Verstand, und dieser formt mich selbst nach diesen Gesetzen, er, der doch mein, des Philosophen seelisches Vermögen ist. Das sich selbst setzende Ich, von dem Fichte spricht, kann es ein anderes sein als das *Fichtes*? Wenn dies keine wirkliche Absurdität sein soll sondern ein auflösbarer Paradox, wie könnte eine andere Methode uns zur Klarheit verhelfen als die der Befragung unserer inneren Erfahrung und eine in ihrem Rahmen erfolgender Analyse?"). One might be inclined to think that Schopenhauer had a few, not perhaps altogether uninteresting things to say about this very problem ('the miracle κατ' 'εξοχηv!') and also about the need of an analysis of inner experience; but, quite characteristically, his name is not even mentioned in the outline given here by Husserl. Now it is surely not our intention to claim Schopenhauer as a precursor of phenomenology — we have already seen too much of that sort of thing in our previous chapter! — and Hübscher is certainly right in observing that phenomenology had starting-points other than Schopenhauer, so that the latter could only with much reservation be called its forerunner. But he also points to two insights of phenomenology (to wit, the vindication of the rights of intuition (*Anschauung*) to underlie abstract concepts, and the analysis of our own body as a different, more pristine kind of 'givenness' than the remainder of — to use a Cartesian term — *extensio*) which indeed are to be found already in Schopenhauer. Some interest in Schopenhauer to come from that side would not be unjust, one would therefore say.

also more lastingly, among artists, with musicians (like Wagner) and writers being in the front line. Nor were these *German* writers only, as the Marxist 'hangover' thesis would have led one to expect. Surely, the first generation of authors with an enthusiasm for Schopenhauer *were* Germans; here one thinks of such typical representatives of what textbooks on the history of German literature call 'poetic realism' as Wilhelm Busch and Wilhelm Raabe. But for the later period one could mention not only Thomas Mann but also, to name but a few, Lev Tolstoy, Marcel Proust and, more recently, Samuel Beckett. As compared to the period of 'poetic realism', however, there is a marked shift in the nature of Schopenhauer's reception, which went indeed in the direction of the 'more esoteric portion' or, more especially, of Schopenhauer's metaphysics of *art* as laid down in the third book of *The World as Will and Representation.* This particular aspect is what exerted its attraction upon Nietzsche, who became the first really important Schopenhauerian — soon to become a heretical one, to be sure. It is characteristic that, starting from the *fin de siècle*, Schopenhauer had to share his popularity with this heretical disciple until, on the eve of the first World War, the disciple's star had definitely overshadowed that of his former master. After that war Schopenhauer moved into the background; and this circumstance makes somewhat questionable Horkheimer's claim that the 20th century has demonstrated the truth of Schopenhauer's pessimistic diagnosis of history and that, therefore, Schopenhauer is of more urgent interest than ever before.[529]

At first sight, much is to be said in favour of such a claim. Had not, indeed, the apogee of European greatness in the *belle époque* come down in the apocalypse of two world wars, with no. I — the first modern war in its display of mechanical and anonymous violence expressing the destructive capacities of industrial civilisation and mass society — striking the big blow to a belief in progress which, until then, had by and large remained unshaken in spite of the warnings of individual prophets like Burckhardt, and also in spite of the paradox that science itself, with Freud and others, had revealed the forces of the irrational beneath the apparently solid surface of civilisation? All this looks like one big vindication of Schopenhauer's philosophy or at least of its general tenor, and yet, we must observe that by the 1920s its hour had ticked away. How can that be?

Perhaps a clue to the riddle is to be found in the final section of that third book of *The World as Will and Representation* just mentioned. There we are informed that, after all, the artist's element is the world as *representation*, with the particularity that he stands on the level of pure intuition, free from the vexations of Will. He halts at the fascinating spectacle of Will's objectivations, which he emulates in his art. But that is precisely why, except in those rare cases when he attains the *seriousness* of the world-renouncing saint, all this is bound to remain a *play*.[530] Did not this very *playfulness* of the artist fit in wonderfully with the *fin de siècle* spirit — a spirit of secret flirtation with decline, yet on the surface still full of trust in progress — rather more than with the period after the First World War, when matters had become *serious*?

Well, it is easy to say that we meanwhile know too well the outcome of all this seriousness. It has led, not only to the philosophical 'seriousness' found back in a man like Kierkegaard (who along the paths of Karl Barth's dialectical theology and of existentialism greatly influenced 20th century thought indeed), but also to the political 'seriousness' of totalitarianism (which still — or again — *did* believe in progress albeit not in the civilized way of the 19th century). And this may perhaps make understandable the new interest in Schopenhauer that arose in our own *fin de siècle,* with its postmodernist rejection of 'grand narratives', especially the political ones. A very relative interest, to be sure; for it does not look as if Schopenhauer is on his way back to fashion. Nor is this deeply to be regretted, in view of the fact that popularity out of fashion is usually popularity for the wrong reasons. Still, this entire study of ours would have somewhat less of a point if we we were not rather firmly convinced that Schopenhauer's philosophy has still — or again — something significant to say to us. This 'something' is to be found, at least in good part, in that third book of *The World as Will and Representation* which contains the 'artists' philosophy' referred to in the above.

b. Schopenhauer's philosophy and the mechanization-cum-aestheticization of the world-picture

Perhaps the reader will object to our sticking upon Schopenhauerianism the label 'artists' philosophy' in that Schopenhauer grants to aesthetic contemplation no more than the power to provide a temporary sedative, whereas only ascesis can bring about final redemption. Still, there are some good (as well as some not-so-good) reasons for doing so. Among the rather bad reasons is surely the one inaugurated by Russell and frequently parroted since — to point to Schopenhauer's conspicuous failure to live up to his own doctrine; which is true enough but also somewhat cheap.[*531] More significantly, the third book of *The World as Will and Representation* tends to strike one as more persuasive overall than the fourth; not so much in view of its invariably interesting yet often bizarre theories but, rather, because of its tone of personal concern, especially where music is concerned.[†532] Still, whatever one's preferences in this regard may be, to

---

[*] In our Introduction we already noted how Russell reproached Schopenhauer's philosophy for being not only "not very consistent", but also "not very sincere". He went on to illustrate the latter feature, not only by means of Schopenhauer's habit of dining well (which we have mentioned there) but also of his having "many trivial love affairs that were sensual, but not passionate" and of his being, not only "exceedingly quarrelsome", but also "unusually avaricious"; with the entire indictment culminating, as might be expected in this formidable list, in the '*obit anus abit onus*' story.

[†] Thomas Mann rightly says that it is not only the high place accorded to aesthetics in Schopenhauer's system that turns it into a typical artists' philosophy, nor even his literary brilliance, but, above all, it is the tension between spirit and sensual urge which runs through it. We saw already that Mann compared *The World as Will and Representation* with a symphony; and, taking the reader's point of view, he postulates quite early on in his

Schopenhauer himself the ethical and the aesthetic spheres were closely linked, as they had been to Kant, whose (however self-willed) disciple he once again proves to be. The latter's *Critique of Judgement*, as we recall, had been meant to provide a bridge between the realms of 'necessity' and 'freedom' — the successive objects of the two previous *Critiques*. However, there can be no doubt that for Kant there was a primacy of the ethical sphere. As Safranski has phrased it, "in art one may warm oneself up to the higher tasks which are commanded by Practical Reason."[*533]

In this regard, what of Schopenhauer's other master, Goethe? For all his special appreciation of the third *Critique*, Goethe held firmly to a primacy of nothing. In his view, science, art, and morals all partake on an equal footing in the universe of culture, and his ideal was the man in whom (cf. Chapter I, p. 35) these are harmoniously united. Hence, no unpoetical science (against Newton), no irrational art (against Romanticism), no unnatural morals (against Christianity and Stoicism). We further recall from Chapter I, p. 27, Goethe's cheerful admission that he lacked "the organ for philosophy proper"; this in contrast to his companion, Schiller, who, among other things, wrote his well-known '*Letters on the aesthetic education of man*' (*Briefe über die ästhetische Erziehung des Menschen*). Compared to Kant there is still a primacy of the ethical here, but also a distinctive shift in the direction of the aesthetic sphere,[†] as need not be overly wondered at in a poet.

Finally, what of Schopenhauer, himself? Though we must reiterate that, to him, the artist is only a half-way house towards the saint, with him, too, we find an unmistakable shift in the direction of the aesthetic sphere. Though not a 'poet' in the proper sense, he surely felt much more natural affinity with it than Kant, whose *Critique of Judgement* he did not in fact value very highly.[‡534] Yet he shared

---

essay that the joy we take from any metaphysical system is always of an aesthetic nature because it is akin to the effect art exerts upon us, to wit, that it makes life's confusions transparent and surveyable.

[*] "In der Kunst darf man sich warmlaufen für die höheren Aufgaben, welche die praktische Vernunft gebietet."

[†] According to Schiller, aesthetic joy gives us a sense of wholeness which transcends the inner division caused in us by the struggle between virtuousness and desire. The *Spieltrieb* is higher (since it is able to grip us more completely) than either the *Stofftrieb* (sensual inclinations which make us belong to the realm of necessity) or the *Formtrieb* (which is the *proprium* of man as a *personality* in the Kantian sense and thus makes him belong to the realm of freedom).

[‡] Cf. his pertinent remarks in the 'Critique of Kantian Philosophy'. He considers it to be Kant's great merit that, other than in all preceding aesthetic theories, here, too, he started from the subject (since here, too, it is first of all necessary thoroughly to know the effect in order to be able to determine the nature of its cause). Kant's principal shortcoming here is that he neglected immediate perception of beauty in favour of our *judgement* about the beautiful, "what, in a very ugly manner, he calls the judgement of taste" ("dem sehr häßlich sogenannten Geschmacksurteil"). Recall from Chapter IV, p. 96, that this neglect of concrete, perceptual knowledge, with its corresponding overestimation of abstract,

Kant's conception, later to be scorned so fiercely by Nietzsche,[535] of the "disinterested pleasure" (*interesseloses Wohlgefallen*) provided by aesthetic experience.*[536] This was indeed quite in line with his 'Apollonian' view of art,

---

conceptual knowledge, was also one of Schopenhauer's principal objections against the first *Critique*! According to Schopenhauer, Kant's treatment of the Sublime was more successful than that of the Beautiful, since in this case not only the right method of the investigation has been indicated but also at least part of the actual road towards the solution of the problem (cf. also footnote on p. 239).

* According to Heidegger, whose lack of appreciation for Schopenhauer's philosophy (also of the 'esoteric' part) we already mentioned, Schopenhauer used for its contents the men he abused, like Schelling and Hegel, while at the same time misunderstanding Kant whom he praised ("Inhaltlich lebte Schopenhauer von denen, die er beschimpft, von Schelling und Hegel. Wen er nicht beschimpft ist Kant; aber stattdessen mißversteht er Kant von Grund aus"). With the latter point, at least, we agree in many a case (as the reader will have noticed), albeit with the proviso that German Idealism misunderstood Kant, too, though in an opposite and, historically, still more fatal way. But is the same true, as Heidegger claims, of Kant's theory of the *interesseloses Wohlgefallen*? Certainly, Schopenhauer's introduction of the Platonic Ideas gave a new dimension to that theory, in that 'disinterestedness' is now, not only a necessary *mark* of aesthetic pleasure but, in a sense, its very *object*, as Julian Young has rightly remarked. Still, for Kant, too, the core of aesthetic pleasure consisted in the experience of a kind of orderliness which differs from scientific orderliness in that it is more 'free' since it does not force the object of perception into any abstract *concept* (cf. Schopenhauer's distinction between Platonic Idea and concept). Heidegger's verdict on Schopenhauer's aesthetics in general is that it cannot even remotely sustain a parallel with Hegel's aesthetics ("Eine Aesthetik, die auch nur aus der Ferne derjenigen Hegel's vergleichbar wäre, kann dies nicht genannt werden"). Well, this is a matter of taste; but more than a matter of taste is Heidegger's wholly unfounded, demonstrably wrongheaded suggestion that Schopenhauer became abusive in so boundless and tasteless a way in order to hide the heavy debt of his own main works to those of Schelling and Hegel, whose main works were already known at the time. There may be some truth in this as far as Schelling's '*On the Essence of Human Freedom*' is concerned (cf. our Chapter I, p. 38); but what about Hegel's *Aesthetics*? Note here that *The World as Will and Representation* came out in 1818, and Hegel's posthumous *Aesthetics* (based on lectures delivered between 1820 and 1829) not until 1835. For surely Heidegger did not mean to say that the rather brief remarks on art at the end of the *Phenomenology of Mind* and in the concluding chapter on the 'Absolute Spirit' in *Encyclopedia 3* were sources of unacknowledged inspiration for Schopenhauer's aesthetics?

It is certainly the case that, just as with colours, here, too, there are interesting parallels between Hegel's and Schopenhauer's respective positions, especially where the hierarchy of the arts is concerned. Thus, to Hegel, too, architecture stands below, for basically the same reasons as with Schopenhauer, and poetry (especially dramatic poetry) above. To Schopenhauer, though, music is unique in that (as we shall see) it transcends the whole hierarchy whereas to Hegel music is just one of the arts, above the plastic arts yet below poetry in his hierarchy. What formal similarities there are, however, cannot conceal the fact that the two works are written in a quite different spirit. Further down in our main text we shall have a few more things to say about this; for here our point is that Schopenhauer's aesthetics, whatever one may think of its 'tenability', is (like Hegel's, as we for our part are which (as noted in the Introduction. p. 4) made him feel more akin to 'classic' Goethe than to the Romantics.

Though it cannot be our purpose to give a detailed account of Schopenhauer's aesthetic doctrines as set forth in Book III of *The World as Will and Representation*, our argument requires some grasp of their core. The central point is that, to Schopenhauer, aesthetic experience is a way of looking at the world such that the knowing subject is no longer the servant of the willing subject and thus has a less distorted view of the world. Since, therefore, in aesthetic experience the 'I' is more objective, indeed, a "clear world-eye" ("klares Weltauge"),[537] the 'World' for its part presents itself to him more directly such as it is in its deepest essence. Thus the two poles of cognition are more closely united than in ordinary, will-governed knowledge. Ordinary man — "this manufactured commodity of nature" ("diese Fabrikwaare der Natur") as he kindly calls us — is not capable of such a view, in that he can direct his attention to things only insofar as these have some (possibly indirect) relationship to his will. The man of genius, by contrast, is characterized by his possession of this very capacity.[538] Like Kant, Schopenhauer reserved the term 'genius' for the great artist. Men of science investigate the world with the principle of sufficient reason as their guide; hence, they can obtain knowledge of relationships only, not of essences. This is why great artistic geniuses lack a talent for mathematics. How often was not Goethe taken to task for this by stupid opponents of his theory of colours? Schopenhauer indignantly asks, going on to point out, of course, that what matters in this field of science is not counting and measuring on the basis of hypothetical data, but immediate knowledge, provided by Understanding, of cause and effect![539] In short, his concept of genius is confined to that category of people who possess this capacity for cognition 'independently from the principle of sufficient reason' to such a degree as to be able not only to see the essences in things, but also to reproduce them and thus to communicate them to others. However, the very same capacity must to a lesser degree and in a different manner be present in all people save those incapable of any aesthetic pleasure whatsoever, since otherwise man could not be moved by the beautiful and the sublime,[*540] whether in nature or in art.[541]

---

quite willing to admit) a highly original piece of thought, *pace* Heidegger's disparaging remarks, which once again seem to us to be characteristic of the general attitude taken towards Schopenhauer by his arch-enemy, the German philosophy professor (cf. also footnote on p. 234).

[*] This Kantian distinction, too, was adopted by Schopenhauer (cf. footnote on p. 237), though of course in his own way: according to him we experience the sensation of the sublime when confronted — but confronted, as in the experience of the beautiful, as a 'pure knowing subject' — with fearful natural phenomena (i.e., with phenomena that stand in a hostile relation to our will). What distinguishes the experience of the sublime from the experience of the beautiful is that in the former case the situation of being a pure knowing subject had first to be attained by consciously compelling ourselves to tear ourselves loose from the relations of the object in question to our will, which relations we know to be

What, then, are these essences? Here the Platonic Ideas appear on the stage. As noted in Chapter VIII, p. 200, Schopenhauer differs from Plato in that not the Ideas but Will is the ultimate reality; however, they represent the *objectivation* of Will on its successive levels. They are therefore the eternal, basic forms, the original and general *types*, in which all moulding (*Gestaltung)* in nature and in art is ultimately grounded.* In the unconscious creation of inorganic nature we already see certain unchanging, typical basic trends at work: the 'forces of nature', like gravity, impenetrability, cohesion, heat, magnetism, electricity, and so on. Changes in time occur *to* them (*an ihnen*), yet as such they transcend all temporariness. This is clearer still in the organic world: whereas the particular specimens of a certain species come into being and perish again, the pure type remains eternally. In man, finally, there is a division between the character of the species and that of the individual, so that in a sense every man represents an Idea of his own.[542] Schopenhauer then goes on to treat the various arts in accordance with this general view, from architecture, which has to do with the lowest stage of objectivation of Will, namely, with gravity, cohesion, rigidity, and hardness, as also with the next stage, light, which is in many respects their opposite, up to those arts whose aim is the representation (*Darstellung*)† of the idea of humanity. These, for the reason mentioned, have for their dual task beauty as characteristic of the species, but also individual character; not, however, as something accidental which is wholly specific for the individual in its uniqueness, but as one aspect of the idea of humanity which is particularly conspicuous in this particular individual. Architecture already, in the struggle between gravity and rigidity as well as in the antithesis of both to light, displays something of the intrinsic dividedness of Will. In its fully developed guise the same dividedness is on display in those arts which have the highest stage of objectivation of Will (Man) for their subject matter, and this is the case most of all in dramatic poetry,[543] which is, therefore, the highest form of art ... but wait! This place really belongs to *music*, or rather, music is an art form *sui generis* in that it stands beyond the hierarchy of the arts. For, whereas all other arts give us an experience of the Platonic Ideas that lie behind the myriads of forms in nature, music gives us a direct experience of the Thing-in-itself, i.e., of "Will".[544] In one sense music might just as well as "Will" be said to be the inner core of the world as a whole. Varying Leibniz' expression, therefore, "music is an unconscious exercise in *metaphysics* [Leibniz: *arithmetic*] in which the mind does not know that it is philosophizing [Leibniz: *counting*]."‡[545]

---

unfavourable. In other words, in the experience of the sublime we *freely* and *consciously* rise above will and the kind of knowledge related to it.

* This makes them quite different from Kant's 'Ideas', as Schopenhauer never tired of emphasising (cp. Chapter VII, p. 179).

† Cf. our remarks in Chapter II, footnote on p. 43, about *Vorstellung* and *Darstellung* and how questionable 'representation' is as an adequate translation of the former expression.

‡ "Musica est exercitium metaphysices occultum nescientis se philosophari animi"; Leibniz' corresponding statement had run: "musica est exercitium arithmeticae occultum

Ideas, Schopenhauer warns us, are not to be conflated with concepts (*Begriffe*). 'Idea' is original unity fallen apart owing to the forms of space and time of our perceptive apprehension (*unitas ante rem*), whereas 'concept' is unity reconstructed from this plurality by the abstracting faculty of our reason (*unitas post rem*).[546] Ordinary empirical thought, because it is confined to the comparison of individual cases, does not discern Ideas, as the synthesizing power of genius alone is capable of accomplishing. It is this capability which makes the genius "understand nature as it were on a cue, and thus express clearly what Nature only stammers".*[547] And that is why he considered it to be "one of the greatest and most

---

nescientis se numerare animi." Schopenhauer finds this true only insofar as music can formally be reduced to quite determinate rules expressed in numbers, from which it cannot deviate without wholly ceasing to be music; but this refers only to the external aspect of music, its shell, so to speak. Were it nothing else, the satisfaction it provides would be of the kind we experience by the correct completion of a task in arithmetic, not that intense joy with which we see the deepest core of our essence voiced by her. Leibniz' view of music is therefore deemed superficial by Schopenhauer; yet one might ask how music could enjoy and console us in that it "flows along as an intimately known yet perpetually remote paradise" ("als ein so ganz vertrautes und doch ewig fernes Paradies an uns vorüberzieht"), if the essence of the world which it is said by Schopenhauer to unveil would only be terrible "Will", which is rather like Hell? Schopenhauer's explanation "that it reflects all stirrings of our innermost nature, though quite devoid of reality and far removed from its pain" ("daß sie alle Regungen unsers innersten Wesens wiedergibt, aber ganz ohne die Wirklichkeit und fern von ihrer Quaal") is not as such too convincing. For does this not suggest once again that a metaphysical dualism which allows for a principle of order and harmony alongside blind "Will" would be more appropriate (if one must at all costs have a metaphysical system in the first place) than that Will-monism to which, as we saw in the preceding chapter, Schopenhauer was not really able to stick in a consistent manner? Iris Murdoch, attracted as we recall her to be by Schopenhauer's ethics (not in the last place because of its link to aesthetics), is more consistent when she writes (not in a context of polemics against Schopenhauer, to be sure) that beauty appears as the visible and accessible aspect of the Good, which itself is not visible since it is transcendent whereas Will is not. But for her avoidance of the word 'God' and preferring (Platonic) 'Good', she is in tune here with Simone Weil, her indebtedness to whom she often professes (cf. Chapter VII, p. 180); compare, e.g. the latter's remark: "En tout ce qui suscite chez nous le sentiment pur et authentique du beau, il y a réellement présence de Dieu. Il y a comme une espèce d'incarnation de Dieu dans le monde, dont la beauté est la marque." ("In everything which provokes in us the pure and authentic sentiment of the beautiful, there is actually the presence of God. There is a kind of incarnation of God in the world, of which beauty is the mark.").

* "Daß er … gleichsam die *Natur auf halbem Worte versteht* und nun rein ausspricht, was sie nur stammelt." True, in a sense beauty exists already in nature according to Schopenhauer, for every natural object *objectivates* the Idea of its species; and if its form is such that it *expresses* this Idea, it is beautiful. But in practice the really existing objects of nature are almost always very imperfect specimens of the Idea they represent; therefore the imaginative power of the genius is required to see in them, not what Nature has really accomplished but what she has sought to accomplish.

well-known mistakes of this great man [Plato]" that he despised and rejected art, especially poetry, as a consequence of his wholly erroneous view (expressed in Book X of the *Republic*) that it is not the Idea but the particular thing which the fine arts intend to depict.[548]

It is true that Plato banished the artist from his Republic; on the other hand, we would add (for Schopenhauer prefers to keep silent about this) he proclaimed: 'ουδεις αγεωμετρητος 'εισιτω μου την στεγην (*oudeis ageômetrètos eisitô mou tèn stegèn*: "let no one uneducated in geometry enter my house", i.e., the Academy)! To Plato, the ideal propaedeutic to the recognition of universal and eternal values was mathematics, because its theses relate to ideal relationships, not to what we see with our eyes. For Schopenhauer the artist, for Plato the mathematician was the half-way house to the saint, so to speak, and this already turns Schopenhauer's 'Platonism' into a Platonism of sorts. Recall those just-cited remarks of his about the lack of mathematical talent in the genius as illustrated by Goethe. Recall further that already as a youngster he had been impressed by Goethe's capacity for 'contemplation' — which Kant, he felt, had lacked — and also that, in looking back upon his short-lived collaboration with Goethe on the theory of colours, he had expressed his admiration for the "amazing objectivity of [Goethe's] mind" as contrasted with Kantian "subjectivism".*

Goethe was not only a poet but also a theoretician of art, in particular, of the plastic arts, which, as the reader may remember (Chapter III, p. 55), had competed until his forties with poetry for the central position in his artistic preoccupations. This makes some comparison highly instructive. Compare, for example, Schopenhauer's remark on the artist who expresses clearly what Nature only stammers, with Goethe's remark to Eckermann: "Once the artist has examined how nature proceeds in shaping its works he is able to lift up real, lesser nature to the height of his own mind, in order to make that real which — whether out of weakness on the inside or of hindrances from the outside — in natural phenomena has only remained an intention."†[549] In Goethe's writings a recurrent theme is that the artist must go into the multifariousness of the phenomena and recognize in them the permanence of the form (*Gestalt*) in order then to go back and make this sensually visible in his work. Not Understanding (*Verstand*), but art is the true mediator between ourselves and the *Urphänomene* (before which, when they reveal themselves to our senses, we feel a kind of awe or even terror, from which

---

* Cf. Chapter III, p. 64.

† "Hat der Künstler studiert, wie die Natur bei Bildung ihrer Werke verfährt, so ist er imstande, die geringere reale Natur zu der Höhe seines Geistes heranzuheben, und dasjenige wirklich zu machen, was in natürlichen Erscheinungen aus innerer Schwäche oder aus äußerem Hindernis nur Intention geblieben ist."

sensual man saves himself in his attitude of simple amazement).*[550] His first meeting with Schiller caused Goethe to go even further, in that (Chapter I, p. 23) not only *Gestalt* and *Urphänomen* but the 'Idea' as well was now allowed to enter his artistic and scientific vocabulary. Listen to Goethe (and compare this with what Schopenhauer taught us): "Concepts are the *sum*, Ideas the *result* of experience; Understanding is required to infer the former, but Reason to grasp the latter."†[551] Or: "What one is wont to call an Idea: that which always comes into appearance, and therefore impresses us as the law of all appearances."‡[552] This realm of Ideas, then, is the very domain of the artist. Not until he has gained an insight into this 'law of all appearances', i.e., into that which constitutes visual reality as such, will he be able to reshape it into a work of art. So, in spite of what some people think who therefore are inclined to detest artists, these are not engaged in simply imitating what one sees with the eyes; rather, they go back to that rational kernel of which nature consists and in accordance with which it acts.§[553] To be sure, Goethe's Ideas, insofar as they are stages in Nature's creative striving for always higher forms by way of polarity and intensification,** are not merely transcendent but also immanent, and this makes them different from Plato's but akin, rather, to the 'forms' of Aristotle, whose concept of 'entelechy' Goethe likewise adopted. Exactly the same, however, may be said of Schopenhauer's 'Ideas', of which it is also as true as it is of Goethe's that they find most perfect expression, not in mathematics, but in works of art. In short, Goethe might have served with as much, or even with more, justice than Plato as the patron saint of the third book of *The World as Will and Representation.* It seems likely, however, that even for the unprofessorial Schopenhauer it would have gone a bit too far to grant such a status in what was, after all, a metaphysical system to a poet who confessed himself to lack the proper organ for philosophy...

---

* "Vor den Urphänomenen, wenn sie unseren Sinnen enthüllt erscheinen, fühlen wir eine Art von Scheu, bis zur Angst. Die sinnlichen Menschen retten sich in's Erstaunen; geschwind aber kommt der thätige Kuppler Verstand und will auf seine Weise das Edelste mit dem Gemeinsten vermitteln. Die wahre Vermittlerin ist die Kunst."

† "Begriff ist Summe, Idee Resultat der Erfahrung; jene zu ziehen, wird Verstand, dieses zu erfassen, Vernunft gefordert." That Goethe here uses 'Reason' (which Schopenhauer would reject, as we know; he would have spoken of 'better consciousness' or of the 'pure knowing subject') suggests the influence of the Kantian Schiller; the way in which he differentiates between 'concept' and 'idea', however, is quite in keeping with Schopenhauer, as might be expected in a poet who, unlike Kant but quite like Schopenhauer, lived from *Anschauung* and distrusted abstractions.

‡ "Was man Idee nennt: das, was immer zur Erscheinung kommt und daher als Gesetz aller Erscheinung uns entgegentritt."

§ "Wollte aber jemand die Künstler verachten, weil sie die Natur nachahmen, so läßt sich darauf antworten, daß die Naturen auch manches andere nachahmen; daß ferner die Künstler nicht das geradezu nachahmen, was man mit Augen siehet, sondern auf jenes Vernünftige zurückgehen, aus welchem die Natur bestehet und wonach sie handelt."

** Cf. Chapter III, p. 57.

True, there is one important difference between Schopenhauer's conception of art and Goethe's. The former ranked music above all other arts, whereas Goethe, a typical *Augenmensch* ('eye person'), had more affinity with the plastic arts, which goes for Kant, too, insofar as with him one can speak of any affinity at all. In this respect, Schopenhauer, for once, appears as a true heir of Romanticism, even though it is no less true that his *taste* in music remained rather classical; for example, he never fell into the Wagner trap, not even temporarily, as Nietzsche, famously, was to do.[*554]

Returning thus to Nietzsche, we first recall our characterization of his philosophy as an effort to overcome the contradictions in Schopenhauer's philosophy which, in the course of this book, we have seen arise out of the unspoken presence in it of two coexistent yet different metaphysics. Just as Nietzsche, as a philosopher of 'Will', remained a disciple of Schopenhauer, in however heretical a fashion, he also remained one in respect of the central place of art in his own philosophy. It would carry us much too far to describe the evolution of Nietzsche's thought in detail; even more than with Schopenhauer's philosophy of art, a brief outline must suffice. In his first period, when still under the spell of Schopenhauerian pessimism, Nietzsche, too, postulated a metaphysical world-will behind the empirical phenomenal world of space, time, and causality. His description of the Apollonian position reminds one of Schopenhauer's Platonic Ideas, as indeed, using Schopenhauer's vocabulary, he calls Apollonian art the "apotheosis of the *principium individuationis*",[555] with the Dionysian "spirit of music", from which tragedy is born, standing for the ecstatic overcoming of the *principium individuationis*.[†] But here, already, there is a difference in that Nietzsche's Will is not a blind, but a creative, will. This is not to be taken in the Biblical sense, for Nietzsche agrees with Schopenhauer that the character of the world is incompatible with a morally perfect Creator. Nietzsche's Creator-God is

[*] We are of course aware that this was easier for a man of Schopenhauer's generation than for Nietzsche, since Schopenhauer did not live to see the apogee of Wagner's fame. A reading of the libretto of the *Ring* (which Wagner sent him in 1854) made him think that the author had perhaps some literary talent; performances of *Tannhäuser* and of the *Flying Dutchman* provoked him to the sole comment that Wagner did not know what music was. However much one may regret that *Tristan and Isolde*, the highest product of Wagner's genius and also his most Schopenhauer-imbued work, never had a chance to be attended by the master, it is certain that Wagner's conception of the *Gesamtkunstwerk* was quite contrary to Schopenhauer's conviction that music is self-sufficient. That, in any case, is how he himself explained his, at first sight, perhaps somewhat astonishing predilection for Rossini (cf. our remarks on the parallel between Schopenhauer and Stendhal in our Introduction, p. 4) — his operas do not actually require words at all and therefore produce their full effect even when rendered by instrumental means alone.

[†] Note that, to Nietzsche, the Dionysian act of Greek tragedy is a symbolic, artistic, and therefore *disciplined* expression of the original Dionysian ecstasy (which we saw him describe in Chapter VIII, p. 190). That is why he considers Greek tragedy as a genius-like *synthesis* of the Dionysian and the Apollonian principles.

the "Dionysian worlds artist" ("der dionysische Weltenkünstler"); even of what we customarily call the world of art we are not the real creators. Rather are we ourselves images and artistic projections to its true creator and we derive our highest value from our importance as works of art; "for only as an *aesthetic* phenomenon are existence and world eternally *justified*."[*556] But only the genius knows something of the eternal essence of art insofar as in the act of artistic begetting he fuses with the original artist of the world; for now he is at the same time subject and object: poet, actor and spectator. Thus what for Schopenhauer was only the preparatory stage on the road to salvation becomes for Nietzsche salvation itself. The escape from individuality by the intoxication of Dionysian art (i.e., the immersion in the Cosmic Will rather than the emancipation from it) is offered as the only remedy for the pain and sufferings inherent in the human condition.

Thus already in the young Nietzsche can we observe a decisive further step taken on the road away from a balance between the ethical and the aesthetic in favour of the latter — a road, to be sure, already taken by Schopenhauer, in whom, however, there was still some sort of primacy of the ethical sphere.[†] This becomes perfectly clear in Nietzsche's later work. After the 'positivist' intermezzo of his second period, when he ranked science much higher than art and was generally playing the role of a freethinker, his third period came under the aegis of a new religious outburst (already foreshadowed in his *Die Fröhliche Wissenschaft* ('*Gay Science*')): that of the Superman (*Übermensch*) and Eternal Recurrence (*Ewige Wiederkehr*). For his valuation of art this meant, in a sense, a return to the first period; only, Dionysian art (still, or again, Nietzsche's ideal) now no longer expresses the pre-conscious world-will, but the self-conscious will-to-power of the creative man (or rather Superman), so that it no longer stands for redemption from the *principium individuationis*, but rather for its jubilant affirmation. A definite turning away, this, from what Nietzsche now called the 'romantic pessimism' of Wagner and of Schopenhauer, and, in a sense, a turning *towards* Goethe, for whom the later Nietzsche, in rejecting Wagner's romantic formlessness in favour of classical severity of form, often expresses his admiration.[‡557]

---

[*] "Denn nur als *ästhetisches* Phänomen ist das Dasein und die Welt ewig *gerechtfertigt*."

[†] An ethics 'of sorts' in that, as we have argued in our Chapter VII, Schopenhauer's world-view is meta-ethical rather than ethical in the strict sense.

[‡] E.g. in the following passage from '*Twilight of the Idols*' (*Götzendämmerung*): "Goethe — not a German event, but a European one: a grandiose attempt to overcome the 18th century by a return to Nature, by a *rising* toward the naturalness of the Renaissance, a kind of self-conquest from the side of this century. He shared in the latter's strongest instincts: the sensitiveness, the idolatry of nature, the anti-historical, the idealist, the unreal and revolutionary (which is just another guise of the unreal). He took advantage of history, of science, of Antiquity as well as of Spinoza and above all of practical activity; he surrounded himself with nothing but closed horizons; he did not cut himself off from life but placed himself inside it; he was not timid and took as much as possible upon himself, above himself, into himself. What he wanted was *Totality*; he fought against the separation

A fundamental difference relative to Goethe nonetheless remains. For all his new-found 'classicism' Nietzsche is 'romantic' (and 'modern') insofar as it has quite rightly been remarked that he starts from the *subjectively* beautiful, i.e., from reality as posited by the creative subject, whereas Goethe had stressed the *objectively* beautiful, i.e., the immanent, aesthetic, lawlike regularity demanded by the beautiful theme itself.*[558] "Will-to-power" is therefore the essence of what Nietzsche considers to be 'classical' art; and one is not quite assured about the nature of Nietzsche's 'classical' taste when he gives as his prime example of 'great style' the Palazzo Pitti, i.e., the most hideous of Florentine Renaissance palaces, which has for its only excuse that behind it lie the Boboli gardens ...† True, there is another Nietzsche, the 'halcyonic' one; and it is this aspect which turns him into one of the heroes of postmodernism; but I agree with Julian Young that this playful, light-footed, ironic artisticity, which reminds one of Nietzsche's near-contemporary Oscar Wilde, was recommended by him as a recipe for the convalescent from Christian–romantic, life-denying pessimism rather than as the ultimate ideal of the fully healthy.[559] So, it seems that 20th-century, post-First World War *seriousness* and 20th-century postmodern *unseriousness* are the two joint, final outcomes of the 'aestheticization of the world-picture'.

Could it really be otherwise? There is, after all, a certain logic in the road that leads from the 18th-century emancipation of the aesthetic as a category in its own

---

between reason, sensuality, sentiment and will (such as preached in a deterrently scholastic way by *Kant*, Goethe's antipode); he disciplined himself into wholeness, he *created* it." ("*Goethe* — kein deutsches Ereignis, sondern ein europäisches: ein großartiger Versuch, das achtzehnte Jahrhundert zu überwinden durch eine Rückkehr zur Natur, durch ein *Hinauf*kommen zur Natürlichkeit der Renaissance, eine Art Selbstüberwindung von seiten dieses Jahrhunderts. — Er trug dessen stärkste Instinkte in sich: die Gefühlsamkeit, die Natur-Idolatrie, das Antihistorische, das Idealistische, das Unreale und Revolutionäre (letzteres ist nur eine Form des Unrealen). Er nahm die Historie, die Naturwissenschaft, die Antike, insgleichen Spinoza zu Hilfe, vor allem die praktische Tätigkeit; er umstellte sich mit lauter geschlossenen Horizonten; er löste sich nicht vom Leben ab, er stellte sich hinein; er war nicht verzagt und nahm so viel als möglich auf sich, über sich, in sich. Was er wollte, das war *Totalität*; er bekämpfte das Auseinander von Vernunft, Sinnlichkeit, Gefühl, Willen (— in abschreckender Scholastik durch *Kant* gepredigt, der Antipode Goethes); er disziplinierte sich zur Ganzheit, er *schuf* sie.").

* Goethe: "The law, in its coming into appearance, produces in greatest freedom, according to its own conditions, the objectively beautiful, which, however, needs to find worthy subjects by whom it is grasped." ("Das Gesetz, das in die Erscheinung tritt, in der größten Freiheit, nach seinen eigenen Bedingungen, bringt das objektiv Schöne hervor, welches freilich würdige Subjekte finden muß, von denen es aufgefaßt wird."). Recall here Schopenhauer's remark (Chapter III, p. 64) on "the amazing objectivity of [Goethe's] mind, which always impresses his poetry with the stamp of genius."

† The 19th-century Dutch author Conrad Busken Huet had already characterized the Palazzo Pitti as a "grin out of Siberian masonry"; nowadays we would say 'style Albert Speer *avant la lettre*'.

right,* through *l'art pour l'art* (which the later Nietzsche rejected as decadent, since it detaches art from life), to 'l'art pour l'artiste' (a formula, to be sure, not used by Nietzsche, but which is the consequence of his conception of art as the self-expression of the strong individual), and, finally, in the course of the democratization of our civilization, to the reduction of art to just another kind of entertainment. In spite of all formal similarities, German *Klassik*, with its ideal of Beauty as an educator, already differed fundamentally from 17th-century French classicism (with its great representatives, Corneille and Racine) precisely because of the intervening emancipation of the Beautiful in the 18th century.† Nor had this emancipation taken place by chance. The 'aestheticization of the world-picture' may be regarded as a logical complement to the 'mechanization of the world-picture', in that the latter confined the domain of objectivity to the sciences.‡ Everything else, i.e., all *values*, then tend to become a matter of taste.§[560] Indeed,

---

* The big landmark was the appearance, in 1750, of Alexander Gottlieb Baumgarten's *Aesthetica* (still written in Latin); here the expression is being used for the first time in its present-day sense of theory of beauty (even Kant, in his 'Transcendental Aesthetics', still used it thirty years later in the etymological sense of theory of perception).

† There is of course still another fundamental difference: whereas the art of the *Grand Siècle* was 'classical' not only in the formal sense, but also in that its 'great style' expressed in idealized form the spirit of a great nation in upheaval (just as the art of Shakespeare, however 'unclassical' in form, expressed the spirit of Elizabethan England; *mutatis mutandis* the same may be said of the great painters of the 'Golden Century' of the Dutch Republic), the classical ideal of Goethe and Schiller was doomed to remain a staged revival of an idealized Hellas performed in a provincial theatre: in fact it was their *personal* solution, which could not found a school, of the problem of how to regain a new universal harmonious world-view which would overcome the chasm between Enlightenment and Romanticism (of which their own *Sturm und Drang* had been the prototype), i.e., between a scientific objectivity which tended to reduce the fullness of life and world to a mechanical device and an artistic sense of life which tended to fly into the dark jungle of irrationalism and subjectivism. This bridge-building had to take place not only in art but in all aspects of life, e.g. in science itself. Hence Goethe's perennial pursuits in the most divergent departments (an unsympathetic spectator might rather call this 'his all-round dilettantism'), not least in (anti-Newtonian) science.

‡ One might go even further, and recall that Kant's critical philosophy even demonstrated the subjective nature of *science*. However, in this case we are concerned with the *transcendental* subject, and, paradoxically, precisely this guarantees the objectivity of science (cf. Chapter I, p. 20). That is why in Chapter 1, p. 33, we distinguished between 'Romantic subjectivism' and 'Enlightenment subjectivism'. It is the former which has given birth in modern times to aesthetics as an independent domain, whereas until then the Beautiful was considered as an illustration of the True and the Good, and thus was subordinated to them.

§ To a reader with an interest in this topic we especially recommend the beautiful study *Homo Aestheticus. L'invention du goût à l'age démocratique* by Luc Ferry, even though we regret that the author, who discusses the positions of Kant, Hegel, and Nietzsche quite thoroughly, mentions Schopenhauer only in passing; neglect of this kind does not, therefore, seem to be confined to German professors! But after all, the once cheerful Gallic

Kant, who in his second *Critique* had sought to rescue the objectivity of ethical judgement and in his third *Critique* sought to rescue at least the intersubjectivity of the aesthetic judgement, could not help but call the latter a judgement of *taste*. But note that this implies a *judgement* of taste! That is to say, although it rests on feeling it can be *argued about*, and this is what makes the 'beautiful' fundamentally different from the 'agreeable', of which (as in culinary matters) it may indeed be said: *de gustibus non est disputandum*. That is why the aesthetic domain is the bridge between man as a rational, intelligible being and man as a natural, sensual being; and we recall that this very aspect of Kantianism was taken up (and elaborated in his own way) by Schiller, who inspired Goethe in his turn.* But, in spite of Goethe's rejection of Romanticism, the Goethe/Schiller concept of 'art as educator' is itself romantic in that art here actually takes the place of religion.† This becomes even more pronounced in Schopenhauer, who in marked contrast to his alleged master Plato, might be said to have expected *too much* of art. It is to be feared that in this respect Hegel showed more realistic sense when he considered art (and also religion) to be a principally superseded form of culture now that, in his own philosophy, World Spirit had finally attained full consciousness. For even those with some scepticism toward the latter claim, and those who have not quite abandoned the hope that great art might still be possible in the future,‡[561] cannot help thinking that the artist, when he no longer recognizes the existence of something greater than himself (as the medieval artist did), is prone, precisely in view of his being a creator, to regard authenticity as the supreme value. That is why Nietzsche scorned Kant's belief (still shared by Schopenhauer) in 'disinterested pleasure', as we have seen. Young, in his critique

---

esprit has in our century, under the auspices of the three H's (Hegel, Husserl, and Heidegger), been soaked quite intensely in the waters of Teutonic profundity and gravity ...

* Cf. chapter I, p. 27, and footnote on p. 237.

† In later life Goethe indeed had his doubts about the classical ideal. These made him stress practical social activity in *Wilhelm Meister*, just as they made him reassess classicism under the influence of the Boisserée brothers (especially Sulpiz) which led him to a renewed appraisal of medieval art; still more important would prove to be the acquaintance with Oriental poetry which resulted in his *West-östlicher Divan*. Cf. our remarks on this topic in Chapter I, footnote on p. 32, and in Section c. below.

‡ Ferry comes to the conclusion that the subjectivization of art has paradoxically led to the end of the *avant garde*, since the latter's attitude of subversion (which attitude is contained in the ideal of absolute innovation by the genius and of originality for originality's sake) has itself become a tradition. Democratized and rendered banal in our own time, it no longer shocks at all, whereas on the other hand the exaltation of human autonomy has done away for good with the old ideal that art should give us a glimpse of the sacred, i.e., of an order superior to ourselves. Therefore, great scientists are still possible, as indeed such have manifested themselves throughout the 20th century, but no contemporary artist could without laughing be compared with Bach, Mozart, and Wagner, or with Jan van Eyck, Velasquez, Rembrandt, and Picasso, or with Brunelleschi, Michelangelo, and Frank Lloyd Wright, or with Shakespeare, Rimbaud, Kafka, and Rilke.

of Nietzsche, nonetheless insists quite rightly that we must distinguish here, first, between the spectator's and the creator's perspective (Nietzsche was interested only in the latter), and, furthermore, *within* the latter category, between a "contemplative" and an "executive" phase, thus revealing that in both cases the notion of 'disinterestedness' is very well fitted to serve as a characteristic of the former part of the dichotomy.[562] Taking this point of view, we may go further and involve Schopenhauer's position in this critique, too. Labels first: the alternative view of the world presented in the third book of *The World as Will and Representation* should not be called another form of 'the world as *representation*', since this implies a manipulating, designing, charting attitude which by definition is linked with a view of the world guided by the principle of sufficient reason. Furthermore, the contemporary reader is likely to be sceptical about the mythology of the Platonic Ideas in which Schopenhauer's theory of art goes attired, even if he may be sympathetic towards its general thesis that aesthetic contemplation (of artistic as well as of natural beauty) can be a way of overcoming the Self and thus of looking more 'objectively' into the world. In that case, however, he will ask whether aesthetic contemplation is not just another guise of that more general, fundamentally *ethical* category of *empathy* which we have also met in Schopenhauer's ethics of compassion. More than that, we recall from Chapter I, p. 26, that in the very domain of *science* such empathy was recommended by Goethe (against Newton's "crucifying" Nature) as a "tender empiricism" which "identifies itself, as intensely as can be, with the object".

c. Schopenhauer's broken world-view: colours and ethics between Kant and Goethe

We appear now to be presenting Goethe as a kind of ethical thinker; is that not something of a joke?

It is certainly true that Goethe was not a man of outstanding moral passions like Schiller, "der Moraltrompeter von Säckingen", as Nietzsche scornfully called him; his indignant remark about Kant's 'radically evil'*[563] demonstrates sufficiently the distance between his own way of thinking and the severely moral thought of his great 'spiritual brother' (as Schopenhauer called the two of them). Still, it has rightly been noted that Epicurus' and Spinoza's ethical ideas were of great significance for Goethe's thought and work.[564] Both were concerned with promoting happiness and suggesting ways of preserving it under adverse

* "By contrast, Kant, too, has feloniously soiled his philosophical robe — after having used a human life-span to clean it from manifold foul prejudices — with the stain of radical evil in order to lure Christians, too, into kissing its hem." ("Dagegen hat aber auch Kant seinen philosophischen Mantel, nachdem er ein langes Menschenalter gebraucht hat, ihn von mancherlei sudelhaften Vorurteile zu reinigen, freventlich mit dem Schandfleck des radikalen Bösen beschlabbert, damit doch auch Christen herbeigelockt werden, den Saum zu küssen.").

circumstances. What Goethe called 'resignation' (*Entsagung*) appears prominently in both Lucretius' *De rerum natura* and Spinoza's *Ethics* as a strategy for maintaining joy and pleasure in life (and also, we would add, in the 'exoterical' part of Schopenhauer's ethics). Both philosophers dealt at length with how to overcome negative thoughts that cause unnecessary suffering and unhappiness. So did Goethe, e.g., in his novel *Die Wahlverwandtschaften* ('*Elective affinities*'). However, this is worldly wisdom rather than ethics in the strict sense of the word. As Prandi, in her book on Goethe's ethics, concedes, he avoided the thing valued most by most religions (and also, we would add, by the 'esoterical' part of Schopenhauer's ethics), namely, self-negation for the benefit of others; just as for Spinoza, too, to act virtuously is to act 'on the basis of seeking what is useful to oneself', not to any other person. Hence, she feels bound to add that Goethe took far less interest in the sufferings of underprivileged humanity than in the sufferings of those in secure possession of life's basic needs: food, clothing, and shelter.[565] To this we would add that, in a sense, Goethe, minister at Weimar, can be seen as a typical representative of *Verantwortungsethik* (an ethics of responsibility at work when, of all people, the author of the 'Gretchentragödie' in *Faust* saw fit to confirm the death-sentence for an unwed child-murderess). Goethe himself hit the mark quite nicely when, in speaking of his relationship to Schiller, the disciple of Kant, he expressed the difference between them in the formula "he preached the gospel of Freedom; I did not want to see the rights of Nature curtailed".* This partisanship for Nature need not cause any wonder, in view of his well-known 'pantheism'. And did not Schopenhauer rightly say: "Pantheists cannot have any serious morals, since for them all is divine and excellent?"†[566]

Now it might be objected that Goethe was not a pantheist *tout court*. "I, for my part, cannot, in view of the manifold aspects of my being, be content with one sole mode of thought: As a poet and artist I am a polytheist; as an investigator of nature, a pantheist; and I am the one as firmly as the other. When I need a God for my personality as a moral being, this has already been fixed, too."‡[567] It is perhaps not too risky an assumption that this statement, which must sound so frivolous (especially as far as the last sentence is concerned!) in the ears of all serious-minded men-of-one-piece, was not only a somewhat coquettish showing of his Protean nature, but also a rather ironic kind of confession on the side of 'The Last

---

* Cf. Chapter I, p. 28.

† "Die Pantheisten können keinen ernsthaften Moral haben; weil bei ihnen Alles göttlich und vortrefflich ist."

‡ "Ich für mich kann, bei den mannigfaltigen Richtungen meines Wesens, nicht an einer Denkweise genug haben: als Dichter und Künstler bin ich Polytheist, Pantheist hingegen als Naturforscher, und eines so entschieden als das andre. Bedarf ich eines Gottes für meine Persönlichkeit als sittlicher Mensch, so ist dafür auch schon gesorgt." Also elsewhere: "We are pantheists when studying nature, polytheists when writing poetry, monotheists in moral respect" ("Wir sind naturforschend Pantheisten, dichtend Polytheisten, sittlich Monotheisten.").

Universal Man' (to use this once-standard epithet for Goethe) that the loss of a unified world-picture had already become irrevocable. But perhaps something more may be said. It is worthwhile observing how Goethe, in the last conversation recorded by Eckermann, which took place only eleven days before his death, performed the *tour de force* of synthesizing nature worship and monotheism:*[568] "If one asks me whether it is congenial to me to pay him [i.e., Christ] adoring reverence, I say: but yes! I bow before him as the divine revelation of the highest

* 'Nature worship' here taken to cover both 'polytheism' and 'pantheism'; for it is evident that the former is compatible with the latter as soon as a civilization (at least in its philosophically minded members) has reached the stage of maturity at which the gods are recognized as the personifications of the various natural phenomena and human propensities. And it is no less evident that polytheism's stressing of the pluralist aspect of the One Divine (in the 'pantheist' sense) makes it certainly much more congenial to the poetic temperament than the 'monotonotheism' scorned by Nietzsche. It is well known that, in the third part of his *Zarathustra*, the latter gave a hilarious description of the death of Greek polytheism: confronted with the claim of an old long-beard of a God to be the only one, the ancient Gods laughed themselves to death, crying: 'is it not precisely divine that there are gods, but no God?'. Indeed, the thesis may be maintained that 'God' is a class-name and as such a 'pagan' residue in Judeo–Christian monotheism; and that the latter's fundamental characteristic of God as the origin of the moral law was bound in the end to lead to the rationalist dissolution of religion into ethics *sec*. But even the Kantian moralist Schiller, who 'preached the gospel of freedom', deplored the death of the Olympian gods in his long nostalgic poem 'Die Götter Griechenlands' ("in order to enrich *one*, the world of the gods had to disappear": "*Einen* zu bereichern unter allen / Mußte diese Götterwelt vergehn"), and its result: "de-deified nature" ("die entgötterte Natur"). It need not wonder, then, that Goethe, 'who did not want to see the rights of Nature curtailed', in one of his more 'pagan' moods deplored that Homer had not remained our Bible and that instead we had made the disastrous acquaintance with the "Jewish mess" and its "sodomite stuff and Egyptian–Babylonian whims" ("Beim erneuerten Studium Homers empfinde ich erst ganz welches unnennbare Unheil der jüdische Praß uns zugefügt hat. Hätten wir die Sodomithereien und ägyptisch-babylonischen Grillen nie kennen lernen, und wäre Homer unsere Bibel geblieben, welch' eine ganz andere Gestalt würde die Menschheit dadurch gewonnen haben!"). This in spite of his great knowledge of, and interest in, the Bible, displayed on many other occasions, and his statement to Eckermann: "May intellectual culture progress always further, may the natural sciences grow ever more both in breadth and depth and the human mind enlarge its scope as it wishes, — it will never surpass the sublimity and the moral culture of Christianity as it shimmers and shines in the Gospels" ("Mag die geistige Cultur nur immer fortschreiten, mögen die Naturwissenschaften in immer breiteren Ausdehnung und Tiefe wachsen und der menschliche Geist sich erweitern, wie er will, — über die Hoheit und sittliche Cultur des Christentums, wie es in den Evangelien schimmert und leuchtet, wird er nicht hinauskommen!"). It has a certain piquancy in this context that even so full-fledged a Christian poet as Camões, who in his *Os Lusíadas* praised the Portuguese for their keeping firm to Catholic faith and for their crusading spirit, could not help but invoke the Olympian Gods as either opponents (Bacchus) or partisans (Venus, Mars) of their heroic enterprise and even (in the ninth and tenth *Canto*) make Venus devise an 'Island of Love', where the Portuguese mariners, as a reward for their hardships, enjoy the charming company of beautiful nymphs ...

principle of morality. When one asks me whether it is congenial to me to worship the sun, I say once more: but yes! For it is likewise a revelation of the Highest, namely, the most powerful of which the perception is granted to us, children of the earth. I adore in it the Light and God's generating power, through which we all live, move and are, and all plants and animals together with us."*[569] To this Goethean worshipping of Light the Newtonian analysis of the latter must really have seemed blasphemous, and it is not by chance that in his *West-östlicher Divan* (book 'Suleika Nameh') we find a poem entitled 'Wiederfinden' ('Finding Back') in which it is suggested that for creatures dwelling in darkness the perception of pure Light would be too powerful: it is the *dawn* which makes the reconciliation between Creator and creation possible.†[570] Who will not be reminded of the doctrine of the 'turbid medium' and the resulting colours? Now in the same *Divan* we find a poem (the one poem of which the 'Parsi Nameh' consists), entitled 'Vermächtnis altpersischen Glaubens' ('Legacy of old-Persian faith'); moreover, in Goethe's 'notes and essays to the *Divan*' a long digression appears under the heading 'Ältere Perser' ('Older Persians'). It is stated there that the old Parsis' religion was based on God's omnipresence in his works, i.e., in the world of the senses: it was a worship of *Light*, of which the Sun, rising above the snow-capped peak of Mount Demavend ('Darnawend', says Goethe) in the Elburz Mountains, and, still more directly, Fire was the visible symbol, which implied, however, reverence towards all other elements as well, since they all proclaim God's existence and power (hence the Parsis' curious way of getting rid of the mortal remains of the deceased). But (and now recall the conversation with Eckermann!) this attitude showed at the same time the path to social culture and personal morality: for it exhorted to diligence, to gratitude for God's gifts and to truthfulness based on respect for reality as God's creation. Not until Zarathustra, Goethe thought, did this 'noble, pure religion' become an elaborate, outer-directed, priest-dominated cult.[571] This was certainly a misunderstanding, due to

---

* "Fragt man mich: ob es in meiner Natur sey, ihm anbetende Ehrfurcht zu erweisen? so sage ich: durchaus! — Ich beuge mich vor ihm, als der göttlichen Offenbarung des höchsten Princips der Sittlichkeit. — Fragt man mich, ob es in meiner Natur sey, die Sonne zu verehren, so sage ich abermals: durchaus! Denn sie ist gleichfalls eine Offenbarung des Höchsten, und zwar die mächtigste die uns Erdenkindern wahrzunehmen vergönnt ist. Ich anbete in ihr das Licht und die zeugende Kraft Gottes, wodurch allein wir leben, weben und sind und alle Pflanzen und Thiere mit uns."

† Upon picturing God's creation of the world and the resulting separation of Light and Darkness and the falling apart of the elements, Goethe in the third stanza arrives at the final vision: "Stumm war alles, still und öde, / Einsam Gott zum erstenmal! / Da erschuf er Morgenröthe, / Die erbarmte sich der Qual, / Sie entwickelte dem Trüben / Ein erklingend Farbenspiel, / Und nun konnte wieder lieben / Was erst auseinander fiel." ("All was dumb, silent and barren / and God lonesome for the first time! / He then created the dawn, / which had mercy, / and produced from turbidity / a ringing play of colours; / and from then on / what had fallen apart could love again.").

the old Persian religion being still little known in the West in Goethe's time;[*][572] yet Goethe was on the right track insofar as the opposition of Light and Darkness (which, like every opposition, was to him polar, with the one pole as necessary as the other; hence, no 'radically evil'!) in Zarathustra's teachings became a sharp dualism between the Good Principle (Ahura Mazda) and the Bad Principle (Ahriman).

Well, Schopenhauer's view on nature was very 'Ahriman'-like indeed; that is why he rejected pantheism and why he said: "Meanwhile an optimist summons me to open my eyes and look into the world: how beautiful it is in the sunshine with its mountains, valleys, streams, plants, animals, etc. But is the world a show-piece? Surely, these things are beautiful to *see*; but to *be* them is quite another matter."[†][573] In sum: "With the pantheists I have their *hen kai pan* [one and all] in common, but not the *pan theos* [all is God]."[‡][574] We have already remarked in the preceding chapter (p. 205) that in Schopenhauer's case one might speak of pan*daemonism* rather than of pan*theism*.[§] The one is, of course, as senseless as the other — Nature is innocently amoral, not immoral. Rather, it is Man's very resentment of being bound to nature, i.e., of not being God, i.e., of not accepting the anthropological paradox,[**] which lies at the root of evil. But it is precisely this Faustian temptation that sheds some new light on Goethe's concern with the 'rights of nature'. In his own time this might have seemed a somewhat superfluous concern; it is less so in our time. "Goethe called our limit — which we need not fear because we can realize ourselves only within it — 'nature', and could still think it superior in power. We have meanwhile experienced that it should be spared so as to protect us from ourselves," the Swiss essayist and literary critic Adolf Muschg has written.[††][575] Certainly in the strict sense one cannot speak of 'rights of nature';[‡‡] and we wish to stress that *ecological concern should not be conflated with vitalist*

---

[*] Zarathustra's role was in fact rather a stressing of the ethical at the expense of the magical aspect of the Persian religion; and it was only much later, at the time of the Sasanid dynasty (AD 224–651), that it became a fully fledged priest-ridden state cult. It is precisely this pronounced ethical character of Zarathustra's teachings which made Nietzsche, in an ironic reversal, choose this name for his own gospel of a life 'beyond good and evil'.

[†] "Inzwischen heißt ein Optimist mich die Augen öffnen und hineinsehen in die Welt, wie sie so schön sei, im Sonnenschein, mit ihren Bergen, Thälern, Strömen, Pflanzen, Thieren usf. — Aber ist denn die Welt ein Guckkasten? Zu sehn sind diese Dinge freilich schön; aber sie zu seyn ist ganz etwas Anderes."

[‡] "Mit den Pantheisten habe ich zwar jenes 'εν και παν gemein, aber nicht das παν θεος."

[§] Cf. also Nietzsche's remark on Schopenhauer's failing to deify Will, which we quoted in Chapter VIII, footnote on p. 223.

[**] Cf. Chapter VIII, p. 225.

[††] "Goethe nannte unsere Grenze, die wir nicht zu fürchten haben, weil wir uns nur innerhalb ihrer verwirklichen können, 'Natur' und durfte es noch für übermächtig halten. Wir haben erfahren daß es geschont werden muß, damit es uns vor uns selber schutzen kann."

[‡‡] Cf. Chapter VII, p. 173.

nature-worship. It is (or at least it ought to be) moved by the vulnerability of the concrete forms of life, not overwhelmed by the tremendous majesty of Life writ large; rather than aiming at an ecstatic submersion of man in nature, it should aim for the rise of all creatures into the Realm of Ends (to use Kant's terminology). In this sense it is compatible with 'monotheism', which, besides much else to be sure, has brought about at least one great liberation: it has done away with the sacral powers that held dominion over man. That is how, in the 'disenchantment of the world' it has in the end come to entail, monotheism has made room, not only for the modern scientific world-view,* but also for a moral attitude centred on individual freedom and responsibility. But that is also how a modern philosopher of Jewish provenance (Cathérine Chalier) could keep firmly to God's "transcendence" while at the same time warning against a one-sided neglect of His "immanent" aspect,†[576] and also conclude her book with a vision of Man as the potential redeemer of nature,‡[577] just as in Chapter VIII (p. 205) we saw

* To avoid a possible misunderstanding here: monotheism, though perhaps a necessary condition, was not of course a sufficient reason for the rise of modern science, which took place only in the context of Western, Latin Christendom (i.e., of one specific variety of monotheism), and even there at an advanced stage of its development; cf. Chapter I, p. 16.

† "Il n'en demeure pas moins que si l'on s'en tenait à l'idée de la transcendance absolue du Créateur, de Son altérité inatteignable par la pensée et le coeur, le risque inéluctable serait celui de l'athéisme ou d'une séparation irrévocable et insurmontable entre Dieu et son oeuvre ... L'homme n'aurait dans cette perspective, ni le besoin ni la nostalgie du divin, il pourrait oublier d'où il vient sans désespérer ni mourir, et la nature serait livrée à la merci de ses désirs souvent promethéens." ("It is no less true that, if one would stick to the idea of the Creator's absolute transcendence and of His otherness which cannot be grasped by either brain or heart, the inevitable risk would be that of atheism in the sense of an irrevocable and insurmountable separation between God and his work ... Seen in this perspective, man would have neither need nor nostalgia for the divine; he might forget where he comes from without despairing or dying, and *nature would be at the mercy of his often Promethean desires*." [our italics. PL]).

‡ By way of illustration we quote here two passages from the final chapter, "La rédemption de la nature", of Chalier's book: "La célébration de la nature en son pouvoir révélant s'éteint en effet d'emblée sur les lèvres de celui qui ne mesure les choses qu'a l'aune étroite et vaine de ses intérêts alors qu'elle éclot dans l'âme de celui qui sait qu'il peut rencontrer le Très-Haut dans les plus simples réalités de la vie, telles les semences de la terre, parce qu'il ne cherche pas à s'évader de sa condition de créature parmi d'autres de créature vouée à elles. L'homme, ... 'est cette créature qui peut et doit amener toutes les autres à leur ultime accomplissement et qui, en temps opportun, à coup sûr le fera'." ("The celebration of nature in its revealing power in fact falls silent instantly on the lips of him who measures things only with the narrow-minded and vain yardstick of his interests, whereas it opens in the soul of him who knows that he may encounter the Most High in the simplest realities of life, like the seeds of the earth, because he does not try to evade his condition of being a creature among the other creatures and devoted to them. Man ... 'is that creature which can and must lead all others to their ultimate accomplishment and will certainly do so when the time has come'."). This is of course an allusion to the Messianic era, but: "Il ne faut pas attendre pour entamer cette oeuvre de rédemption que les temps ultimes soient enfin

Schopenhauer do — though not of course in the latter's quasi-Buddhist sense of world-annihilation, but rather in the Biblical sense of sanctification of the world.* Schopenhauer would certainly have dismissed this as 'optimism'; yet it was this presentation of man as "the *secret hope* of the world and of all creatures" ("die *geheime Hoffnung* der Welt und aller Kreatur") which made Thomas Mann nonetheless speak of Schopenhauer as a great *humanist*, albeit a pessimistic one.[578] As Mann admitted, it is unusual to juxtapose these two words; and certainly we are concerned here with a 'humanism of sorts'. This is the case in particular when we compare Schopenhauer's variety with that of his great teacher, Kant. On this score we have already remarked (Chapter VIII, p. 219) that in fact, though not in name, little is left in Schopenhauer of Kant's 'intelligible character' as the 'proud castle of human freedom'.

Now this 'intelligible character' should remind us of history's course. With Descartes' *cogito*, modern philosophy, i.e., the process of *subjectivization* of the world-picture, is rightly said to have started. For all that, his 'thinking substance' is still an unbroken, metaphysical subject, and God's existence is still held to be demonstrable. One century and a few decades later, in Kant a cleavage of the subject has taken place into 'transcendental' (the subject proper), 'empirical' (the individual), and 'intelligible' (personality), and God has been reduced to a subordinate postulate of Practical Reason. Consider that, whereas Descartes still needed God even for the question 'What can I know?' (as God, not being a malign spirit, cannot deceive us as to the existence of the extended world), Kant did not even need him any longer for the question 'What ought I do?' but only for the remaining question 'What may I hope for?'. Thus, Kant's world was already a good deal more of a broken world than than that of Descartes, owing to Hume's preceding, *psychologist* assault on *classic* rationalism. With Schopenhauer, then, Kant's new, *critical* rationalism (and *a fortiori* its elaboration into a new scholasticism by Hegel) came under psychologist assault in its turn. But this new assault went far deeper. That it did, was due to Schopenhauer's deepening of the 'empirical subject' so as to arrive at the discovery of the Unconscious for which the way had been paved by Romanticism. Clearly, this was to the detriment both of

---

advenus, que les terribles méfaits de l'histoire aient partout fait place à la paix. Elle commence en effet, dès maintenant, dans le regard de compassion porté sur tout existant et dans le refus d'en mépriser aucun sous prétexte qu'il ne serait là que pour servir les appétits humains." ("To enter upon this work of redemption it is not necessary to wait until the end of times has finally arrived when the terrible misdeeds of history have everywhere given way to peace. It begins in fact from now on in a compassionate look at all existing beings and in the refusal to disdain any of them on the pretext that it is only there to serve human appetites.").

* Buddhism in its original guise can be said to have performed likewise a 'disenchantment of the world', but in an inner-directed sense, so to speak; therefore it aims, not at bringing the world under God's commandment (in fact 'God' is denied, and in this sense Buddhism may be said to be even more radical in the application of the Second Commandment than monotheism itself!), but at world-negation.

the 'transcendental subject' (which he turned into nothing but a 'knowing subject', thus robbing it of what is specifically human about it),* and of the 'intelligible character' (which he saved in hardly anything but the name).† Therefore, the chasm between the domain of science and the domain in which our world at large is being made sense of became much deeper in Schopenhauer's *professed* metaphysics ('Will' vs. 'Representation') than had been the case with Kant, in whose work these domains were still subsumed under the closely related categories of theoretical and practical *reason*, respectively. It remains true, nonetheless, that Schopenhauer's *hidden* metaphysics ('Nirvana' vs. 'Will') still left a place, if not for 'practical reason', then at least for 'better consciousness': Intellect (albeit in a somewhat peculiar guise) could still in the end vanquish Will! It was Schopenhauer's *humanism* indeed which prevented him from himself drawing the consequences of his Will-metaphysics — this it was left to Nietzsche to do; yet his own deconstruction of Kantianism left room for nothing but this *pessimistic* humanism.

This is why Schopenhauer, for all his anti-Hegelian scepticism about the scope of 'Reason' and all his contempt for vulgar ('Pelagian') rationalism, always, and quite rightly, considered himself to be an heir of the Enlightenment; he may not, therefore, *pace* Lukács, be claimed as a simple protagonist of 20th-century irrationalism. Here, for example, is what he had to say on the subject in *On Will in Nature:* "The respectable word 'Enlightenment' has grown into almost a term of abuse; the great men of the past century, Voltaire, Rousseau, Locke, Hume, are scoffed at, these heroes, these ornaments and benefactors of mankind, whose fame in both hemispheres can, if by anything, still be glorified only by the fact that always and everywhere that obscurantists appear, these are their bitter foes — as they have every reason to be."‡[579] He used to single out Voltaire for his praise, significantly doing so, among other things, because the latter was fully alive to the predominance of evil and to the misery of human existence.[580] Indeed, the author of *Le désastre de Lisbonne* and of *Candide* has almost as little claim as Schopenhauer himself to be called an optimistic believer in Progress. But his example shows that, to vary Thomas Mann ("one need not be a rhetor and flatterer of humanity to be a humanist"),§[581] one need not be an obscurantist to reject belief in Progress. For this concept, which is nothing but the secularization of the Judeo–

---

* Cf. Chapter IV, p. 107.

† Cf. Chapter VII, p. 194.

‡ "Das ehrwürdige Wort Aufklärung ist eine Art Schimpfwort geworden, die großen Männer des vorigen Jahrhunderts, Voltaire, Rousseau, Locke, Hume, werden verunglimpft, diese Heroen, diese Zierden und Wohlthäter der Menschheit, deren über beide Hemisphären verbreiteter Ruhm, wenn durch irgend etwas, nur noch dadurch verherrlicht werden kann, daß jederzeit und überall, wo Obskuranten auftreten, solche ihre erbitterten Feinde sind — und Ursache dazu haben."

§ "Man braucht kein Schönredner und Menschheitsschmeichler zu sein um Humanist zu sein."

Christian concept of Providence (God to be found no longer — as in Paganism — in Nature, but in History), was in fact a kind of sugar-coating of the message of Enlightenment. That message was not so joyous after all ('there is a Creator-God, but once having done His work he has withdrawn, so that man must make shift with what he has — which is not all that much', so one might paraphrase it), and its sugar-coating was intended for the 'Lampes' (the 'enlightened' Lampes, this time) of this world.

Even in its still idyllic, 19th-century guise the message of unrelenting progress was ethically dubious, because it fails in the last resort to take seriously what happens to concrete man.[*582] Our 20th-century experiences with what may be called 'optimistic anti-humanism' ('a brilliant future lies before us if only we have first exterminated a few — 'class' or 'race' — enemies')[†] gives indeed a new, contemporary significance to what we saw Thomas Mann call Schopenhauer's 'pessimistic humanism'. For Schopenhauer, for his part, always took concrete individual suffering seriously, and this, in spite of his political stance against revolutions of any kind, made him both sensitive to social problems like slavery in the Southern states of the US or the state of workers in the factories and sympathetic towards attempts at improvement, even though he would always add that, as soon as man has escaped from indigence, boredom announces itself as a new way of suffering. We reiterate that this interest of his in the concrete, living subject has more than a little to do with that emphasis on concrete *Anschauung* Schopenhauer shared with Goethe.

Certainly, the very same emphasis seduced Goethe (and also, in his wake, Schopenhauer, albeit in his own, self-willed manner) into his ill-fated effort to destroy Newton, in favour of an alternative theory of colours to be grounded in the same directly intuitive experience which is proper to the aesthetic world-view. One would wish that Goethe, who understood so well the art of resignation, had

---

[*] One may think here of Ivan Karamazov's protest against a 'future that was built on unrevenged child's tears'. Although, as we have already said, 19th-century belief in progress was still idyllic as compared to some of its 20th-century guises (see below), these are already announced in Hegelianism/Marxism and in Darwinism. On the latter we mention Chesterton's telling comment: "When scientific evolution was announced, some feared that it would encourage mere animality. It did worse: it encouraged mere spirituality. It taught men to think that so long as they were passing away from the ape they were going to the angel. But you can pass from the ape and go to the devil."

[†] Against those who see in the Enlightenment the source of all evil, especially of 20th-century totalitarianism, it should be stressed, first, that the totalitarian malady in its communist guise was certainly characterized by an obsession with science and technology, and can to that extent be regarded as a (perverted) offspring of the Enlightenment, but that it could attain power only in backward countries with a despotic tradition. Second, it should be stressed that, in its fascist guise, it adopted only 'instrumental reason' (in order to attain complete control of society) joined, however, to a cult of irrationalism which made it a (once again: perverted) heir to Romanticism rather than to the Enlightenment (cf. our remarks in the Introduction, p. 5).

resigned here, too, in his own vein saying something rather like: 'As scientists, we are (alas!) Newtonians (and therefore to that extent Kantians: to wit, partisans of the Kant who, in the *Critique of Pure Reason*, acutely defined the legitimate scope of Newtonian science), but as living beings with a capacity to confer ethical–aesthetic meaning to life we are and remain Goetheans (and are pleasantly surprised to find ourselves here to a large extent in agreement with the Kant of the *Critique of Judgement*, though *pace* our friend Schiller we still have some difficulties with the *Critique of Practical Reason*).' To ask for such a display of modesty might perhaps have been asking a bit too much ... But in all seriousness, it is of course quite legitimate for the scientist to claim that colours are 'really' wave-lengths. It is no less legitimate for the humanist, starting from the other end, to claim that those poor, quantitative wave-lengths reach their real destiny only in the human susceptibility for their qualitative richness as colour-phenomena. What we need, therefore, is certainly not some esoteric, 'higher' knowledge ('the sixth sense of the bats' scorned by Schopenhauer): science itself is really the best esoteric knowledge that ever was. For it may rightly be called 'esoteric' in view of its distrust of the testimony of 'common sense' and of its search, in its own way, for the 'true' world behind it, using a language of its own (mathematics) which is accessible only to the initiated; and it may be called the 'best one', since it is the only one which leads to something other than mere speculative fantasy. What we *do* need, it seems, is rather a vindication of that *concrete realism* which has its roots in what, in our fourth chapter (p. 100) we have described as the young child's primordial experience of resistance which makes all conscious experience and knowledge possible in the first place, and which on the ethical plane consists of the humble awareness that there is something other than our own self which deserves our respectful attention. In this sense we whole-heartedly agree with Iris Murdoch when she writes: "We need to be enabled to think in terms of degrees of freedom, and to picture, in a non-metaphysical, non-totalitarian and non-religious sense, the transcendence of reality. A simple-minded faith in science, together with the assumption that we are all rational and totally free, engenders a dangerous lack of curiosity about the real world, a failure to appreciate the difficulties of knowing it."[583]

But let there be no misunderstanding: This comes down, in effect, to a reappraisal of the role of freedom in 'theoretical' and 'practical' Reason, respectively, as compared to how Kant saw it. Surely, just as Kant taught, the domain of Theoretical Reason is that of 'necessity' and the domain of Practical Reason that of 'freedom', as far as their *subject matter* is concerned. But are not science and, in its wake, technology, in that they are also an *enterprise* (a *Faustian* enterprise), not at least as much an expression of human freedom from nature as, nay more so than, an ethics in which (like in the *Westöstlicher Divan*) man cannot but feel respect for, and therefore exert, self-constraint towards other human beings in the first place, but also, by analogy, towards animals and even towards nature as a whole (the analogy being suggested by the animal capacity for suffering, and also by organic nature's purposefulness and nature's beauty in general)? Should we not then distinguish between *two kinds* of 'heteronomy': the 'bad' one (rightly rejected by Kant) which makes man dependent either on fear of God (if not just fear of social conventions) or on his own passions, and a 'good' one (not to be found in Kant) due to which respect and compassion are enjoined upon him by the confrontation with the reality of the 'Other'? 'Respect' and 'compassion', not because we are all part of one huge, irrational 'Will' (as Schopenhauer thought), but because we see our own, always relative freedom to cut ourselves loose from 'Will' reflected in other men and, by analogy, in all other creatures?

Elements of such 'concrete realism' are surely to be found in Schopenhauer: in his stressing of *Anschauung* against bloodless abstractions; in his sense of natural as well as artistic beauty, and, above all, in his experience-rooted ethics of compassion leading to the rejection of a belief in remorseless progress. He might have inaugurated it ... but for his πρωτον ψευδος (*prôton pseudos*).* But let there once again be no misunderstanding. Although in that case he might have avoided his metaphysical paradox, his world-view would no less have remained a broken one, in that the two kinds of 'representation' described in Books I and III of *The World as Will and Representation* would still stand unconnected beside each other. And precisely that is what makes him so modern. For the ineluctable problem remains that 'concrete realism' (for which Schopenhauer's expression 'representation independent of the principle of sufficient reason' is not such a bad description after all) † is doomed to share the world with the wholly different-natured realism of science and technology ('representation subject to the principle of sufficient reason'). This *abstract* realism, inaugurated by the Scientific Revolution, has come to occupy a dominant portion of the modern world without being able in itself to provide 'a sense of life' or to be integrated into a unified world-view (such as was still possible in the synthesis of the Scholastics, but was already doomed to failure in the final attempt at synthesis undertaken by the German Idealists). The great modern schizophrenia lucidly observed by Patočka and quoted as such at the start of our first chapter cannot be healed. We cannot hope for any new 'synthesis', but perhaps we may acquire the ability to find the ever-unstable balance of the 'compromise'.

---

* Chapter VIII, p. 226.

† One may call it thus, with a proviso derived from the end of the previous section, to wit, that this alternative view of the world had better not be taken in terms of aesthetic contemplation only.

ENDNOTES

SSW: Arthur Schopenhauer, *Sämtliche Werke*, hrsg. von Arthur Hübscher (Wiesbaden: Eberhard Brockhaus Verlag, 1948-52).

KAT: Immanuel Kant, *Werke* (Akademie Textausgabe, Berlin: Walter de Gruyter, 1968).

KdRV: Immanuel Kant, *Kritik der reinen Vernunft*, nach der ersten und zweiten Original-Ausgabe neu herausgegeben von Raymund Schmidt (Hamburg: Felix Meiner, 1956).

GW: *Goethes Werke*. (Weimar: Hermann Böhlau, 1887 – 1912).

[1] Bertrand Russell, *A History of Western Philosophy* (9th impression, London 1961), p. 726.

[2] Arthur Schopenhauer, *Philosophie in Briefen* (hrsg. von Angelika Hübscher und Michael Fleiter), Frankfurt a.M.: Insel Verlag, 1989, p. 351.

[3] *ibid.*, p. 179.

[4] Bertrand Russell (*op. cit.* in n. 1), p. 726.

[5] Thomas Mann, 'Schopenhauer', in: *Adel des Geistes. Sechzehn Versuche zum Problem der Humanität* (Stockholm, 1945), p. 339.

[6] Wilhelm Gwinner, *Arthur Schopenhauer aus persönlichem Umgang dargestellt. Ein Blick auf sein Leben, seinen Charakter und seine Lehre* (Leipzig, 1862), p. 52.

[7] Arthur Schopenhauer, *Handschriftlicher Nachlaß* I (Frankfurt a.M., 1966), p. 13.

[8] Arthur Schopenhauer, *Gespräche. Neue, stark erweiterte Ausgabe, hrsg. von Arthur Hübscher* (Stuttgart-Bad Canstatt, 1971), p. 324. (Conversation with Frédéric Morin in March 1858).

[9] Russell, *op. cit.* in n. 1, p. 722.

[10] The text quoted in the appended footnote is from Edward Conze, *Der Buddhismus. Wesen und Entwicklung* (original title: *Buddhism, its Essence and Development*), 3. Auflage, Stuttgart 1962, p. 92.

[11] J. Aler, 'Schopenhauer als metafysicus', in *De waarheid en de kunst* (Amsterdam/ Meppel, 1996), p. 35.

[12] Rüdiger Safranski, *Schopenhauer und die wilden Jahre der Philosophie* (2. Auflage, München/Wien 1988), p. 101 [NB the same circumstance is stressed by Aler in his essay mentioned in the previous endnote].

[13] Thomas Mann, *op. cit.* in n. 5, p. 397.

[14] Jean-François Revel, *Histoire de la philosophie occidentale. La philosophie classique. Humanistes et Cartésiens*, Chapitre deuxième: 'L'abandon de l'idéal contemplatif' (Editions Stock, 1970), p. 39-74.

[15] Jan Patočka, *Le monde naturel comme problème philosophique*. Den Haag, 1976, p. 1.

[16] cf. E.J.Dijksterhuis, *De mechanisering van het wereldbeeld* (Amsterdam, 1950).

[17] The expression 'deviation from the general human pattern' is due to the Dutch historian Jan Romein, 'De Europese geschiedenis als afwijking van het Algemeen Menselijk

Patroon' (1954), in: idem, *Historische lijnen en patronen*. Amsterdam, 1971. For the view that Europe's 'deviance' both came to the fore in, and was consolidated by, the Scientific Revolution of the 17th century, see H. Floris Cohen, *The Scientific Revolution. A Historiographical Inquiry* (University of Chicago Press 1994), p. 4.

18 Edwin Arthur Burtt, *The Metaphysical Foundations of Modern Physical Science. A Historical and Critical Essay* (London 1972; reprint of second revised edition of 1932), p. 300f.

19 The reference in the appended footnote to Dijksterhuis is to his *op. cit.* in n. 16, p. 3, and to Cohen, to his *op. cit.* in n. 17, p. 17.

20 Patočka, *op. cit.* in n. 15, p. 8f. The reference to Meyerson in the appended footnote is to his *Identité et réalité* (Paris, 1907); to Dijksterhuis, his *op.cit.* in n. 16, p.7.

21 Alexandre Koyré, 'Du monde de l'"à-peu-près" à l'univers de la précision' in: *Etudes d'histoire de la pensée philosophique*, ed. Gallimard 1971, p.344.

22 *ibid.*, p.150ff.

23 *ibid.*, p.349. The reference in the appended footnote to Dijksterhuis is to his *Val en worp: Een bijdrage tot de geschiedenis van de mechanica van Aristoteles tot Newton*. Groningen: Noordhoff, p. 171.

24 cf. Lewis Mumford, *Technics and Civilization* (Harbinger Books Edition, New York and Burlingame 1963; first edition 1934).

25 Koyré, *op. cit.* in n. 21, p. 352.

26 Revel, *op. cit.* in n. 14, p. 193-262 (Chapitre sixième: 'Le premier philosophe moderne').

27 Blaise Pascal, *Pensées*, 887/78, in *Oeuvres complètes* (Editions du Seuil, 1963: p. 615).

28 This is the final sentence of a chapter on Descartes in a book-in-progress by H.F. Cohen, provisionally entitled 'How Modern Science Came Into the World'.

29 Georges Sorel, *Les préoccupations métaphysiques des physiciens modernes* (Paris, 1907). The reference in the appended footnote to J. de Kadt is to his *Georges Sorel: Het einde van een mythe* (Amsterdam, 1938; reprinted in the same author's collection of essays *Verkeerde voorkeur*, Amsterdam 1948).

30 The references to Martin Heidegger in the appended footnote are to his *Nietzsche*. Zweiter Band, in: *Gesamtausgabe*, Bd. 6.2, Frankfurt a.M., 1997, p. 168-171); his *Holzwege*, in *Gesamtausgabe*, Bd. 5, Frankfurt a.M., 1977, p. 294), and his *Nietzsche*. Erster Band, *Gesamtausgabe*, Bd. 6.1, Frankfurt a.M., 1996, p.33. The reference to Victor Farias is to his *Heidegger et le nazisme* (éd. Verdier, Lagrasse, 1987).

31 Martin Heidegger, 'Die Frage nach der Technik', in *Die Technik und die Kehre* (Pfullingen 1962), p.12ff.

32 The reference to Ernst Cassirer in the appended footnote is to his *An essay on man. An introduction to a philosophy of human culture* (Yale UP, 29th printing, 1978), p. 211f.

33 The references in the appended footnote are to: Lynn White, *Medieval Technology and Social Change* (Oxford University Press, 1962); Joseph Needham, *The Grand Titration: Science and Society in East and West* (London, 1969); Fernand Braudel, third volume (*Le temps du monde*. Paris 1979) of his monumental *Civilisation matérielle, économie et capitalisme, XVe-XVIIIe siècle*, and Immanuel Wallerstein, *The Modern World System*, vol. I: *Capitalist Agriculture and the Origins of the European World-Economy in the Sixteenth Century* (New York, San Francisco, London, 1974; two other volumes have followed since).

[34] See previous endnote.

[35] The reference to R. Hooykaas in the appended footnote is to his article 'The Portuguese Discoveries and the Rise of Modern Science' (in *Selected Studies in History of Science*, Coimbra, 1983).

[36] Bertrand Russell, *op. cit.* in n. 1, p. 678.

[37] KdRV, B XII-XIII.

[38] *ibid.*, A 51 - B 75.

[39] GW, II. Abt. (*Naturwissenschaftliche Schriften*), 2. Bd. (*Farbenlehre*, polemical part), p. 69.

[40] *ibid.*, 1. Bd. (Introduction, p. XXVI).

[41] GW, 11. Bd., p. 17f. [NB Reported by Goethe in his essay 'Glückliches Ereignis'].

[42] GW, *op. cit.* in n. 39, p. XXXI.

[43] A. Koyré, *Etudes Galiléennes*, 2. ed. (Paris, 1966; the first edition, dated 1939, came out in 1940), p. 213.

[44] C.I. Dessaur, *De droom der rede* (The Hague 1982), p. 61. For the significance of the hermetic tradition in European thought, cf. also Frances A. Yates, *Giordano Bruno and the Hermetic Tradition* (Chicago, 1964).

[45] GW, I. Abt., 27. Bd., p. 204f. (*Dichtung und Wahrheit,* second part, eighth book).

[46] GW, I. Abt., 14. Bd., p. 28 (*Faust*) [lines translated as in: Charles Coulston Gillispie, *The Edge of Objectivity* (Princeton University Press, 3d.ed., 1990), p. 196].

[47] Ronald D. Gray, *Goethe the Alchemist. A Study of Alchemical Symbolism in Goethe's literary and scientific works* (Cambridge University Press, 1952), p. 258.

[48] GW, I. Abt., 29. Bd., p. 11.

[49] *ibid.*, p. 9f.

[50] GW, I. Abt., 28. Bd., p. 288f.

[51] GW, II. Abt., 11.Bd., p. 128 ('Maximen und Reflexionen').

[52] Johann Peter Eckermann, *Gespräche mit Goethe in den letzten Jahren seines Lebens*, hrsg. von H.H. Houben (Leipzig, 1909), p. 196f. (conversation of 11 April 1827).

[53] GW, II. Abt., 11.Bd., p. 21-37.

[54] *ibid.*, p. 47 ('Einwirkung der neueren Philosophie').

[55] *ibid.*, p. 50.

[56] *ibid.*, p. 52.

[57] *ibid.*, p. 54-56 ('Anschauende Urteilskraft').

[58] The reference in the appended footnote to Alain Finkielkraut is to his *La défaite de la pensée*, ed. Gallimard 1987, p. 107.

[59] Eckermann, *op.cit.* in n. 52, p. 583 (conversation of 14 March 1830).

[60] *ibid.*, p. 180f. (conversation of 31 January 1827).

[61] *ibid.*, p. 263 (conversation of 2 May 1829), and p. 578 (conversation of 14 March 1830).

[62] Heinrich Heine, 'Die Romantische Schule', in: *Heinrich Heines Sämtliche Werke, mit Einleitungen, erläuternden Anmerkungen und Verzeichnissen sämtlicher Lesearten von Dr. Ernst Elster*, Fünfter Band (Leipzig 1887-1890), p. 240.

[63] GW, II. Abt., 11. Bd., p. 159.

[64] GW, I. Abt., 2. Bd., p. 216 ('Gereimte Distichen').

[65] Cf. for this Alfred Schmidt, *Goethes herrlich leuchtende Natur. Philosophische Studie zur deutschen Spätaufklärung* (München/Wien, 1984), especially p.133-134.

[66] The reference in the appended footnote to the conversation with Eckermann is to *op. cit.* in n. 52, p. 294f.

[67] A most enlightening exposition of this fundamental difference between Goethe's conception of science and that of mechanical-mathematical science, to which I am much indebted, has been given by Ernst Cassirer in his thoughtful essay 'Goethe und die mathematische Physik. Eine erkenntnistheoretische Betrachtung' (*Idee und Gestalt. Goethe-Schiller-Hölderlin-Kleist*, Darmstadt 1971, p. 35-80).

[68] GW, I. Abt, 2. Bd., p. 215 ('Gedichte.Zweiter Teil', section 'Gott, Gemüth und Welt').

[69] Heinrich Heine, *Deutschland. Ein Wintermärchen*; in *Sämtliche Werke*, Zweiter Band (Leipzig o.J.), p. 432.

[70] *ibid.*, p. 445.

[71] As quoted by H.A.M. Snelders, *Wetenschap en intuitie. Duits romantisch en speculatief natuuronderzoek rond 1800* (Baarn, 1994), p.177.

[72] For a more detailed account we refer the reader to Safranski's biography (n. 12).

[73] *ibid.*, p. 239.

[74] SSW VII, p . 91.

[75] *ibid.*

[76] *ibid.*, p. 3; also in the second version (SSW I, first part: *Über die vierfache Wurzel des Satzes vom zureichenden Grunde*, p. 1).

[77] *ibid.*, p. 7; and *ibid.*, p. 4.

[78] The reference in the appended footnote to Stephen Toulmin is to volume 1 of his *Human Understanding* (Oxford, 1972), section 2.4: 'A Digression on 'Representation', p. 192 f.); the one to Christopher Janaway is to his *Self and World in Schopenhauer's Philosophy* (Oxford, 1989), p. 38).

[79] The reference in the appended footnote to Kant is to his KdRV, B 93; to Schopenhauer, to SSW II, p. 618 ('Critique of Kantian Philosophy').

[80] SSW VII, p. 70, and SSW I, p.142.

[81] SSW VII, p. 70.

[82] *ibid.*, p.71.

[83] *ibid.*, p.72.

[84] *ibid.*, p.74.

[85] SSW I, p. 145.

[86] SSW VII, p. 51; also in the second version (SSW I, p. 101).

[87] SSW VII, p. 56.

[88] SSW VII, p. 91. The references made in the appended footnote to other places in Schopenhauer's work are to SSW I, p. 110f., and to SSW VII, p. 56.

[89] KdRV, A 163-166, B 203-207.

[90] SSW VII, p. 67, and SSW I, p.138.

[91] *ibid.*, p. 34f.; and *ibid.*, p. 87f.

[92] *ibid.*, p. 38; and *ibid.*, p. 90.

[93] *ibid.*, p. 33f.; and *ibid.*, p. 86.

[94] *ibid.*, p. 36 [NB In the second version, the digression on the 'immediate object' has been moved to a new section 22 with the same heading 'Vom unmittelbaren Objekt': SSW I, p. 84].

[95] To wit, Section 21 of the second version ('The *a priori* nature of the concept of causality. The intellectual nature of empirical intuition. Understanding': '*Apriorität des Kausalitätsbegriffes. Intellektualität der empirischen Anschauung. Der Verstand*'). SSW I, p. 51-84.

[96] I have previously put forward this particular thesis (even though, as I would say now, in a somewhat overstated manner) in my article: 'Five decisive years: Schopenhauer's epistemology as reflected in his theory of colour'. *Studies in the History and Philosophy of Science*, Vol. 18, No. 3, pp. 271-291, 1987.

[97] For a more detailed account of this episode, and more generally for the biographical data in this paragraph we once again refer the reader to Safranski's biography cited in n. 12.

[98] *ibid.*, p. 267.

[99] Eckermann, *op. cit.* in n. 52, p. 261 (conversation of 20 February 1829).

[100] Gray, *op. cit.* in n. 47, p. 101.

[101] GW, II. Abt., 3. Bd., p. 358f.

[102] GW, II. Abt., 4. Bd., p. 296 ('Confession des Verfassers' in the historical part of the *Theory of Colours*).

[103] Cf. for this Dennis L. Sepper, *Goethe contra Newton: Polemics and the Project for a New Science of Color* (New York, 1988), p. 27. For a critical evaluation of Goethe's place in the history of science, and of his theory of colours in particular, see also George A. Wells, *Goethe and the Development of Science 1750-1900* (Alphen a/d Rijn, 1978); Frederick Burwick, *The Damnation of Newton: Goethe's Color Theory and Romantic Perception* (New York, 1986); Frederick Amrine, Francis J. Zucker, and Harvey Wheeler (eds.), *Goethe and the Sciences: A Reappraisal* (Dordrecht, 1987); and, last but not least, Albrecht Schöne, *Goethes Farbentheologie* (München, 1987).

[104] Sir Isaac Newton, *Opticks* (Dover Publications, New York, 1952), Book One, Part I, Prop. II, Theor. II, p. 26-63.

[105] *ibid.*, Prop.I.

[106] GW, II. Abt., 2. Bd., p. 14.

[107] The reference in the appended footnote to Rudolf Laemmel is to his *Isaac Newton* (Zürich, 1957), as cited in Maurice Martin, *Die Kontroverse um die Farbenlehre*, Novalis Verlag, Schaffhausen 1979, p. 30; further to Richard Westfall in his *Never at Rest. A Biography of Isaac Newton* (Cambridge University Press, 1980), p. 279.

[108] GW, II. Abt., 4. Bd., p. 203-6.

[109] GW, II. Abt., 3. Bd., p. 281f.

[110] GW, II. Abt., 1. Bd., p. 64.

[111] For the fragment by Georg Christoph Tobler discussed in the appended footnote, see Alfred Schmidt, *op. cit.* in n. 65, p.101, and GW, II. Abt., 11. Bd., p. 10-12 (the fragment itself is to be found there on p. 5-9).

[112] GW, II. Abt., 1. Bd., p. 364

[113] *ibid.*, p. 202.

[114] *ibid.*, p. 203f.

[115] *ibid.*, p. 225.
[116] GW, II. Abt., 11. Bd., p. 147 ('Über Naturwissenschaft im Allgemeinen. Einzelne Betrachtungen und Aphorismen').
[117] GW, II. Abt., 5.2. Bd., p.366 ('Paralipomena' CIV).
[118] GW, I. Abt., 3. Bd. ('Gott und Welt'), p. 103 ('Was es gilt. Dem Chromatiker').
[119] The reference in the appended footnote to Malus is to his review of the *Theory of Colours* in *Annales de Chimie* 79 (p. 199-219), as quoted in Goethe's *Die Schriften zur Naturwissenschaft. Vollständige mit Erläuterungen versehene Ausgabe im Auftrag der Deutschen Akademie der Naturforscher Leopoldina*, Bd. 5A (Weimar 1992), p. 55ff.
[120] GW, II. Abt., 1. Bd., p. 443.
[121] Quoted by Schopenhauer himself in the second volume of his *Parerga and Paralipomena* (SSW VI, p. 92).
[122] *ibid.* See also the introduction to the second edition of *On Vision and Colours* (SSW I, p. iii-vi).
[123] In view of the two versions of *On Vision and Colours* mentioned in the appended footnote, and the relatively small difference between them, we have consulted the first version of 1816 in Arthur Schopenhauer, *Sämtliche Werke*, edited by Franz Mockrauer in Munich in 1923 (sixth volume, p. 5-56), but refer here throughout to the second version as printed in the Hübscher edition (SSW) used by us for all his other works.
[124] SSW II, p. 618.
[125] SSW I, p. 8f.
[126] Wilhelm Ostwald, *Schopenhauer und die Farbenlehre* (Leipzig, 1918), p. 108. The reference in the appended footnote to Schopenhauer's mentioning Müller in another context is to SSW I, p. 89.
[127] SSW I, p. 10.
[128] *ibid.*, p. 10 ff.
[129] *ibid.*, p. 19.
[130] GW, II. Abt., 1. Bd.,p .ix.
[131] SSW I, p. 25.
[132] SSW I, p. 38.
[133] *ibid.*, p. 33.
[134] *ibid.*, p. 38.
[135] *ibid.*, p. 45f.
[136] *ibid.*
[137] *ibid.*, p. 53.
[138] *op. cit.* in n. 2 (letter of 11 November 1815), p. 121f.
[139] *ibid.*, p. 124.
[140] GW, II. Abt., 1. Bd., p. 1.
[141] *ibid.*, p. 28f.
[142] SSW I, p. 4.
[143] Werner Heisenberg, 'Die Goethesche und die Newtonsche Farbenlehre im Lichte der modernen Physik', in *Wandlungen in den Grundlagen der Naturwissenschaft*, 4. Ausg. (Leipzig, 1943), p. 58-76.

[144] Section 14 of the first version is entitled 'Von den äußern Reizen, welche die qualitative Theilung der Thätigkeit der Retina bestimmen'. Section 13 of the second version is an enlarged combination of the original Section 14 with the two sections that follow it in the first version: 'Unterschied der physischen und chemischen Farben' and 'Scheinbare Inkongruenz und wirkliche Übereinstimmung des physiologischen und physischen Gegensatzes der Farben'.

[145] SSW I, p. 39.

[146] Arthur Hübscher, *Denker gegen den Strom. Schopenhauer gestern – heute – morgen.* 2. Auflage: Bonn 1982, p. 70.

[147] SSW I, p. 73.

[148] SSW I, p. 90; p. 87f.; p. 91.

[149] Cf. for this Paul D. Sherman, *Colour vision in the nineteenth century: the Young-Helmholtz-Maxwell-theory* (Bristol,1981). For Schopenhauer's anticipating some notions of modern physiological optics: Joachim Gerlach, 'Schopenhauers Farbenlehre und die moderne Sinnesphysiologie', in: W. Schirmacher (Hrsg.), *Zeit der Ernte: Studien zum Stand der Schopenhauer-Forschung. Festschrift für Arthur Hübscher zum 85. Geburtstag* (1982).

[150] SSW I, p .48.

[151] The reference in the appended footnote is to Eckermann, *op. cit.* in n. 52, p. 260.

[152] SSW I., p. 3 (*Theoria colorum physiologica, eademque primaria*).

[153] Sir Isaac Newton, *Opticks*, Book One, Part II, Prop. III, Prob. I, and Prop.VII, Theor.V.

[154] cf. John Gage, *Colour and Culture. Practice and Meaning from Antiquity to Abstraction* (Thames and Hudson, 1993), p. 203ff.

[155] Hermann von Helmholtz, 'Über das Sehen des Menschen', in: *Vorträge und Reden*, volume 1 (Braunschweig, 1903), p. 115f.

[156] *op. cit.* in n. 2, p. 149f. (letter of 20 January 1856).

[157] Hermann von Helmholtz, 'Die Thatsachen in der Wahrnehmung', in: *op. cit.* in n. 155, volume 2, p. 27.

[158] SSW II, p. 36.

[159] The reference in the appended footnote to Schopenhauer's views on Lamarck is to SSW II, p. 169. See also Arthur Hübscher, *op. cit.* in n. 146, p. 151.

[160] SSW III, p. 307ff.

[161] The passage quoted in the appended footnote is to be found *ibid.*, p. 4.

[162] *ibid.*, p. 22.

[163] cf. Karl Popper, *Objective Knowledge*, rev. ed. (Oxford, 1979); especially chapter 3 ('Epistemology without a knowing subject') and 4 ('On the Theory of an Objective Mind'), passim.

[164] Ludwig Wittgenstein, *Tractatus Logico-Philosophicus* 5.631, 5.632 and 5.64 (in Edition Suhrkamp on p. 90/91).

[165] The passages referred to in the appended footnote are: *op. cit.* in n.2, p. 91f. (letter to Julius Frauenstädt of 12 July 1852), and SSW III, p. 153.

[166] SSW III, p. 17.

[167] SSW III, p. 314f.

[168] *op. cit.* in n. 2, p. 81 (Karl Rosenkranz to Schopenhauer, 9 September 1837), and p. 84f. (response, 25 September).
[169] KdRV, A 367.
[170] *ibid.* A 368-373.
[171] *ibid.* A 378-389.
[172] *ibid.* B 409-413.
[173] KAT IV, p. 372ff.
[174] Kant's passage referred to in the appended footnote is in KdRV, B 519ff.
[175] KdRV, B 275-279
[176] SSW I, p. 80f.
[177] SSW II, p. 15.
[178] SSW I, p. 52f.
[179] SSW III, p. 309.
[180] *ibid.* p. 310.
[181] SSW II, p. 11f.
[182] For a searching analysis of this particular problem of Schopenhauer's epistemology from a different starting point cf. Joachim Aul, *Schopenhauer's Begründungstheorie im Lichte der Ergebnisse der modernen Wissenschaftstheorie* (Göttingen, 1983), especially Chapter VIII ('Das Problem der Gegenstandskonstitution'), p. 122-137.
[183] SSW II, p. 497.
[184] *ibid.*, p.21.
[185] Russell, *op.cit.* in n. 1, p. 634.
[186] *ibid.*, p. 640f.
[187] KdRV, A 122.
[188] Kant's proof appears in the second edition of the *Critique of Pure Reason* as an introduction to a lengthy digression (KdRV, A 190-211 and B 233-256) on the second analogy of experience, entitled 'Grundsatz der Erzeugung' in the first edition and 'Grundsatz der Zeitfolge nach dem Gesetze der Kausalität' in the second edition. [NB The first analogy is 'Grundsatz der Beharrlichkeit der Substanz'; the third 'Grundsatz des Zugleichseins nach dem Gesetze der Wechselwirkung oder Gemeinschaft'; all to be found in Section 3 of Chapter 2 of Book 2 of the 'Transcendental Analytic' (itself Part 1 of the 'Transcendental Logic', with the 'Transcendental Dialectic' as Part 2)].
[189] Bertrand Russell, 'On the Notion of Cause', in *Mysticism and Logic* (ed. Unwin Books, London 1963), p. 141.
[190] SSW VII, p. 36f.
[191] SSW II, p. 519.
[192] *ibid.*
[193] The references in the appended footnote to Popper and Arnheim are to K.R. Popper, *op.cit.* in n. 163, p. 71f., and to Rudolf Arnheim, *Visual Thinking* (California University Press, first edition 1969).
[194] KdRV, A 120. The reference in the appended footnote to Martin Heidegger is to his *Kant und das Problem der Metaphysik*, Bonn 1929, reprint in *Gesamtausgabe*, Bd. 3 (Frankfurt a.M. 1951), p. 168ff.); to Ernst Cassirer, to his part in a debate with Heidegger which he published as a review of Heidegger's book in *Kantstudien*, 1931,

but which we have examined in: Ernst Cassirer & Martin Heidegger, *Débat sur le kantisme et la philosophie [Davos, mars 1929] et autres textes de 1929-1931*, présentés par Pierre Aubenque, éd. Beauchesne, Paris 1972, p. 78-79.

195 cf. SSW II, p. 605ff.

196 SSW II, p. 118.

197 The reference in the appended footnote to Arnold Gehlen is to his 'Die Resultate Schopenhauers', in *Gedächtnisschrift für Arthur Schopenhauer zur 150. Wiederkehr seines Geburtstages*, hrsg. von C.A. Emge und Otto v. Schweinichen (Berlin, 1938), p. 104; to Hübscher, to his *op. cit.* in n. 146, p. 268f.

198 SSW III, p. 41f.

199 SSW I, p. 58-70; p. 54; p. 56; p. 57.

200 *ibid.*, p. 76.

201 KdRV, A 373.

202 SSW II, p. 596.

203 KdRV, B 275.

204 SSW II, p. 526f.

205 *ibid.*

206 *ibid.*, p. 520

207 SSW II, p. 595.

208 *ibid.*

209 The Schopenhauer passage quoted in the appended footnote refers to *ibid.*, p. 9.

210 SSW II, p. 120

211 The reference in the appended footnote to Edmund Husserl is to his 'On intentional experiences and their 'contents'' ('Über intentionale Erlebnisse und ihre 'Inhalte'', in: *Logische Untersuchungen*, Zweiter Band, Erster Teil: 'Untersuchungen zur Phänomenologie und Theorie der Erkenntnis' (*Husserliana* XIX/1, Martinus Nijhoff Publishers 1984, p. 312-528), especially p. 438f.

212 SSW IV (*Die beiden Grundprobleme der Ethik*), p. 31.

213 KdRV, *op. cit.* B 132.

214 SSW VII, p. 69; and SSW I, p. 141.

215 *ibid.*, p. 72; and *ibid.*, p. 143.

216 *ibid.*, p. 68 and p. 141.

217 Wittgenstein, *op.cit.* in n. 164, no. 2.1 (p. 16).

218 The reference in the appended footnote to Janaway is to his *op. cit.* in n. 78, p. 148.

219 SSW III, p. 70.

220 The reference in the appended footnote to Volker Spierling is to his dissertation *Schopenhauers transzendentalphilosophisches Selbstmißverständnis. Prolegomena zu einer vergessenen Dialektik* (München, 1977).

221 Helmuth Plessner, *Conditio humana* (Pfullingen, 1964), p. 16. The reference in the appended footnote to Konrad Lorenz is to his *Behind the Mirror. A search for a natural history of human knowledge*, London 1977 (originally published as *Die Rückseite des Spiegels*, München 1973), p. 7; p. 9.

222 Helmuth Plessner, *op. cit.* in previous endnote, p. 25ff.

[223] The reference in the appended footnote to *Über den Willen in der Natur* is to SSW IV, p. 36.
[224] GW, I. Abt., 14. Bd., p.163 (monologue 'Wald und Höhle' in *Faust*).
[225] Eckermann, *op.cit.* in n. 52, p. 253; conversation of 17 February 1829.
[226] As reported by Ernst Otto Lindner and Julius Frauenstädt in *Arthur Schopenhauer. Von ihm. Über ihn* (Berlin, 1863); quoted from *op. cit.* in n. 8, p. 31.
[227] GW, II. Abt., 12. Bd., p. 406 ('Maximen und Reflexionen').
[228] Spierling, *op. cit.* in n. 220, p. 65f.
[229] GW, I. Abt., 2. Bd., p. 230.
[230] GW, I. Abt., 36. Bd., p. 112.
[231] GW, I. Abt., 14. Bd., p. 87.
[232] GW, IV. Abt. (Briefe), 23. Bd., p. 180f. (Letter to F.J. Seebeck, 28 November 1812; translation taken from *Hegel's Philosophy of Nature* I, ed. and transl. by M.J. Petry (London, 1970), p. 83).
[233] *ibid.*
[234] Eckermann, *op. cit.* in n. 52, p. 532.
[235] To be found in *Briefe von und an Hegel*, ed. Johannes Hoffmann (Hamburg, 1952-54), *II* and *III*; and in *Goethe-Hegel Briefwechsel*, ed. Hermann Bauer (Stuttgart, 1970).
[236] P.F.H. Lauxtermann, 'Hegel and Schopenhauer as Partisans of Goethe's Theory of Color' in *Journal of the History of Ideas* (1990, 4, p. 599-624).
[237] G.W.F. Hegel, *op. cit.* in n. 232, p. 242.
[238] *ibid.*, II, p. 19.
[239] *ibid.*, p. 142.
[240] *ibid.*, p. 133. For he reference in the appended footnote to 'entoptic colours' see Burwick, *op. cit.* in n. 103, p. 59.
[241] *ibid.*, p.135.
[242] *ibid.*, p. 140.
[243] The reference in the appended footnote is to Arthur Schopenhauer, *Metaphysik der Natur*, hrsg. u. eingel. von Volker Spierling (München-Zürich 1984), p. 19ff.
[244] G.W.F. Hegel, Werke in zwanzig Bänden III (Frankfurt a.M., 1971), p. 22f.
[245] *ibid.*, p. 22
[246] *ibid.*, p. 64.
[247] G.W.F.Hegel, *Vorlesungen über die Geschichte der Philosophie* III (Leipzig, 1971), p. 206ff.
[248] *op. cit.* in n. 244, p.120.
[249] The reference in the appended footnote to Thomas S. Kuhn is to his 'Mathematical versus Experimental Traditions in the Development of Physical Science' in *The Essential Tension: Selected Studies in Scientific Tradition and Change* (Chicago, 1977), p. 31-66.
[250] SSW VI, p. 144.
[251] G.W.F. Hegel, *op. cit.* in n. 232, I, p. 264f.
[252] *ibid.* Cf. also Frans H. van Lunteren, 'Hegel and Gravitation' in *Hegels Philosophie der Natur*, ed. by Rolf-Peter Horstmann and Michael John Petry (Stuttgart, 1986), p. 45-53.

[253] *The Mathematical Principles of Natural Philosophy* by Sir Isaac Newton, translated by Andrew Motte 1729; with an introduction by I. Bernard Cohen, London 1968, p. 392.

[254] Arthur Schopenhauer, *Philosophische Vorlesungen*, Teil II, hrsg. von Volker Spierling (München, 1984), p. 133f.

[255] SSW II, p.117.

[256] Alfred Schmidt, *Idee und Weltwille. Schopenhauer als Kritiker Hegels* (Carl Hanser Verlag, München/Wien, 1988), p. 24.

[257] G.W.F. Hegel, *op. cit.* in n. 232, p. 272.

[258] The reference in the appended footnote to Schopenhauer's habilitation is to *op. cit.* in n. 8, p. 47.

[259] Emile Meyerson, *De l'explication dans les sciences*, tome 2 (Paris, 1921), p. 122-138; and F.A.Hayek, *The Counter-Revolution of Science* (Indianapolis, 2d. ed. 1979), p. 367-400.

[260] SSW IV, p. 86.

[261] The reference in the appended footnote to Joseph Needham is to his 'Integrative Levels: A Revaluation of the Idea of Progress' in idem, *Moulds of Understanding. A Pattern of Natural Philosophy* (London 1976, p. 131-172), in particular to p. 140f.; further to volume 2 of his monumental *Science and Civilisation in China*, and to the highly instructive sections on Needham in Chapter 6 of H. Floris Cohen, *op. cit.* in n. 17, p. 418-482.

[262] SSW III, p. 508.

[263] Hübscher, *op. cit.* in n. 146, p. 74.

[264] GW, I. Abt., 14. Bd., p. 14. The conversation with Luden is included in Kurt Rossmann, *Deutsche Geschichtsphilosophie. Ausgewählte Texte von Lessing bis Jaspers* (DTV, München 1969), p. 187.

[265] GW, I. Abt., 7. Bd., p. 157.

[266] SSW III, p. 509f.

[267] Karl R. Popper, *The poverty of historicism* (London,1957).

[268] K. Rossmann, *op. cit.* in n. 264, p. 180.

[269] Alfred Schmidt, *op. cit.* in n. 256, p. 98.

[270] The reference in the appended footnote to Franz Rosenzweig is to his *Der Stern der Erlösung*, 4. Auflage (Den Haag, 1976), p. 8; for Schopenhauer's conversation with Wieland, see *op. cit.* in n. 8, p. 22.

[271] The reference in the appended footnote to Max Scheler is to his 'Die transzendentale und die psychologische Methode', in: *Frühe Schriften*: *Gesammelte Werke* I, Bern, 1971, p. 217f.

[272] KAT VIII, p. 22f.

[273] *ibid.*, p. 64f.

[274] J.de Kadt, 'De sociologische plaats van het existentialisme'. In: *Verkeerde Voorkeur* (Amsterdam 1948), p. 334.

[275] Meyerson, *op. cit.* in n. 259.

[276] Francis Fukuyama, Chapter 5 ('An Idea for a Universal History') of his *The End of History and the Last Man* (London, 1992).

[277] SSW VI, p. 258 ('Parerga und Paralipomena' II).

[278] Thomas Mann, *op. cit.* in n. 5, p. 381.
[279] *op. cit.* in n. 8, p. 326 (Conversation with Frédéric Morin, March 1858).
[280] *op. cit.* in n. 62, vol. IV, p. 259.
[281] The reference in the appended footnote to Kant is to KAT IV, p. 356,
[282] *ibid.*, p. 350f.
[283] KdRV, B XXX.
[284] KAT V, p. 4.
[285] KdRV, B 433-434.
[286] *ibid.*, B 472-473, A 444-445. For Schopenhauer's comment in the appended footnote see SSW II, p. 585.
[287] KdRV, B 364, A 307-308.
[288] SSW II, p. 571.
[289] *ibid.*, p. 573.
[290] *ibid.*, p. 585.
[291] *ibid.*, p. 599.
[292] *ibid.* p. 598.
[293] Max Horkheimer, *Zur Kritik der instrumentellen Vernunft*. Frankfurt a.M., 1967.
[294] KdRV, B 536-537, A 508-509. The line quoted from Schopenhauer in the appended footnote is in SSW II, p. 575.
[295] KAT IV, p. 392.
[296] *ibid.*, p. 395f.
[297] *ibid.*, p. 398.
[298] Quoted from 'Die Philosophen' (in: *Schiller's sämtliche Werke in zwölf Bänden*, Bd. I, Stuttgart 1881, p. 346).
[299] KAT IV, p. 400.
[300] *ibid.*, p. 407f.
[301] *ibid.*, p. 413.
[302] *ibid.*, p. 414.
[303] *ibid.*, p. 419f.
[304] *ibid.*, p. 421.
[305] *ibid.*, p. 429.
[306] *ibid.*, p. 432f.
[307] *ibid.*, p. 433.
[308] *ibid.*, p. 446f.
[309] *ibid.*, p. 450.
[310] *ibid.*, p. 94f.
[311] *ibid.*, p. 96f.
[312] *ibid.*, p. 98f.
[313] *ibid.*, p. 99.
[314] SSW IV, p. 176 (*Die beiden Grundprobleme der Ethik*).
[315] *ibid.*, p. 117 f.
[316] *ibid.*, p. 120.

[317] *ibid.*, p. 121.
[318] *ibid.*, p. 124.
[319] *ibid.*, p. 129.
[320] *ibid.*, p. 132; the reference in the appended footnote to Cassirer is to his *op. cit.* in n. 194, p. 72.
[321] SSW IV, p. 133.
[322] KAT V, p. 84.
[323] SSW IV, p. 134.
[324] *ibid.*, p. 142.
[325] *ibid.*, p. 143.
[326] *ibid.*
[327] *ibid.*, p. 148.
[328] *ibid.*, p. 157.
[329] *ibid.*
[330] *ibid.*, p. 162f.
[331] *ibid.*, p. 162.
[332] *ibid.*, p. 137.
[333] *ibid.*, p. 136.
[334] *ibid.*, p. 158.
[335] *ibid.*, p. 205f.
[336] *ibid.*, p. 208.
[337] *ibid.*, p. 210.
[338] *ibid.*, p. 238f.
[339] *ibid.*, p. 271.
[340] *ibid.*, p. 267.
[341] *ibid.*, p. 251.
[342] *ibid.*, p. 3.
[343] *ibid.*, p. 6f.
[344] *ibid.*, p. 9.
[345] *ibid.*, p. 19.
[346] *ibid.*, p. 26f.
[347] *ibid.*, p. 29f.
[348] *ibid.*, p. 36.
[349] *ibid.*, p. 46f.
[350] *ibid.*, p. 49.
[351] *ibid.*, p. 55.
[352] *ibid.*, p. 93.
[353] *ibid.*, p. 95.
[354] *ibid.*, p. 97.
[355] As quoted by Julien Benda, *La trahison des clercs* (éd. 'Libertés', dirigée par Jean-François Revel, 25, J.J. Pauvret éditeur, 1965), p. 8 (original edition Bernard Grasset 1927).

[356] David Hume, *A Treatise of Human Nature*, Book III, Part I, Section 1.
[357] *ibid.*, Book II, Part III, Section 5; and Book III, Part I, Section 1.
[358] Iris Murdoch, *Metaphysics as a guide to morals* (London, 1992), p. 25.
[359] SSW IV, p. 147 (*Die beiden Grundprobleme der Ethik*).
[360] Murdoch, *op. cit.* in n. 358, p. 223
[361] Emmanuel Lévinas, *Difficile liberté* (Editions Albin Michel, 3e édition, 1990), p. 279f.
[362] Cathérine Chalier, *Judaisme et altérité* (Editions Verdier, 1982), p. 263.
[363] The references in the appended footnote to Kant are to KdRV, A 362, and to KAT IV, p. 428, and VI, p. 223.
[364] Michel Foucault, *Les mots et les choses. Une archéologie des sciences humaines* (ed. Gallimard,1966), p. 398
[365] Richard Rorty, *Contingency, Irony and Solidarity* (Cambridge University Press, 1989), especially Part III ('Cruelty and Solidarity').
[366] Emmanuel Lévinas, *Humanisme de l'autre homme* (Paris, Fata Morgana, 1972), p. 60. The reference in the appended footnote to Alain Finkielkraut is to his books *La défaite de la pensée* (Paris, 1980) and *La sagesse de l'amour* (Paris, 1987), with the passage quoted deriving from the latter book, p. 151f.
[367] KAT V, p. 86.
[368] SSW IV, p. 260f. (*Die beiden Grundprobleme der Ethik*).
[369] *ibid.*, p. 213.
[370] The reference in the appended footnote to Lévinas is to his *Difficile liberté*, p. 21 and p. 23.
[371] Samuel P. Huntington, *The Clash of Civilizations and the Remaking of World Order* (New York, 1996).
[372] Leszek Kolakowski, *Religion* (Fontana Paperbacks 1982), p. 190f.
[373] Murdoch, *op. cit.* in n. 358, p. 438.
[374] The reference in the appended footnote to Schopenhauer is to SSW IV, p. 144 (*Über den Willen in der Natur*, chapter 'Hinweisung auf die Ethik'); to Thomas Mann, to his *op. cit.* in n. 5, p. 378.
[375] The reference in the appended footnote to Schopenhauer is to SSW IV, p. 222f.; to Constant and Kant's attempted refutation, to KAT VIII, p. 425f.
[376] The reference in the appended footnote to G.K.Chesterton is to his *Orthodoxy* (London, 1907), p. 50f.; to Pascal, to his *Pensées*, 681/353 (*op. cit.* in n. 27, p. 590).
[377] Max Weber, 'Politik als Beruf', in *Gesammelte politische Schriften*, 3.Aufl., Tübingen 1971, p. 549f.
[378] G.K.Chesterton, *Saint Thomas Aquinas* (Hodder and Stoughton, London 1956), p. 85.
[379] The reference in the appended footnote to Schopenhauer on women is to his SSW IV, p. 215; to Carol Gilligan, to her *In a different voice. Psychological theory and women's development* (Cambridge, 1982).
[380] SSW IV, p. 242 (*Die beiden Grundprobleme der Ethik*).
[381] Jeremy Bentham, *The Principles of Morals and Legislation*, Chapter XVII, Section 1 (1789).
[382] Luc Ferry, *Le nouvel ordre écologique. L'arbre, l'animal et l'homme* (éd. Grasset, 1992), p.64 and 193.

[383] KAT VI, p. 443.
[384] KAT V, p. 464.
[385] This has been rightly pointed out by Ferry in his *op. cit.* in n. 382, p. 209.
[386] SSW III, p. 405 (note).
[387] Ferry, *op.cit.* in n. 382, p.206 f.
[388] KAT IV, p. 462f.
[389] SSW II, p. 505f.
[390] *ibid.*
[391] *op. cit.* in n. 2, p. 170f. (Letter to Julius Frauenstädt of 21 August 1852).
[392] SSW II, p. 599.
[393] GW, I. Abt., 4. Bd., p.15 (From a poem Goethe wrote for the jubilee of his Weimar fellow-minister von Voigt on 27 September 1816).
[394] Immanuel Kant, *Die drei Kritiken in ihrem Zusammenhang mit dem Gesamtwerk*, mit verbindendem Text zusammengefaßt von Raymund Schmidt (Stuttgart, 1975), p. 487f.
[395] The reference in the appended footnote to Schmidt is to *ibid.*, p. 388f.
[396] SSW II, p. 379.
[397] The reference in the appended footnote to Kant is to KAT IV, p. 443. Cf. Günther Patzig, "Platons Ideenlehre, kritisch betrachtet" (p.119-143 of his collection of essays *Tatsachen, Normen, Sätze* (Stuttgart 1980)
[398] Murdoch, *op. cit.* in n. 358, p. 438f.
[399] SSW II, p. 606.
[400] Simone Weil, *La pesanteur et la grâce* (Librairie Plon, édition 1988), p. 196.
[401] *ibid.*, p. 116.
[402] As quoted by Murdoch in her *op. cit.* in n. 358, p. 394.
[403] *ibid.*, p. 396.
[404] *ibid.*, p. 411.
[405] *ibid.*, p. 508.
[406] *ibid.*, p. 412.
[407] The Sartre passage in the appended footnote has been quoted from Finkielkraut, *La sagesse de l'amour* (cited before in n. 366), p. 114f.
[408] Weil, *op. cit.* in n. 400, p. 132.
[409] Lévinas, *Difficile liberté*, p. 24-42.
[410] Murdoch, *op. cit.* in n. 358, p. 507.
[411] Murdoch, *ibid.*, p. 438.
[412] KAT V, p. 125.
[413] KdRV, A 634, B 662.
[414] *Critique of Practical Reason*, Part I, Book II, Chapter II, section 'Von der Dialektik der reinen Vernunft in Bestimmung des Begriffs vom höchsten Gut', in: KAT V, p. 110f. (on the 'highest good'); p. 113f. (on the antinomy); p. 114-119 (on the solution of the antinomy); p. 122-124 (on immortality); and p. 124-132 (on God).
[415] The reference in the appended footnote to Schopenhauer is to SSW II, p. 591.
[416] KAT VI, p. 36.
[417] *ibid.*, p. 26f.

[418] *ibid.*, p. 21.
[419] *ibid.*, p. 40f.
[420] *ibid.*, p. 43. The reference in the appended footnote to Günther Patzig is to his *Ethik ohne Metaphysik* (Göttingen, 1971), p. 58.
[421] Chesterton, *op. cit.* in n. 376, p. 242.
[422] The reference in the appended footnote to Kant is to KAT VI, p. 449.
[423] The reference in the appended footnote to Patzig is to his *op. cit.* in n. 420, p. 46f.
[424] Friedrich Nietzsche, *Werke in drei Bänden*, hrsg. von Karl Schlechta, Erster Band (9. Aufl., Darmstadt, 1982), p. 27 (*Die Geburt der Tragödie oder Griechentum und Pessimismus* 2). Cf. Julian Young, *Nietzsche's Philosophy of Art* (Cambridge University Press, 1992), p. 28.
[425] SSW II, p. 449.
[426] KAT VI, p. 469 and 449.
[427] Eckermann, *op. cit.* in n. 52, p. 606 (Conversation of 2 August 1831).
[428] Rosenzweig, *op. cit.* in n. 270, p. 11.
[429] Ernst Cassirer, *Kant's Life and Thought* (Yale University Press, 9181), p. 257 (original title: *Kants Leben und Lehre*, first ed. 1918, reprint Darmstadt 1977, p. 273f.).
[430] The reference in the appended footnote to Milan Kundera is to his *The unbearable lightness of being* (London/Boston 1984), p. 253.
[431] The reference in the appended footnote to Schopenhauer is to SSW II, p. 480f.
[432] As quoted by Alain Finkielkraut in his essay on Péguy: *Le mécontemporain* (éditions Gallimard 1991), p. 37.
[433] Arthur Schopenhauer, *op. cit.* in n. 7, p. 23.
[434] SSW II, p. 219.
[435] Thomas Mann, *op. cit.* in n. 5, p. 397.
[436] SSW II, p. 476f.
[437] SSW VII, p. 73.
[438] SSW I, p. 143.
[439] *ibid.*
[440] SSW II, p. 121.
[441] SSW III, p. 220.
[442] SSW II, p. 127.
[443] SSW III, p. 219.
[444] SSW II, p. 124.
[445] *ibid.*, p. 179.
[446] The reference in the appended footnote to Schopenhauer is to SSW III, p. 315; to Ernst Cassirer, to his *Das Erkenntnisproblem in der Philosophie und Wissenschaft der neueren Zeit*, 3. Bd. (Darmstadt, 1974), p. 433f.
[447] Max Scheler, *Die Stellung des Menschen im Kosmos* (7.Aufl., Francke Verlag Bern-München, p. 56-71.
[448] Nietzsche, *op. cit.* in n. 424, Zweiter Band, p. 226 (*Die fröhliche Wissenschaft* 357).
[449] *ibid.*, p. 576 (*Jenseits von Gut und Böse* 11).
[450] Nietzsche, *op. cit.* in n. 424, Erster Band, p. 479f. (*Menschliches Allzumenschliches* 39).

[451] SSW II, p. 487.
[452] *op. cit.* in n. 2, p. 170.
[453] *ibid.*, p.172.
[454] The reference in the appended footnote to Russell is to his *op. cit.* in n. 1, p. 726; to Schopenhauer, to SSW II, p. 471.
[455] *op. cit.* in n. 2, p. 376.
[456] The reference in the appended footnote to Schopenhauer's letter to Frauenstädt is to *op. cit.* in n. 2, p. 173; to Schopenhauer's letter to Adam von Doss: *ibid.*, p. 95; for Schopenhauer's influence upon Wittgenstein, see especially Allan Janik & Stephen Toulmin, *Wittgenstein's Vienna* (New York 1973).
[457] SSW II, p. 450.
[458] *ibid.*
[459] *ibid.*
[460] *ibid.*
[461] *ibid.*
[462] The reference in the appended footnote to Richard F. Gombrich is to his *How Buddhism Began: The Conditioned Genesis of the Early Teachings* (London 1996).
[463] The reference in the appended footnote to Schopenhauer quoting Aristotle is to SSW III, p. 398.
[464] SSW III, p. 186 and 723f.
[465] For a description of this philosophical current cf. Bertrand Russell, *The Philosophy of Logical Atomism* (London, 1918); for a comparison with Buddhism cf. Chris Gudmunsen, *Wittgenstein and Buddhism* (London, 1977).
[466] According to his own account in SSW III, p. 186.
[467] Helmuth von Glasenapp, 'Schopenhauer und Indien', in: *Schopenhauer Jahrbuch* 36 (1955), p. 32-49. and idem, 'Die Weisheit Indiens bei Schopenhauer und in der neueren Forschung', in: *Schopenhauer Jahrbuch* 42 (1961), p. 52-61.
[468] The reference in the appended footnote to Schopenhauer is to SSW III, p. 697.
[469] von Glasenapp, *op. cit.* in n. 467: 'Schopenhauer und Indien', p. 39.
[470] von Glasenapp, *op. cit.* in n. 467: 'Die Weisheit Indiens ...', p. 58.
[471] SSW II, p. 535; the passage referred to in the appended footnote is in SSW III, p. 314.
[472] The reference in the appended footnote to Wittgenstein is to his *Tractatus logico-philosophicus*, 5.633.
[473] The reference in the appended footnote to Schopenhauer is to *op. cit.* in n. 2, p. 345.
[474] The reference in the appended footnote to Schopenhauer is to *op. cit.* in n. 2, p. 172.
[475] Bryan Magee, *The Philosophy of Schopenhauer* (Oxford-New York, 1983), p. 143f.
[476] SSW III, p. 55.
[477] SSW I, p. 42f., and SSW II, p. 571.
[478] SSW III, p. 52.
[479] SSW II, p. 10.
[480] Ernst Cassirer, *Substanzbegriff und Funktionsbegriff. Untersuchungen über die Grenzfragen der Erkenntnistheorie* (Berlin, 1910; repro: Darmstadt 1969), passim. and E.J. Dijksterhuis, *op. cit.* in n. 16, p. 550.

[481] Magee, *op. cit.* in n. 475, p. 112.
[482] *ibid.*
[483] SSW III, p. 51.
[484] Magee, *op. cit.* in n. 475, p. 145.
[485] The reference in the appended footnote to Schopenhauer is to SSW III, p. 55; to Wilhelm Ostwald, to his *Vorlesungen über Naturphilosophie*, Leipzig 1902, p. 413f. Cf. for all this C. Hakfoort, 'Science Deified: Wilhelm Ostwald's Energeticist World-View and the History of Scientism', *Annals of Science* 49, 1992, p. 525-544.
[486] A lucid analysis of this peculiar kind of scientism is given in Tomas Vanheste, *Copernicus is ziek. Een geschiedenis van het New Age denken over natuurwetenschap* (Delft, 1996).
[487] Murdoch, *op. cit.* in n. 358, p. 511.
[488] SSW III, p. 219.
[489] GW, I. Abt., 3. Bd., p. 106 ('Ultimatum', in section 'Gott und Welt' of 'Gedichte. Dritter Band').
[490] SSW III, p. 739.
[491] *op. cit.* in n. 2, p. 307 (Letter of 17 April 1853)
[492] *ibid.*, p. 313f. (Letter of 11 September 1853).
[493] *ibid.*, p. 319f. (Letter of 30 November 1854).
[494] Magee, *op. cit.* in n. 475, p. 283.
[495] Sigmund Freud, *Studienausgabe* Bd.I, Frankfurt a.M. 1969, p. 540 ('Angst und Triebleben').
[496] *ibid.*, Bd. V, p. 46 (*Drei Abhandlungen zur Sexualtheorie.* Vorwort zur vierten Auflage).
[497] SSW III, p. 458.
[498] SSW III, p. 243. Especially interesting in this respect are Chapters 19 ('Vom Primat des Willens im Selbstbewußtseyn') and 32 ('Über den Wahnsinn').
[499] SSW III, p. 456f.
[500] Freud, *op. cit.* in n. 495, Bd. IX, p. 181 (*Die Zukunft einer Illusion*).
[501] Janaway, *op. cit.* in n. 78,p. 102.
[502] Freud, *op. cit.* in n. 495, Bd. III, p. 322 (*Das Ich und das Es*).
[503] *ibid.*, p. 295.
[504] *ibid.*, p. 304.
[505] *ibid.*, p. 320.
[506] *ibid.*, p. 303.
[507] *ibid.*, Bd. I, p.516 (*Die Zerlegung der psychischen Persönlichkeit*).
[508] For those interested in these problems and their historical background, see Henri F. Ellenberger, *The Discovery of the Unconscious. The History and Evolution of Dynamic Psychiatry* (London-New York,1970).
[509] The reference in the appended footnote to Scheler is to his *op. cit.* in n. 447, p. 55.
[510] Nietzsche, *op. cit.* in n. 424, Dritter Band, p. 917 ('Aus dem Nachlaß der achtziger Jahre').
[511] *ibid.*, p. 556.

[512] Jean-Paul Sartre, *L'existentialisme est un humanisme* (Les éditions Nagel, Paris 1967), p. 17; p. 21f.; also (on Gide, as cited in the appended footnote) p. 74f.

[513] The reference in the appended footnote to Pascal is to his *Pensées*, 199-*72*, Ed. du Seuil, p. 528 (translated here as by A.J. Krailsheimer in the Penguin edition, 1966, p. 94); to St Augustine, to his *De civitate Dei* xxi, 10.

[514] The references in the appended footnote to Kant are to KAT V, p. 400. p. 410, p. 415.

[515] The references in the appended footnote to Schopenhauer are to SSW II, p. 519 and to SSW IV, p. 120.

[516] SSW II, p. 517.

[517] *op. cit.* in n. 2, p. 331 (Letter of 28 November 1851).

[518] Alfred Schmidt, *Emanzipatorische Sinnlichkeit. Feuerbachs anthropologischer Materialismus* (Reihe Hanser, Carl Hanser Verlag, München 1973), p. 110.

[519] Arthur Schopenhauer, *Briefwechsel und andere Dokumente* ausgewählt und herausgegeben von Max Brahn. Leipzig, 1911.

[520] Franz Mehring, 'Arthur Schopenhauer', in *Philosophische Aufsätze* (*Gesammelte Schriften*, Bd. 13, p. 150).

[521] Georg Lukács, *Die Zerstörung der Vernunft* (Berlin, 1954), chapter 'Schopenhauer'; p. 156-198.

[522] Karl R. Popper, *The open society and its enemies*, vol. 2 ('The High Tide of Prophecy: Hegel, Marx and the Aftermath'), Princeton University Press, fifth revised edition 1966, p. 306.

[523] Max Horkheimer, 'Die Aktualität Schopenhauers' in *Kritik der instrumentellen Vernunft*, p. 252f.

[524] Safranski, *op. cit.* in n. 12, p. 488f.

[525] SSW V, p. 183; SSW IV, p. 87.

[526] The *Parerga* are to be found in SSW V (first volume) and VI (second volume).

[527] Martin Heidegger, *Nietzsche*. Erster Band, in: *Gesamtausgabe*, Bd. 6.1 (Frankfurt a.M. 1996), p. 60.

[528] The reference in the appended footnote to Edmund Husserl is to his 'Die Krise der europäischen Wissenschaften und die transzendentale Phänomenologie', in *Husserliana*, Bd. VI (The Hague, 1962), p. 205; to Hübscher, to his *op. cit.* in n. 146, p. 264.

[529] Horkheimer, *op. cit.* in n. 523, p. 251.

[530] SSW II, p. 315f.

[531] Russell, *op. cit.* in n. 1,p. 726.

[532] The reference in the appended footnote to Thomas Mann is to his *op. cit.* in n. 5, p. 339 and p. 337.

[533] Safranski, *op. cit.* in n. 12, p. 328.

[534] The reference in the appended footnote to Schopenhauer is to SSW II, p. 627f.

[535] Nietzsche, *op. cit.* in n. 424, Band II, p. 845f. (*Zur Genealogie der Moral*, Chapter 3: 'Was bedeuten asketische Ideale?'; Section 6).

[536] The reference in the appended footnote to Heidegger is to his study on Nietzsche cited in n. 30 (Erster Band), p. 107 and p. 32; to Young, to his *op. cit.* in n. 424, p. 118f.

[537] SSW II, p. 218.

[538] *ibid.*, p. 220f.
[539] *ibid.*, p. 223.
[540] *ibid.*, p. 238
[541] *ibid.*, p. 229f.
[542] *ibid.*, p. 265.
[543] *ibid.*, p. 301.
[544] *ibid.*, p. 304.
[545] The reference in the appended footnote to Schopenhauer is to *ibid.*, p. 312 and p. 302; to Iris Murdoch, to her 'On God and Good', in: *Existentialists and Mystics. Writings on Philosophy and Literature* (London, 1977), p. 356f., and to Simone Weil, to her *op. cit.* in n. 400, p. 171.
[546] SSW II, p. 276.
[547] *ibid.*, p. 222.
[548] *ibid.*, p. 250
[549] Eckermann, *op. cit.* in n. 52, p. 237 (Conversation of 20 October 1828).
[550] GW, I. Abt., Bd. 48, p. 200 ('Aus den Heften zur Morphologie').
[551] *Goethe's Werke* (Hamburger Ausgabe) Bd. XII (Hamburg, 1953), p. 438 ('Maximen und Reflexionen').
[552] GW, I. Abt., Bd. 42-2, p. 256 ('Maximen und Reflexionen über Kunst, Natur und Wissenschaft').
[553] GW, I. Abt., Bd. 48, p. 198 ('Maximen und Reflexionen über Kunst. Aus Wilhelm Meisters Wanderjahren').
[554] The reference in the appended footnote to Schopenhauer is to SSW II, p. 262.
[555] Nietzsche, *op. cit.* in n. 424, Erster Band, p. 22 (*Die Geburt der Tragödie oder Griechentum und Pessimismus* 4). At first publication in 1872 the title was *Die Geburt der Tragödie aus dem Geiste der Musik* ('*The Birth of Tragedy from the Spirit of Music*'). The new edition, published in 1886 under the name just cited, contained a new introduction by the author under the title 'Versuch einer Selbstkritik' ('An attempt at self-criticism') in which he made up an unsentimental balance of the juvenile work.
[556] *ibid.*, p. 40.
[557] *ibid.*, Zweiter Band, p. 1024.
[558] The reference in the main text is to Theo Meyer, *Nietzsche und die Kunst* (Tübingen-Basel, 1993), p. 91; the passage quoted in the appended footnote is from GW, I. Abt., Bd. 48, p. 205 ('Maximen und Reflexionen über Kunst. Aus Wilhelm Meisters Wanderjahre').
[559] Julian Young, *op. cit.* in n. 424, p. 100f.
[560] The reference in the appended footnote to Luc Ferry is to his *Homo Aestheticus. L'invention du goût à l'age démocratique* (Paris, 1990).
[561] *ibid.*, p. 319 f.
[562] Young, *op. cit.* in n. 424, p. 120.
[563] GW, IV. Abt., Bd. 10, p.75 (Letter to Herder and his wife, 7 June 1793).
[564] E.g, by Julie D. Prandi in her book *Dare to be happy. A Study of Goethe's Ethics* (Lanham, New York, London 1993), p. 6-27.
[565] *ibid.*, p. 43.

[566] SSW V, p. 141.

[567] Goethes Briefe, Bd. III. Hamburg, 1965 (Letter to Jacobi, 16 January 1813). The passage quoted in the appended footnote only is from GW, I. Abt., Bd. 42-2, p. 211 ('Maximen und Reflexionen uber Literatur und Ethik. Aus dem Nachlaß').

[568] The references in the appended footnote are to Nietzsche, *op. cit.* in n. 424, Zweiter Band, p. 43 (Section 2 of the chapter 'Von den Abtrünnigen' ('On renegades'); to *Schillers Sämtliche Werke in zwölf Bänden*, 1. Bd. (Stuttgart 1881), p. 66f.; to *Goethes Gespräche. Gesamtausgabe, neu hrsg. von Flodoard Freiherr von Biedermann* (Leipzig 1909), Bd. I, p. 202, and to Eckermann, *op. cit.* in n. 52, p. 614 (Conversation of 11 March 1832).

[569] Eckermann, *op. cit.* in n. 52, p. 613.

[570] GW, I. Abt., Bd. 6, p. 188f.

[571] GW, I. Abt., Bd. 7, p. 19f.

[572] For the statements about Persian religion in the appended footnote, cf. Richard H. Grutzmacher, *Die Religionen in der Anschauung Goethes* (Baden-Baden 1950), p. 45.

[573] SSW III, p. 667 (Section 'Von der Nichtigkeit und dem Leiden des Lebens' in the second volume of *The World as Will and Representation*).

[574] SSW III, p. 739 (Section 'Epiphilosophie' in the second volume of *The World as Will and Representation*).

[575] Adolf Muschg, *Goethe als Emigrant. Auf der Suche nach dem Grünen bei einem alten Dichter* (Insel Verlag, Frankfurt a.M., 1996; originally Suhrkampf Verlag 1986), p. 12.

[576] Cathérine Chalier, *L'Alliance avec la nature* (Les Editions du Cerf, Paris 1989), p. 54.

[577] *ibid.*, p. 184; p. 191.

[578] Thomas Mann, *op. cit.* in n. 5, p. 397.

[579] SSW IV, p. 16.

[580] Cf. SSW III, p. 671.

[581] Thomas Mann, *op. cit.* in n. 5, p.387.

[582] The reference in the appended footnote to G.K. Chesterton is to his *op. cit.* in n. 376, p. 282.

[583] Iris Murdoch, 'Against Dryness', in her *op. cit.* in n. 545, p. 293.

Index

Science and Philosophy

Series Editor:

Nancy J. Nersessian, *Program in Cognitive Science, Georgia Institute of Technology, Atlanta*

1. N.J. Nersessian: *Faraday to Einstein: Constructing Meaning in Scientific Theories.* 1984 ISBN Hb 90-247-2997-1 / Pb (1990) 0-7923-0950-2
2. W. Bechtel (ed.): *Integrating Scientific Disciplines.* 1986 ISBN 90-247-3242-5
3. N.J. Nersessian (ed.): *The Process of Science.* Contemporary Philosophical Approaches to Understanding Scientific Practice. 1987 ISBN 90-247-3425-8
4. K. Gavroglu and Y. Goudaroulis: *Methodological Aspects of the Development of Low Temperature Physics 1881–1956.* Concepts out of Context(s). 1989 ISBN 90-247-3699-4
5. D. Gooding: *Experiment and the Making of Meaning.* Human Agency in Scientific Observation and Experiment. 1990 ISBN 0-7923-0719-4
6. J. Faye: *Niels Bohr: His Heritage and Legacy.* An Anti-realist View of Quantum Mechanics. 1991 ISBN 0-7923-1294-5
7. D.A. Anapolitanos: *Leibniz: Representation, Continuity and the Spatiotemporal.* 1999 ISBN 0-7923-5476-1
8. G. Châtelet: *Figuring Space.* Philosophy, Mathematics, and Physics. 2000 ISBN 0-7923-5880-5
9. X. Chen: *Instrumental Traditions and Theories of Light.* The Uses of Instruments in the Optical Revolution. 2000 ISBN 0-7923-6349-3
10. P.F.H. Lauxtermann: *Schopenhauer's Broken World-View.* Colours and Ethics between Kant and Goethe. 2000 ISBN 0-7923-6585-2

Kluwer Academic Publishers – Dordrecht / Boston / London